COMET HALLEY
Investigations, Results, Interpretations
Volume 2: Dust, Nucleus, Evolution

THE ELLIS HORWOOD LIBRARY OF SPACE SCIENCE AND SPACE TECHNOLOGY

SERIES IN ASTRONOMY

Series Editor: JOHN MASON
Consultant Editor: PATRICK MOORE

This series aims to coordinate a team of international authors of the highest reputation, integrity and expertise in all aspects of astronomy. It makes valuable contributions to the existing literature, encompassing all areas of astronomical research. The titles will be illustrated with both black and white and colour photographs, and will include many line drawings and diagrams, with tabular data and extensive bibliographies. Aimed at a wide readership, the books will appeal to the professional astronomer, undergraduate students, the high-flying 'A' level student, and the non-scientist with a keen interest in astronomy.

PLANETARY VOLCANISM: A Study of Volcanic Activity in the Solar System
Peter Cattermole, formerly Lecturer in Geology, Department of Geology, Sheffield University, UK, now Freelance Writer and Consultant and Principal Investigator with NASA's Planetary Geology and Geophysics Programme

DIVIDING THE CIRCLE: The development of critical angular measurement in astronomy 1500–1850
Allan Chapman, Centre for Medieval and Renaissance Studies, Oxford, UK

SATELLITE ASTRONOMY: The Principles and Practice of Astronomy from Space
John K. Davies, Royal Observatory, Edinburgh, UK

THE ORIGIN OF THE SOLAR SYSTEM: The Capture Theory
John R. Dormand, Department of Mathematics and Statistics, Teesside Polytechnic, Middlesbrough, UK, and Michael M. Woolfson, Department of Physics, University of York, UK

THE DUSTY UNIVERSE
Aneurin Evans, Department of Physics, University of Keele, UK

SPACE-TIME AND THEORETICAL COSMOLOGY
Michel Heller, Department of Philosophy, University of Cracow, Poland

ASTEROIDS: Their Nature and Utilization
Charles T. Kowal, Space Telescope Institute, Baltimore, Maryland, USA

COMET HALLEY — Investigations, Results, Interpretations
Volume 1: Organization, Plasma, Gas
Volume 2: Dust, Nucleus, Evolution
Editor: J. W. Mason, B.Sc., Ph.D.

ELECTRONIC AND COMPUTER-AIDED ASTRONOMY: From Eyes to Electronic Sensors
Ian S. McLean, Joint Astronomy Centre, Hilo, Hawaii, USA

URANUS: The Planet, Rings and Satellites
Ellis D. Miner, Jet Propulsion Laboratory, Pasadena, California, USA

THE PLANET NEPTUNE
Patrick Moore, CBE

ACTIVE GALACTIC NUCLEI
Ian Robson, Director of Observatories, Lancashire Polytechnic, Preston, UK

ASTRONOMICAL OBSERVATIONS FROM THE ANCIENT ORIENT
Richard F. Stephenson, Department of Physics, Durham University, Durham, UK

EXPLORATION OF TERRESTRIAL PLANETS FROM SPACECRAFT: Instrumentation, Investigation, Interpretation
Yuri A. Surkov, Chief of the Laboratory of Geochemistry of Planets, Vernandsky Institute of Geochemistry, USSR Academy of Sciences, Moscow, USSR

THE HIDDEN UNIVERSE
Roger J. Taylor, Astronomy Centre, University of Sussex, Brighton, UK

AT THE EDGE OF THE UNIVERSE
Alan Wright, Australian National Radio Astronomy Observatory, Parkes, New South Wales, Australia, and Hilary Wright

COMET HALLEY
Investigations, Results, Interpretations
Volume 2:
Dust, Nucleus, Evolution

Editor
J. W. MASON B.Sc., Ph.D.

ELLIS HORWOOD
NEW YORK LONDON TORONTO SYDNEY TOKYO SINGAPORE

First published in 1990 by
ELLIS HORWOOD LIMITED
Market Cross House, Cooper Street,
Chichester, West Sussex, PO19 1EB, England

A division of
Simon & Schuster International Group
A Paramount Communications Company

© Ellis Horwood Limited, 1990

Printed and bound in Great Britain
by Hartnolls, Bodmin, Cornwall

British Library Cataloguing in Publication Data

Comet, Halley: Investigations, results, interpretations.
1. Halley's Comet
I. Mason, J. W.
523.642
ISBN 0–13–171083–4

Library of Congress Cataloging-in-Publication Data

Comet Halley: Investigations, results, interpretations / editor. J. W. Mason
p. cm. — (The Ellis Horwood library of space science and space technology. Series in
astronomy)
Includes bibliographical references and index.
Contents: v. 2. Dust, nucleus, evolution
ISBN 0–13–171083–4
1. Halley's comet. I. Mason, John (John W.) II. Series.
QB723.H2C63   1990
523.6'42–dc20

90–4789
CIP

Directory of Contributors

Brownlee, D.E. Dept. of Astronomy, University of Washington, Seattle, Washington WA 98195, USA.

Greenberg, J.M. Dept. of Laboratory Astrophysics, University of Leiden, Niels Bohrweg 2, 2333CA Leiden, The Netherlands.

Houpis, H.L.F. Mission Research Corp., 2300 Garden Road, Suite 2, Monterey, California CA 93940-5326, USA.

Keller, H.U. Max-Planck-Institut für Aeronomie, D-3411 Katlenburg-Lindau, Federal Republic of Germany.

Kissel, J. Max-Planck-Institut für Kernphysik, Postfach 103980, D-6900 Heidelberg 1, Federal Republic of Germany.

Kresák, L. Astronomical Institute, Slovak Academy of Sciences, 84228 Bratislava, Czechoslovakia.

Lamy, P.L. Laboratoire d'Astronomie Spatiale, Traverse du Siphon, Les Trois Lucs, 13012 Marseille, France.

Larson, S.M. Lunar and Planetary Laboratory, Dept. of Planetary Sciences, University of Arizona, Tucson, Arizona AZ 85721, USA.

McDonnell, J.A.M. Unit for Space Sciences, University of Kent at Canterbury, Canterbury, Kent CT2 7NR, England.

McIntosh, B.A. Herzberg Institute of Astrophysics, National Research Council of Canada, Ottawa, Canada KIA 0R6.

Massonne, L. ESOC Darmstadt and Max-Planck-Institut für Kernphysik, Heidelberg, now with mbp Software & Systems GmbH, Geschäftsstelle Bremen, Am Wall 140, D-2800 Bremen 1, Federal Republic of Germany.

Pankiewicz, G.S.A. Unit for Space Sciences, University of Kent at Canterbury, Canterbury, Kent, CT2 7NR, England.

Rickman, H. Astronomiska Observatoriet, Box 515, S-75120 Uppsala, Sweden.

Sagdeev, R.Z. Space Research Institute, USSR Academy of Sciences, Profsoyuznaya 84/32, SU-117810, Moscow GSP-7, USSR.

Sekanina, Z. MS 183-401, Jet Propulsion Laboratory, California Institute of Technology, 4800 Oak Grove Drive, Pasadena, California CA 91109, USA.

Sitarski, G.	Space Research Centre, Bartycka 18, PL-00-716 Warsaw, Poland.
Stephenson, F.R.	Dept. of Physics, University of Durham, Science Laboratories, South Road, Durham, DH1 3LE, England.
Szegö, K.	Central Research Institute for Physics, PO Box 49, H-1525 Budapest, Hungary.
Vaisberg, O.L.	Space Research Institute, USSR Academy of Sciences, Profsoyuznaya 84/32, SU-117810, Moscow GSP-7, USSR.
Watanabe, J.-i.	Dept. of Astronomy, Faculty of Science, University of Tokyo, Bunkyo-ku, Tokyo, 113 Japan.
Weissman, P.R.	Earth and Space Sciences Division, Jet Propulsion Laboratory, California Institute of Technology, 4800 Oak Grove Drive, Pasadena, California CA 91109, USA.
Yeomans, D.K.	MS 264-664, Jet Propulsion Laboratory, California Institute of Technology, 4800 Oak Grove Drive, Pasadena, California CA 91109, USA.
Zarnecki, J.C.	Unit for Space Sciences, University of Kent at Canterbury, Canterbury, Kent, CT2 7NR, England.
Ziolkowski, K.	Space Research Centre, Bartycka 18, PL-00-716 Warsaw, Poland.

Table of Contents

List of names,
acronyms and abbreviations

AIAA	American Institute of Aeronautics and Astronautics
AMPTE	Active Magnetospheric Particle Trace Experiment
amu	Atomic mass unit
APV-N	Wave and plasma analyzer (Vega)
APV-V	Wave and plasma analyzer (Vega)
ASP-G	Automatic stabilized platform (Vega)
ASTRO-1	Three UV telescopes carried in Shuttle payload bay
ASTRON	Franco-Soviet UV space observatory
AU	Astronomical unit (150 million km)
BD	Bonner Durchmersterung
BRL	High data rate telemetry channel (Vega)
BS	Bow shock
BTM	Low data rate telemetry channel (Vega)
CA	Closest approach
CCD	Charge-coupled device
CD	Compact disk
CHON	Carbon-hydrogen-oxygen-nitrogen
CIS	Capacitive impact sensor (Giotto)
CNES	Centre National des Etudes Spatiales
CNRS	Centre National de la Recherche Scientifique
C/P	Coronagraph/polarimeter (SMM)
CRA	Cometary ram analyzer (Vega)
CRAF	Comet Rendezvous and Asteroid Flyby
CSE	Cometocentric solar ecliptic (coordinate system)
CSIRO	Commonwealth Scientific and Industrial Research Organization
CSG	Centre Spatial Guyanais (Kourou, French Guyana)

DE	Disconnection event
DE-1	Dynamics Explorer-1
D/H	Deuterium to hydrogen isotope ratio
DID (DIDSY)	Dust impact detection (system) (Giotto)
DSN	Deep Space Network
DUCMA	Dust particle detector (Vega)
EA	Electrostatic analyzer (Vega)
EESA	Electron electrostatic analyzer (Giotto)
EPA	Energetic particle analyzer (Giotto)
EPONA	Energetic particle experiment (Giotto)
ESA	European Space Agency
ESO	European Southern Observatory
ESP	Energy spectrum of particles plasma experiment (Suisei)
EUV	Extreme ultraviolet
FES	Fine error sensor (IUE)
FFS	Fast forward shock
FIS	Fast ion sensor (Giotto)
FITS	Flexible image transport system
FLD	Field line draping (coordinate system)
FRS	Fast reverse shock
FTS	Fourier transform spectrometer
GRE	Radio science experiment (Giotto)
GRT	Ground received time
GSFC	Goddard Space Flight Center
GSRT	Ground station received time
HERS	High-energy range spectrometer (Giotto)
HIS	High intensity spectrometer (Giotto)
HMC	Halley multicolour camera (Giotto)
HSE	Halley-centred solar ecliptic (coordinate system)
IACG	Inter-Agency Consultative Group
IAU	International Astronomical Union
ICE	International Cometary Explorer (formerly ISEE-3)
IGY	International Geophysical Year
IHW	International Halley Watch
IIS	Implanted ion sensor (Giotto)
IKS	Infrared spectrometer (Vega)
IMC	Interstellar molecular cloud
IMF	Interplanetary magnetic field (also experiment on Sakigake)

IMP-8	Magnetometer experiment on ICE
IMS	Ion mass spectrometer (Giotto)
ING	Neutral gas mass spectrometer (Vega)
IPM-M, -P	Impact momentum and plasma sensor (Giotto)
IPS	Interplanetary scintillation
IR	Infrared
IRAM	30-metre millimetre wave telescope
IRAS	Infrared Astronomy Satellite
ISAS	Institute of Space and Astronautical Sciences
ISEE-3	International Sun-Earth Explorer-3
ISTP	International solar-terrestrial phenomena
IUE	International Ultraviolet Explorer
JPA	Johnstone plasma analyzer
JPL	Jet Propulsion Laboratory
KAO	Kuiper Airborne Observatory
KSC	Kagoshima Space Center
LHS	Left-hand side (of equation)
LJO	Lear-Jet Observatory
L-SPN	Large-scale phenomena network (of IHW)
LWP	Long-wavelength primary camera (IUE)
Ly-α	Lyman-α line of hydrogen
MAG	Magnetometer experiment (Giotto)
MHD	Magnetohydrodynamic
MISCHA	Magnetometer experiment (Vega)
MIT	Massachusetts Institute of Technology
MPAE	Max-Planck-Institut für Aeronomie
MPB	Magnetic pile-up boundary
MSM/RSM	Meteoroid shield momentum sensor (Giotto)
MS-T5	Name for Sakigake before launch
MVA	Minimum variance analysis
NASA	National Aeronautics and Space Administration
NMS	Neutral mass spectrometer (Giotto)
N-NN	Near-nucleus network (of IHW)
NS	Neutral (current) sheet
NSF	National Science Foundation
OAO	Orbiting Astronomical Observatory
OB	Outer boundary
OGO	Orbiting Geophysical Observatory

OPE	Optical probe experiment (Giotto)
PB	Pile-up boundary (cometopause)
PDS	Planetary data system
PHOTON	Shield penetrator detector (Vega)
PIA	Dust mass spectrometer (Vega)
PICCA	Positive ion cluster composition analyzer (Giotto)
Planet-A	Name of Suisei before launch
PLASMAG	Plasma energy analyzer (Vega)
POM	Polyoxymethylene
PUMA	Dust mass spectrometer (Vega)
PVO	Pioneer Venus Orbiter
PWP	Plasma wave probe (Sakigake)
RFC	Ram faraday cup (Vega)
RHS	Right-hand side (of equation)
rms	Root mean square
RPA	Plasma analyzer (Vega)
SCET	Spacecraft event time
SDA	Solar direction analyzer (Vega)
SDFC	Solar direction faraday cup (Vega)
SFS	Slow forward shock
SMM	Solar Maximum Mission
SOFA	Second-order fermi acceleration
SOW	Solar wind experiment (Sakigake)
SP-1, -2	Dust particle detector (Vega)
SPARTAN	University of Colorado experiment on *Challenger*
SR	Sub-region
SRS	Slow reverse shock
TKS	Three-channel spectrometer (Vega)
TÜNDE-M	Energetic particle analyzer (Vega)
TVS	Television camera system (Vega)
UT	Universal time
UV	Ultraviolet
UVI	Ultraviolet imager (Suisei)
VLA	Very Large Array
VLBI	Very-long baseline interferometry
WG	Working group (of IHW or IACG)
WG$^+$	Water group ions

Editor's Preface

In 1985/86, thousands of scientists around the world turned their attention to perhaps the most famous of all cosmic visitors — Halley's comet. Returning to the inner Solar System once every 76 years or so, Halley's comet is the brightest periodical comet, and the return of 1985/86 was the thirtieth consecutive apparition to be recorded since that of 240 BC. The comet is named after the second Astronomer Royal for England, Edmond Halley, not because he discovered it but because he was the first person to calculate its path around the Sun.

Halley saw the comet which now bears his name in 1682, five apparitions ago and, believing it to be the same comet that had been seen previously in 1531 and 1607, predicted it would return in 1758. The comet was duly recovered by an amateur astronomer Johann Palitzsch on Christmas Night 1758, and it passed perihelion in March of the following year. Halley's correct prediction and in particular his use of Isaac Newton's laws of motion incorporating the newly formulated theory of universal gravitation were major scientific achievements — all the more so when one considers the superstitious fears associated with comets in the 17th century.

Studies of Halley's comet at successive returns have mirrored the rapid progress in technological developments during the last 150 years. In 1835, scientific studies were based on telescopic drawings made of the comet: in 1910 the newly discovered tools of photography and spectroscopy were used to secure a permanent record of the comet's appearance at visual wavelengths and determine its chemical composition.

In 1986, a flotilla of six spacecraft from four space agencies — nicknamed the 'Halley Armada' — encountered the comet and conducted *in situ* measurements. Few people alive in 1910 (less than seven years after the Wright brothers had made their first flight) would have believed that such a feat would be possible just 76 years later.

The 1985/86 return has turned out to be the most significant apparition ever of Halley's comet. A wealth of new and important data were obtained by the spacecraft which examined the comet at close range, including the first ever images of a cometary nucleus — the tiny dust- and ice-ball at the heart of the comet. In addition, many exciting and intriguing discoveries were made by observers using spacecraft in orbit around the Earth and Venus, and some of the world's largest ground-based telescopes in conjunction with the most sensitive electronic detectors, working in the infrared and ultraviolet wavebands as well as at visual wavelengths. When combined with the other remote observations carried out from rockets and high-flying aircraft, this was, at the time, the largest and most intensive observing campaign ever mounted on any astronomical object. Only the attention given to supernova SN1987A in the Large Magellanic Cloud, since February 1987, is in any way comparable.

Coordinating the vast comet Halley observing campaign — which began with the comet's recovery in October 1982 and continues even as this book goes to press, nearly eight years later — presented enormous difficulties. It was vital to standardize observing techniques and instrumentation, and help ensure that all data and results were carefully collated and properly documented and archived for future reference. This was especially important in view of some of the problems that were encountered in 1910. At that time, although the project had well-defined aims, many observatories did not cooperate with the central committee and, unfortunately, sufficient money and manpower were not available to utilize efficiently the enormous quantity of data which were collected. Indeed, the only comprehensive study dealing with the 1910 apparition of Halley's comet appeared in 1931 — some 21 years after the comet's perihelion passage — and the most detailed series of photographs, taken at the Lick Observatory, could not be included.

For the 1985/86 return, several thousand professional astronomers from more than 50 countries and tens of thousands of amateur astronomers worldwide, all coordinated by a body called the International Halley Watch (IHW), combined their efforts to monitor the comet's activity almost continuously at a wide range of wavelengths. The activities of the nations involved in spacecraft studies of the comet were coordinated by a separate body called the Inter-Agency Consultative Group (IACG). It is a testimony to the success of this organizational structure that such a wealth of scientific data has been obtained, collated and archived from the comet's most recent apparition.

As usual, when an important, worldwide scientific investigation, such as that of Halley's comet, is carried out, major conferences and symposia around the world provide a forum for the many scientists involved to present their results, identify areas of agreement and conflicting opinion, and discuss the conclusions which can be drawn from them. Conference proceedings containing the multitude of contributed oral and poster papers presented at these meetings are one way in which this information is published and disseminated. Indeed, several of these have already appeared, the most important being those from the conferences held in Heidelberg (October 1986), Brussels (April 1987) and Bamberg, FRG (April 1989), Discussion of observations of Halley's comet was also a major item on the agenda at the 20th General Assembly of the International Astronomical Union (IAU), held in Baltimore, USA in August 1988.

However, with an observational programme as complex, diverse and of such importance as that conducted during the 1985/86 return of Halley's comet, it was felt that something more than just conference publications and papers in the learned journals was needed to review the wealth of data obtained, and summarize in a shortened, more manageable form the most important results of the research carried out. It is with this in mind that the present work has been produced.

Comet Halley — Investigations, Results, Interpretations, has been published in two volumes, each dealing with three specific aspects of cometary research conducted during the recent Halley apparition. Volume 1 deals with Organization, Plasma and Gas and Volume 2 with Dust, Nucleus and Evolution. Here, the wide range of investigative techniques involved in the comprehensive studies of Halley's comet and the results obtained are carefully reviewed and discussed by 54 of the world's top cometary scientists, representing 17 different countries. This is not a collection of scientific papers, but a selection of review articles, each constituting a separate chapter in one of the two volumes. In general, the reviews contained in the present work are based on the published material available prior to October 1989 and include relevant results from IAU Colloqium No. 116, 'Comets in the Post-Halley Era', held in Bamberg, FRG during April 1989.

Although every effort has been made to minimize the degree of overlap between the various chapters in each section, some repetition has been unavoidable. This is particularly so where investigators involved in each of the five principal spacecraft missions, Giotto, Sakigake, Suisei, Vega-1 and Vega-2 have reviewed the results obtained by the instruments on that spacecraft within a particular subject area and compared them with those obtained elsewhere.

Finally a note on consistency. I have attempted to impose some uniformity of style between the chapters, while retaining the flavour of the original contributions. The meanings of all symbols are explained in the chapters as they appear, and, although some standard symbols appear throughout, there are minor differences in style from one chapter to another. No attempt has been made to settle differences of opinion between expert contributors, when to do so would be to fly in the face of common sense. Such differences demonstrate only too well that even after an investigation as intensive as that conducted during the recent apparition of comet Halley, a great many questions remain unanswered. Some of these may remain so until the comet next returns in 2060/61. Other problems may be resolved sooner, when planned future spacecraft missions to comets take place.

This project would not have succeeded without the enormous help that I received from the contributing authors. The task of drawing together all the material for the two volumes took considerably longer than originally anticipated, and I would like to thank all of the contributors for their patience and enthusiasm, and for responding promptly and helpfully to requests for information.

I am also pleased to acknowledge the assistance provided by the staff of Ellis Horwood; special thanks are due to Felicity Horwood for her guidance and encouragement, and to Beverley Ford for dealing with correspondence and secretarial matters.

John Mason

Chichester, England, 1990

Editor's Introduction

Our knowledge of the underlying physical nature of comets took a giant leap forward in March 1986 when the Vega and Giotto spacecraft imaged the nucleus of Halley's comet from close range. This was the first time that conclusive proof had been obtained of Whipple's dust and ice-ball theory for the cometary nucleus, first proposed in the 1950s. Great advances have been made in our understanding of the surface activity of the nucleus, which was irregular in shape as expected. The icy surface was covered in a dusty crust, which was extremely dark and warm: Vega measurements indicated a surface temperature in the range 320-400 K. Only a minor part of the illuminated surface of the nucleus was active. Dust emission from the surface was structured in large jets, some of which could be followed out to distances of several hundred kilometres from the surface. New data was also obtained on the dust particles, including their masses, spatial distribution and elemental abundance. All these results will be of significance in studies of the rate of mass loss by periodic comets, their long term activity, ageing processes and eventual decay.

This book is divided into three sections, entitled Dust, Nucleus and Evolution and orbital motion. Part I, Dust, is the largest section and contains 10 chapters altogether. In chapter 1, P.L. Lamy discusses the detailed structure of comet Halley's dust tail, including streamers, anti-tails, spikes, and striae. The observational circumstances of the dust tail during the 1985/86 apparition are shown, and examples of dust tail features observed between February and May 1986 are presented and discussed. In chapter 2, J.A.M. McDonnell and G.S. Pankiewicz describe the data obtained on the cometary dust environment by the dust impact detectors carried aboard the Vega-1, Vega-2 and Giotto spacecraft. The flux of dust particles with masses in the range $10^{-20} < m < 10^{-3}$ kg measured by the three spacecraft, as a function of cometocentric distance, is presented and two models for dust particle emission from the nucleus are compared. Chapter 3, by O.L. Vaisberg considers the main structural features of the dust coma of comet Halley, the motion and distribution of dust particles, and discusses some of the processes controlling the dust distribution, using results obtained by the SP-1 plasma impact detector on the Vega spacecraft. In chapter 4, S.M. Larson shows how dust jets may be studied by careful enhancement of ground-based images, and gives examples of coma jet features seen during the 1910 and 1985/86 apparitions of comet Halley, demonstrating how these may be used to infer the rotation period of the nucleus. Infrared studies of Halley's comet are reviewed by J.C. Zarnecki in chapter 5. These are particularly useful for the investigation of scattered sunlight and thermal emission from dust grains. Spectroscopic infrared observations may prove to be most fruitful in determining the nature of the cometary dust. The significance of various spectral features at wavelengths between 1.4 and 20 microns is discussed, together with a few at longer wavelengths. Infrared imaging results are also briefly described. L. Massonne describes the coma model, a method used for computing the spatial density of cometary dust at all locations in the coma, in chapter 6. A comparison of the results of the coma model with *in situ* measurements by spacecraft can be used to improve the model parameters. In chapter 7, J.-i. Watanabe shows how quantitative analysis of coma morphology has been carried out by applying the moments method. Contour maps of near nucleus CCD images of comet Halley have been analyzed to obtain information on asymmetric mass ejection from the nucleus. Chapter 8 by D.E. Brownlee and J. Kissel discusses the dust particle composition data obtained by similar instruments flown on three spacecraft near comet Halley. The bulk elemental composition of the dust is consistent with chondritic material, except that Halley dust has a higher carbon and nitrogen content than any meteorite; it is also more heterogeneous than carbon-rich chondrites. Roughly a third of the micrometre and smaller particles are dominated by carbon and other low atomic weight elements. Some of the observations of Halley dust support Greenberg's model, described in chapter 9 by J.M. Greenberg himself. Here it is proposed that comets are mixtures of elongated interstellar grains, having a high carbon content, assembled in a 'bird-nest' structure. This possibility is discussed on the basis of evidence so far obtained, and the question of whether comet nuclei maintain their initial composition and structure or not is addressed. In the last chapter of the Dust section, chapter 10, B.A. McIntosh describes the structure and evolution of the comet Halley meteoroid stream, which gives rise to the Eta Aquarid and Orionid meteor showers every year. The release and dispersion of dust particles around the comet's orbit, and the long-term effects of various perturbing influences on the orbits of the stream particles are discussed. A shell structure

for the stream is proposed, and estimates of the total mass and age of the stream are derived.

Part II, Nucleus, contains five chapters, beginning with a review by H.U. Keller, chapter 11, in which results obtained from high resolution images of the dust jets and the nucleus by Giotto's Halley Multicolour Camera are presented. A detailed discussion of the nucleus shape, dimensions, surface features, albedo, surface activity and density is included. In the following chapter. R.Z. Sagdeev and K. Szegö present the results obtained from analysis of images of the near-nuclear region of the comet acquired by Vega-1 and Vega-2. Measurements of the nucleus albedo, surface temperature, variability of the dust jets on a time-scale of days and dust composition are described. The most challenging task has been the three-dimensional reconstruction of the nucleus shape, its orientation in space and the distribution of dust jets on the surface from the Vega imaging data. In chapter 13, H. Rickman shows how the mass of the nucleus of comet Halley may be estimated from nongravitational effects on its orbital motion. The theoretical basis and observational material used for such mass determinations are reviewed and the probable uncertainties assessed. The most likely range for the mean density of the nucleus is given. Chapter 14 by H.L.F. Houpis reviews the many techniques employed for modelling the internal structure of the cometary nucleus. The relative merits and deficiencies of the various models are assessed in the light of the major results from the comet Halley spacecraft encounters. Chapter 15, is a review by Z. Sekanina of the very complex problem of establishing the rotation vector of Halley's nucleus. The interaction between activity of a comet and its rotation, techniques for determining the spin vector of comets, and the 2.2-day and 7.4-day periodicities noted during the 1985/86 apparition of comet Halley are all discussed. The results of various investigations are presented, but no specific model is widely accepted and the true character of the rotational motion is uncertain.

Part III, Evolution and orbital motion, comprises six chapters. In chapter 16, F.R. Stephenson discusses how early returns of Halley's comet may be identified by careful scrutiny and interpretation of ancient astronomical records. These data have been extremely useful in establishing probable dates of perihelion passage for the comet as far back as 240 BC. In chapter 17, L.G. Sitarski shows how nongravitational forces play an important rôle in the orbital evolution of comets. Studies of variations in the nongravitational parameters may be used to infer changes in the activity of the nucleus or of precessional motion of the spin axis. In chapter 18, D.K. Yeomans presents the results of integrating the equations of motion of comet Halley backwards in time from 2134 AD to 466 BC, and questions whether the comet's motion can be reasonably extrapolated more than a few perihelion passages beyond the period for which observational data is available. In chapter 19, K. Ziolkowski examines old and present-day methods of calculating the long-term motion of comet Halley, and compares the results of such investigations, assessing the accuracy of calculations of the comet's perihelion passage data by various techniques. The final two chapters consider the long-term evolution of cometary nuclei from their formation to eventual decay. In chapter 20, P.R. Weissman discusses cometary formation, storage in the Oort Cloud, their return to the planetary region and the dynamical evolution of their orbits from long- to short-

period. The degree of processing of the cometary material which is likely to have occurred at or near the nucleus surface as a consequence of these processes is assessed. Finally, chapter 21 by L. Kresak looks at the ageing processes occurring in periodic comets. The problems of estimation of the secular brightness decrease, the mass loss per revolution and consequent survival time of comets are discussed. The active lifetime of a periodic comet may be only a few hundred revolutions, but with intermittent dormant phases their total lifetime could be much longer.

Because cometary nuclei may contain the most pristine, primitive material in the Solar System, this has proved to be a strong motivation for spacecraft missions to comets. However, the close-range observations of Halley's nucleus in March 1986 revealed not the sublimating surface itself but a dark, dusty crust that largely insulates the ices lying underneath. Similarly, little knowledge has been gained about a comet's internal physical structure. Some evidence provided by the comet Halley measurements supports the view that cometary material consists, at least in part, of interstellar grains, although a number of unsolved problems remain. We may have to wait for future space missions which can either penetrate the dark insulating crust and reveal what lies beneath, or recover samples of cometary material more pristine than we can infer from flyby or rendezvous missions. In this respect NASA's Comet Rendezvous and Asteroid Flyby (CRAF) mission and ESA's Comet Nucleus Sample Return (CNSR) mission — now renamed 'Rosetta' — are eagerly awaited.

Part I
Dust

1

The dust tail of comet Halley

P. L. Lamy

1 FROM EARLY AGES TO THE PRESENT DAY: AN HISTORICAL PERSPECTIVE OF COMETARY DUST TAILS

If comets, and in particular their everchanging tails, have been a source of fascination for man, their curiosity did not go beyond a mere classification according to their forms (swords . . .) until the 15th century. It seems that Frascator and Apianus were the first, almost simultaneously and independently in the years 1530–1540, to characterize the fundamental behaviour of cometary tails, namely that their direction is opposite to that of the Sun. Festou *et al.* (1985) say that a remark in the annals of the T'ang dynasty in the AD 837 apparition of comet Halley, stating that when a comet rises in the morning, its tail extends toward the west and vice versa, and that is a constant rule, reveals that the Chinese had noticed earlier a property of the tails. The 16th century also witnessed the first explanations of tails, and Cardan proposed an optical theory, according to which they were produced by refraction of solar light off the head of the comet, indeed satisfying the anti-solar orientation. Kepler (1571–1630) must probably be credited with the concept of a repulsive force due to solar light, explaining not only the general direction of the tails but also their curvature and their concavity directed toward the place the comet has left. Newton realized that a 'mechanical theory' describing the motion of particles ejected from the head and accelerated in the anti-solar direction was the most convincing explanation. This idea was pursued by Olbers, who formulated such a theory based on a repulsive force inversely proportional to the square of the heliocentric distance (r^{-2}) which he thought was electrical. After having observed the 1835 apparition of comet Halley, Bessel (1836) worked out the analytical development in an effort to explain the form of its tail. He went one step further than Olbers, introducing two competing forces varying as r^{-2}, a repulsive one (as Olbers did) and an attractive one, the solar gravity. He attributed the origin of the repulsive force to the surrounding ether. Bredikkin (1831–1904; his work was published by Jaegermann 1903) improved the work of Bessel and introduced the key concepts of synchrones and syndynes (or syndynames) which are fundamental, and now commonly used, in the understanding of cometary dust tails. Two parameters are needed to properly define them, namely the ratio β of the radiation pressure force to the solar gravitational force for a given grain (the notation $1-\mu$ has often been used in the past) and the time τ elapsed between its ejection and observation. A synchrone is the geometrical locus of the grains ejected from the nucleus at the same time and having any value of β, while a syndyne is the locus of the grains leaving the nucleus continuously and having a particular value of β. Bredikkin classified cometary tails in three groups according to their curvature. The straight ones were of type I and are now identified with the plasma tails.

Type II tails had intermediate curvature and little structure, and type III had larger curvatures and exhibited structures known as streamers, which Bredikkin incorrectly identified with syndynes. He further linked this geometrical classification with a chemical composition ranging from hydrogen (type I) to hydrocarbon (II) and metallic vapours (III). The evidence for the presence of dust in comets came from spectroscopic observations performed after 1860, and Arrhénius in 1900 proposed that tails are indeed made of dust particles repelled by radiation pressure exerted by the solar light. The second misinterpretation made by Bredikkin was corrected by Moiseyev (1925), who identified streamers with synchrones (and not syndynes) based on his analysis of the tail of Comet 1901 I. This has been largely confirmed by the study of the subsequent bright comets, the streamers being associated with short periods of enhanced dust ejection, i.e., outbursts. It was not until 1966 that the next breakthrough came with the dynamical/photometric model devised by Finson & Probstein (1966, 1968) who linked the mechanical theory with the calculation of the spatial density of dust and the reflected light, providing for the first time a powerful tool for the quantitative study of dust tails. They treated the motion of dust particles as a hypersonic, collision-free source flow and expressed the intensity of light scattered by the dust in terms of three parametric functions: the size–density distribution, the time-dependent dust production rate, and the time-(via the properties of the gas) and size-dependent dust ejection velocity. These functions are determined by fitting the calculated surface-density distribution with the observed isophotes of the tail by trial and error. The method has proven itself extraordinarily powerful, and has been successfully applied to several comets, e.g., Arend–Roland 1957 III, Bennett 1970 II, Seki–Lines 1962 III. In fact, the predicted size distribution function used for the early models of Comet Halley were derived from these results to a large extent. However, the Finson–Probstein model presents some limitations inherent to its approximations. The ejection velocity of the dust which is the terminal velocity for the interaction with the expanding gas in the coma is not rigorously introduced in the theory. Its effect is simulated by uniformly expanding, spherical shells with the same velocity, and the so-called

'hypersonic approximation' further requires that this expansion is slow compared to the motion of the dust in the tail. Avoiding further technical details and summarizing, the method does not apply near the coma (i.e. inner part of tails), for large grains, (present in anti-tails), or for outbursts. This prompted Kimura & Liu (1977) to undertake a general three-dimensional treatment where the trajectory of each grain is rigorously calculated, using Kepler's equations with the appropriate initial velocity which is the terminal velocity already mentioned. By considering a large number of dust particles properly sampling the required parametric functions (e.g., dust production rate, size distribution, spatial direction of the initial velocity . . .), these authors were able to generate brightness maps to be compared with observed isophotes in the case of Arend–Roland 1957 III, Bennett 1970 II, Kohoutek 1973 XII, and IRAS–Araki–Alcock 1983 VII. Perhaps the most interesting aspect of the Kimura–Liu treatment is the prediction of the formation of a new structure in cometary tails, which they called 'neck-line', appearing as a sharp spike, either solar or anti-solar, when the Earth is near the orbital plane of a comet. This structure has very likely been detected in Comets Arend–Roland 1957 III, Bennett 1970 II, and Halley in its 1986 apparition.

2 FROM STREAMERS TO STRIAE: A PERSPECTIVE ON STRUCTURES IN COMETARY TAILS

Plasma and dust tails are often distinguished and opposed by their structures, the latter being usually qualified as structureless. It may often be the case, but dust tails do display a wealth of structures, and the present apparition of Comet Halley certainly confirms this. Beside their aesthetic value, structures are very informative on the physical properties of dust in comets. A brief description of the various structures is now offered to clarify a sometimes confused situation, and to help understanding of those present in Comet Halley.

2.1 Streamers

Streamers appear as discrete reinforcements in or outside the main body of a dust tail; they originate from the nucleus and follow very well theoretical synchrones. They are correctly associated with short periods of enhanced dust ejection from the nucleus, either outbursts or explosive phenomena such as splitting of the nucleus (e.g., Comet West 1976 VI). Streamers have been displayed by a relatively large number of comets.

2.2 Anti-tail (or anomalous tail)

Although the anti-solar direction of dust tails has been constantly emphasized, it happens occasionally that a comet displays a tail orientated in the opposite direction, named for that reason an anti-tail. These structures are perfectly understandable in the synchrones–syndynes framework, and result from grains subjected to very low radiation pressure forces (practically 'large' submillimetre grains). Consequently, their trajectories are strongly curved, and considerably separated from the anti-solar direction. A simple perspective effect allows these grains to be seen, in projection onto the plane of the sky, 'trailing' the coma, that is, in the solar direction. There is a tendency for the old synchrones to pile up toward the earliest ejection times, resulting in a sharp edge or even an apparent reinforcement (particularly, if the Earth is close to the orbital plane of the comet) sometimes called a spike (we shall, however, reserve this name for a different structure). Here again, this is entirely due to the geometry of projection; there is a continuity from this sharp edge, through the fuzzy anti-tail to the normal tail.

Since an unusual geometry is required, the anti-tail phenomenon is quite rare. Comets Arend–Roland 1957 III (Fig. 1), Tago–Sato–Kosaka 1969 IX, and Kohoutek 1973 XII are among the best recent

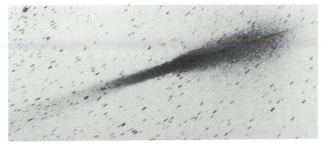

Fig. 1 — Comet Arend–Roland 1957 III photograph by R. Forgelquist (Uppsala) on 25 April 1957 when the Earth crossed the cometary orbital plane. Three phenomena are well illustrated, the tail, the anti-tail, and superimposed on it, the sunward spike (reproduced from *Sky and Telescope* **47** 218, 1974).

examples. Periodic comets do also occasionally exhibit anti-tails, and this has been used by Sekanina & Schuster (1978a, b) in the case of P/d'Arrest 1976 XI and P/Encke 1977 XI, to investigate their large (larger than 10 μm) particles component.

2.3 Spikes

When the motion of dust grains is considered in three-dimensional space, it is a well-known fact of celestial mechanics that a particular grain, ejected at the anomaly f, will again cross the orbital plane of the nucleus at the anomaly $(f + \pi,)$i.e., at the second node of its orbit with respect to that of the nucleus. In their general treatment, Kimura & Liu (1977) observed that a dust shell formed preperihelion will collapse into a flat ellipse postperihelion, at the second node. The size distribution of particles leads to a set of overlapping ellipses whose distance to the nucleus depends upon the radiation pressure force. Large grains, which are subjected to a weak force, form an ellipse approximately centred on the nucleus, half of it toward the Sun, the other half away from the Sun. Smaller grains subjected to this force form ellipses further away, in the anti-solar direction. Kimura & Liu used the term 'neck-line' for the locus of the centres of all the ellipses. Now, when the Earth is close to the comet orbital plane, the set of these ellipses will be seen edge-on, sometimes gaining enough contrast to be seen against the surrounding tail or anti-tail. This results in a sharp spike appearing either in the solar or the anti-solar directions or both (e.g., Bennett 1970 II; see Pansecchi *et al.* 1987). I propose to use the following terminology for the manifestation of this common neck-line, 'sunward spike' and 'anti-solar spike'. As a result of the above mechanism of formation, the sunward spike is shorter than the anti-solar one, and is composed of large, submillimetric particles.

2.4 Striae

At first glance, striae may appear like streamers and have often been misinterpreted as such. They are systems of discrete bands or rays whose direction distinctly differs from synchrones and which do not converge to the nucleus; they usually do not last very long, a fact which renders their observation quite difficult. They have been positively identified as such

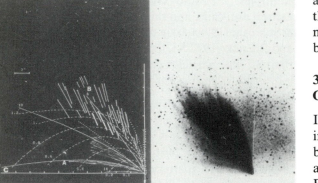

Fig. 2 — Comet West 1976 VI offers a perfect example of a striated tail. Several striae are reproduced on a synchrone–syndynes graph (from Lamy & Koutchmy 1979).

in five comets. Comet West 1976 VI represents by far the best documented example (Fig. 2) and has been studied in detail by Koutchmy & Lamy (1976), Lamy & Kóutchmy (1979), and Sekanina & Farrell (1979, 1980). While the two pairs of authors agree on the basic characteristics of the striae, Lamy and Koutchmy advocate an electromagnetic interaction

and Sekanina & Farrell a fragmentation of grains in the tail. Probably, the question will not be settled until new comets exhibit striae which, we must hope, will be well observed.

3 A GOOD PERSPECTIVE ON THE DUST TAIL OF COMET HALLEY

If tails are intrinsic cometary phenomena, it is important to realize that our perception of them is biased, even spoiled, by the way we actually see them as projected on the plane of sky, presently from the Earth. Indeed, the geometric conditions of projection have already been emphasized, being responsible for the very appearance of a particular tail. A key parameter is the position of the Earth with respect to the orbital plane of the comet, as the tail mostly spreads over it. The cometocentric latitude of the Earth, that is the angle between the comet–Earth vector and the comet orbital plane, gives a very good feeling of how a tail is seen. It is plotted as a function of time in Fig. 3 as well as the phase angle (i.e., the Sun–nucleus–Earth angle) which plays a role in

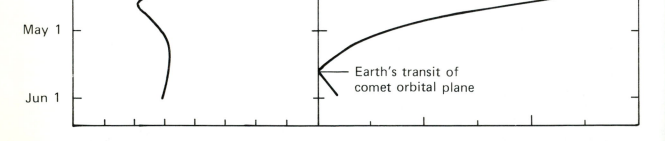

Fig. 3 — The cometocentric latitude of the Earth and the phase angle of the nucleus of Comet Halley as functions of calendar date.

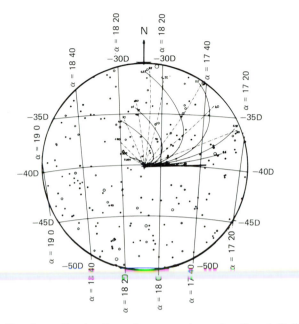

Fig. 5 — Graph of synchrones–syndynes for Comet Halley superimposed on the sky field for 2 April 1986. This is an example of a catalogue of daily maps (prepared by P. Malburet and Y. Valerio, Laboratoire d'Astronomie Spatiale).

connection with the photopolarimetric properties of the dust particles. The latitude was about 12° when the first postperihelion observations started, increasing steadily to reach a maximum of 28° in early April 1986, then decreasing rapidly to zero on 18 May when the Earth crossed the orbital plane of the comet, resulting in an edge-on view of the tail. The phase angle followed approximately the same trend, reaching a maximum of 66°.2 on 20 March, then decreasing to a minimum of 20°.9 on 18 April and slowly increasing again. This is only one part of the story. The appearance of a tail is also a question of contrast above the local background. This involves both the intrinsic brightness of the tail (via the level of dust production) and the brightness of the background: atmospheric (twilight), interplanetary (Zodiacal Light), galactic (Milky Way), and, not to be forgotten, moonlight. The Moon came close to the comet, within a few degrees on 7 March and within 11° on 1 April. After passage at perihelion, twilight is always a problem, and it limits an early detection of the tail; in the case of Comet Halley, the first photographs showing the dust tail were obtained on 19 February, some 10 days after the comet passed its perihelion and

being separated from the Sun by about 35°. Granting that it is impossible to avoid the twilight and the Moon, the observing conditions for the dust tail of Comet Halley were quite favourable, the main interference coming from the Milky Way. In the synoptic view (Fig. 4) covering the period 1 March to 30 April 1986, the tail of Comet Halley is sketched, based on the actual photographic observations and the drawing of a few synchrones. Incidentally, daily synchrone–syndyne maps were calculated in advance and superimposed on the appropriate stellar fields (an example is given in Fig. 5). The resulting catalogue was extremely useful during the observations, for instance for correctly centring the field and orientating the polarizer. The comet spent the first half of April 1986 in the southern Milky Way, crossing the projected galactic plane on 6 April. The damaging effect for the tail is dramatically illustrated in a photograph taken on April 7.3. (Fig. 6): the diffuse clouds interfere with the tail, distorting it and burying its faint external part while the incredibly rich stellar field complicates the analysis and the photometry.

4 THE DUST TAIL OF COMET HALLEY IN 1986

If the comet displayed a modest plasma tail in its preperihelion approach, no large-scale manifestation of the dust was displayed as expected. But as soon as it

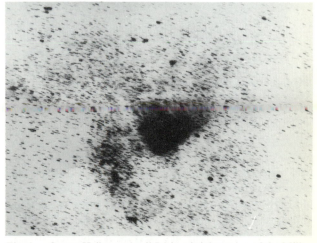

Fig. 6 — Comet Halley on April 7.3 in a bright region of the Milky Way (S. Koutchmy and R. Verseau, Institut d'Astrophysique de Paris). The field is 7°.6 × 11°.5 ; north is up.

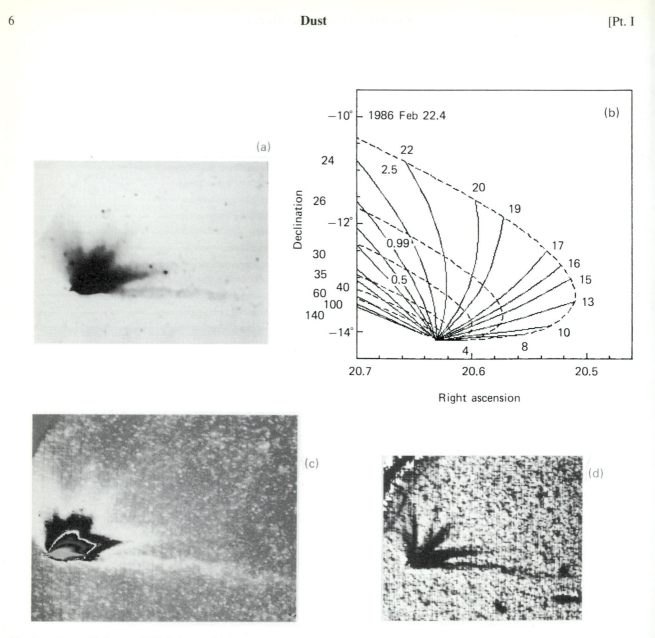

Fig. 7 — Comet Halley on 1986 February 22.4,
(a) an image obtained with the ESO-Wide Field CCD Camera (H. Pedersen) and at the same scale and orientation, (b) the synchrones-syndynes graph (P. Malburet). The false colour isophotes (c)*, * to be found in the colour illustration section in the centre of the book. and the processed image (b) are not strictly to the same scale.

was recovered after perihelion passage it did reward the observers by a fantastic show, a broad fan-shaped tail highly structured by bright streamers 1° to 2° long and spread over a sector angle of more than 120°! Dramatic illustrations are offered by the two images obtained on February 22.4 in Chile (Fig. 7)* and on February 22.8, some ten hours later, in Australia (Fig.

8). The pattern persisted during the first part of March—the photograph in Fig. 9 was obtained on March 8.4—until approximately the 10th, and faded rapidly; thereafter, the tail looked practically feature-less. On April 12.3, the Earth was close to its highest latitude with respect to the cometary orbital plane and at its minimum distance from the nucleus (0.42 AU): the dust tail appeared as a broad fan extending over a few degrees. From then on, its shape was controlled by

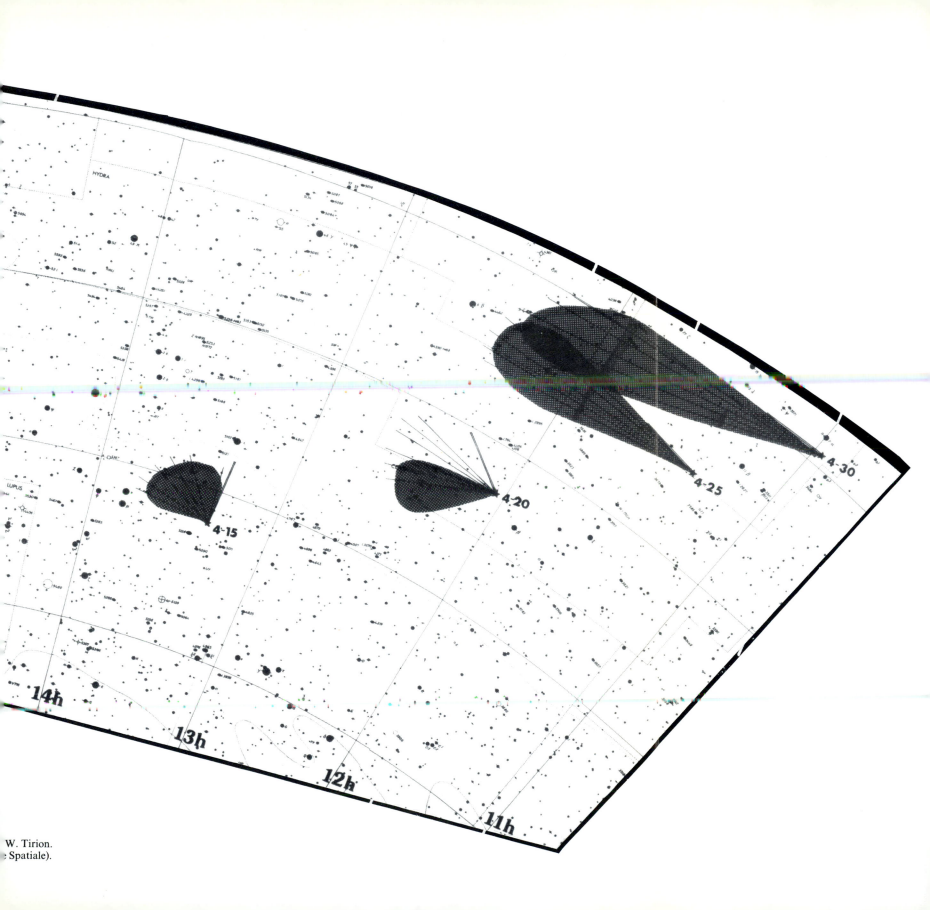

HYDRA

LUPUS

4-15

4-20

4-25

4-30

14h

13h

12h

11h

W. Tirion.
Spatiale).

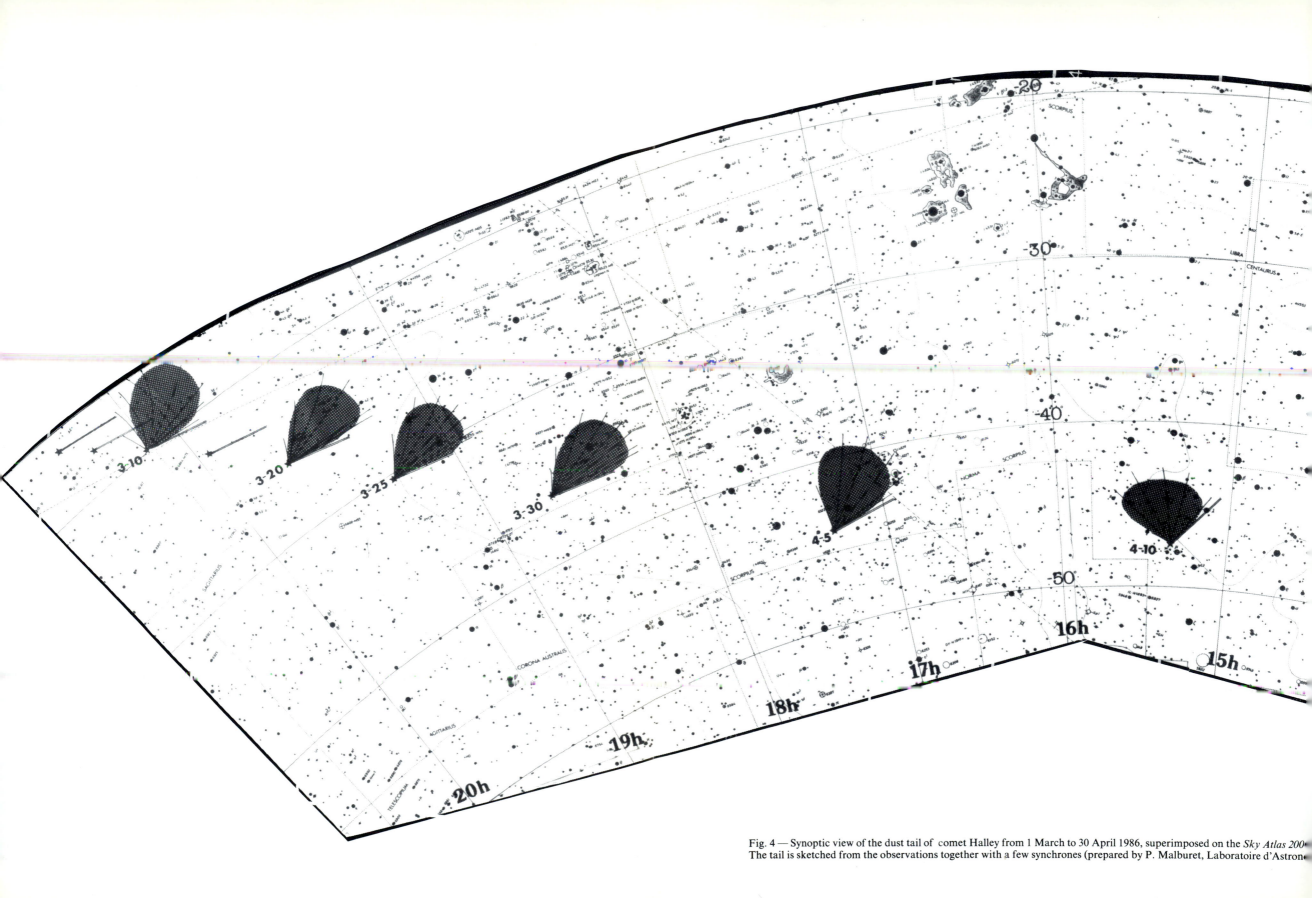

Fig. 4 — Synoptic view of the dust tail of comet Halley from 1 March to 30 April 1986, superimposed on the *Sky Atlas 200*
The tail is sketched from the observations together with a few synchrones (prepared by P. Malburet, Laboratoire d'Astron

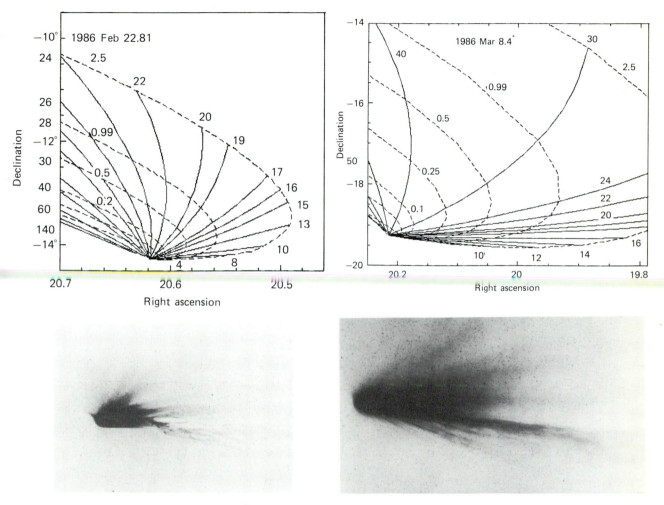

Fig. 8 — Comet Halley on 1986 February 22.81, : a photograph obtained with the UK Schmidt telescope in Australia (K.S. Russel), and the graph of the synchrones-syndynes at the same scale and orientation.

Fig. 9 — Comet Halley on 1986 March 8.4, : a photograph obtained with the ESO Schmidt telescope in Chile (H.E. Schuster), and the graph of the synchrones-syndynes at the same scale and orientation.

perspective effects as the Earth was approaching the orbital plane. This is clearly visible in the sequence of images of April 12.3 (Fig. 10), 18.3 (Fig. 11), 29 (Fig. 12) and May 10 (Fig. 13) as the sector angle decreased steadily from 80° to 5° approximately. The resulting edge-on effect led to a considerable enhancement of the tail and its increasing apparent length projected onto the sky (the decreasing background as the comet moved away from the Milky Way also played a role in increasing the contrast). On Fig. 13, the tail extends clearly over the 13° allowed by the field of the instrument, suggesting a length of the order of 40 million kilometres. This is purely indicative, as the numerous biases involved really preclude a meaningful value of the tail length and its variation with time.

The graphs of the syndynes which are displayed along with the images (Figs 7 to 13) already allow some insight into the properties of the dust grains by analysing the values of β, the ratio of the radiation pressure force to the gravitational force. If β reaches about 2.5, one should notice that the bulk of the tail lies within the syndyne $\beta \simeq 1$. The values of β for various materials given by Burns et al. (1979) indicate a maximum value for silicates of 0.6. As there is

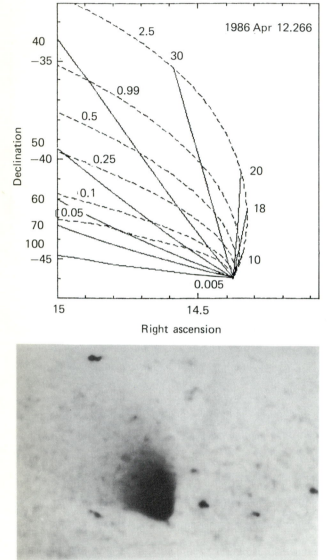

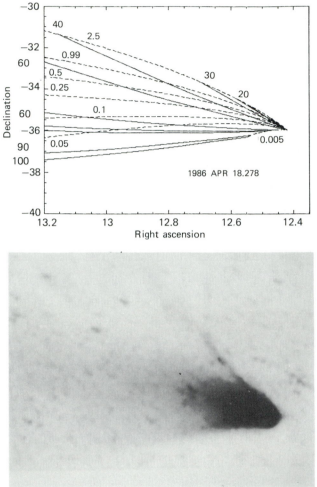

Fig. 11 — Comet Halley on 1986 April 18.278, : same as Fig. 10.

Fig. 10 — Comet Halley on 1986 April 12.266, : an image obtained with the ESO-Wide Field CCD Camera (H. Pedersen), and the graph of the synchrones–syndynes at the same scale and orientation (P. Malburet).

evidence that the density of cometary material is smaller than that of the bulk materials (the grains are porous), the above maximum should be increased, possibly up to 0.8–1. However, an additional absorbing material is required to explain the values of up to 2.5, for instance amorphous carbon or an absorbing

organic material. The qualitative picture which emerges from this straightforward analysis of the dust tail of a dominating silicate population and an additional population of an absorbing material is in agreement with the infrared observations.

4.1 The streamers

The graphs of the synchrones now allow study of the multi-tail structure present from 22 February to 10 May. For the purpose of comparison, each component is characterized by a profile of maximum brightness. Their breadth, which reflects the duration of the events as well as the dispersion in terminal velocities, obviously limits the accuracy of the method, as does

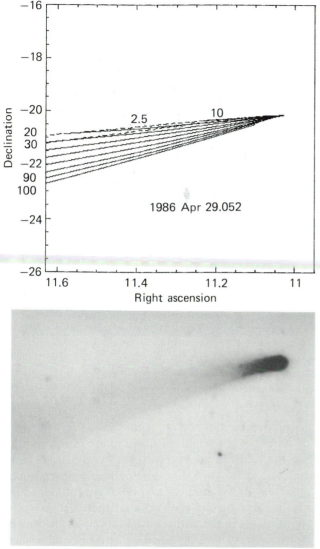

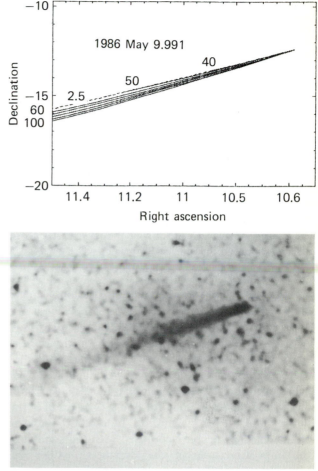

Fig. 13 — Comet Halley on 1986 May 9.991 : same as Fig. 10.

Fig. 12 — Comet Halley on 1986 April 29.052 : same as Fig. 10.

the presence of stars which distort the brightness distribution; finally the spatial resolution of the synchrones decreases with increasing age (they tend to pile up). Before attempting a general photometric modelling, an intermediate step is underway in which a first filtering removes the stars while the second enhances the structures; this is illustrated in Fig. 7 for the February 22.4 observation. The present result is, however, based on previous work (Lamy 1986). The various components are all of synchronic nature, i.e., they are true streamers, and therefore reflect the activity of the nucleus as a succession of outbursts and 'quiet' periods. The time of emission of each component is taken as the synchrone fitting the line of maximum brightness. The times expressed in days referred to perihelion for the observations of Feb. 22.4 and 22.8 and March 8.4 are given in Table 1. There is an overall excellent agreement. The component $T = -0.6$ day, difficult to distinguish on the Feb. 22.4 image, is supported by conspicuous and copious dust emission. The regularity of the pattern of the streamers suggests some periodic recurrence which is compared in Table 1 with one of the proposed rotation periods of the nucleus, namely 52 hours ($\simeq 2.2$ days). The coincidence is indeed striking, suggesting a strong (although not perfect as expected) modulation by the rotation or the precession of the nucleus.

Table 1. Times (in days referred to perihelion) of major dust outbursts. The column 'periodicity' give multiples of the roation period of the nucleus

Feb. 22.4	Feb. 22.81	Mar. 8.4	Periodicity
		+9	8.7
+2.5	+2.4	+2.5	2.2
(≃0.0)?	−0.6	0.0	0
−4.0	−4.1	−4.2	−4.3
−6.3	−6.6	−6.3	−6.5
−10.2	−10.1	−10.0	−10.8
−13	−13.1	−13.0	−13.0

Curiously enough, the other proposed period of 7.5 days does not appear in the present chronology of the streamers. But there is no doubt that the streamers are the manifestations in the tail of the jet activity from discrete sources on the rotating nucleus.

4.2 The anti-tail

The very broad angular extent of the tail conspicuous on the Feb. 22.4 and 22.8 images implies that the northernmost part must be viewed as a true anti-tail. As there is a perfect continuity between the normal tail and the anti-tail, the role of the geometry in creating such a structure is here well illustrated. The external edge is strongly reinforced and is conspicuous on the high-contrast image of Feb. 22.8, a result of the piling-up of the old synchrones. Indeed, although the old synchrones become confused with the syndynes, the impressive role of the streamers justifies the identification of the reinforcement with a synchrone timing the onset of substantial dust production. This corresponds to some 110 days before perihelion when the comet was at a heliocentric distance of 2 AU. Although the overall increase of the dust production was probably progressive, both the ultraviolet (Feldman et al. 1987) and the infrared (Tokunaga et al. 1986) observations indicate a substantial jump in the dust production at about the same distance of 2 AU. Now the corresponding syndyne yields values of β of the order of 0.005 indicating the presence of large, submillimetre grains. As noted by Sekanina (1986), we witnessed the formation of the meteor stream of Comet Halley.

4.3 The spikes

The observation of a structure during the period of late April, early May 1986 forming a sunward spike

Scale (10^6 km)

0   1   2

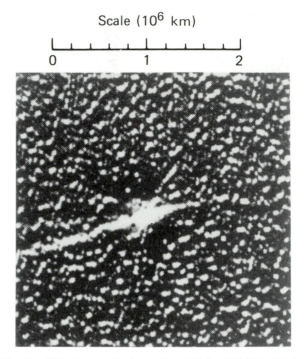

Fig. 14 — This photograph obtained on 1986 April 30.201, with the Schmidt telescope of the Barnard Observatory (G. Emerson) illustrates both the sunward (right extension) and anti-solar (left extension) spikes. North is up (reproduced from Sekanina et al. 1986).

oriented to within 20° of the solar direction and extending over a length of some 20 arc min from the nucleus adds an unexpected plus to an already rich dust tail. The April 30.2 photographic observation processed by Sekanina et al. (1986) is reproduced in Fig. 14. One further remarks that the spike extends considerably in the anti-solar direction, as illustrated by two observations obtained on 28 April and 9 May (Fig. 15) and processed by using Laplacian filtering. To study the sunward extension, Sekanina et al. (1986) have reduced eleven observations, derived its position angle, and tentatively explained it as a combination of an important production of very small, submicrometre grains from one or several areas in the equatorial zone of the nucleus and the subsequent transit of the Earth across this equatorial plane. As the dust ejecta develops in a flat sheet, the transit results in an edge-on view and therefore a brightness enhancement allowing detection above the local background. Very small, dielectric or slightly absorbing grains are advocated, since they feel almost no

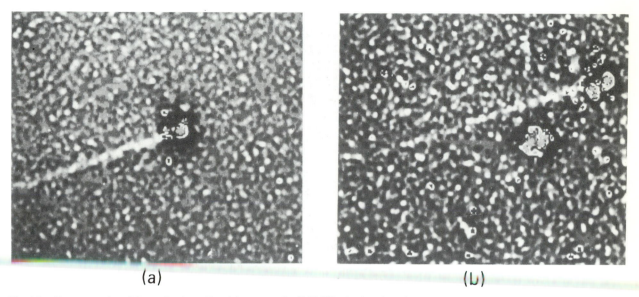

(a)   (b)

Fig. 15 — Two examples of the anti-solar spike: (a) on 1986 April 28.085, the field is 5.5° × 6.3°, (b) on 1986 May 4.183, the field is 11° × 12.6°. The two images were obtained with the ESO-Wide Field CCD Camera (H. Pedersen). North is up.

radiation pressure while they can reach high terminal velocities (> 0.5 km s^{-1}) through the interaction with the expanding gas in the solar direction. These authors calculated that the emission started in mid-April to reach the projected length of some 700 000 km in late April. Incidentally, the time of the Earth's transit across the equatorial plane yielded an independent determination of the orientation of the spin axis of the nucleus. Recently, Fulle (1987) produced a radically different interpretation in terms of a neck-line structure resulting in a sunward spike as defined here in section 2.3. This is indeed supported by the fact that the Earth was getting close to the orbital plane of the comet at the end of April. The calculated position angles are also in good agreement with those derived by Sekanina *et al.* (1986) from the observations. Fulle noted that the ejection velocities required to explain the length of the sunward spike, of the order of 0.05 km s^{-1}, are appropriate to millimetric grains which are indeed insensitive to the radiation pressure, and that the ejection took place between 5 and 20 January 1986. Fulle's interpretation presents two apparent advantages. First, it does not require a new special mechanism since it is based on a well understood dynamical behaviour; second, the neck-line structure is able to simultaneously explain the anti-solar spike observed at the same time (Fig. 15).

However, the low surface brightness of the sunward spike makes the measurement of its direction and its length a very difficult task, not to mention its photometry. More work is clearly required before a convincing interpretation of the two spikes is reached.

5 PHOTOPOLARIMETRIC PROPERTIES OF THE TAIL AND ITS MODELLING

The photopolarimetric analysis of the images of the dust tail represents a formidable task, and only limited results on a few selected images have been obtained so far (Lamy *et al.* 1987b). They are based on images obtained with the ESO wide-field CCD camera on April 6.3 and 11.3 with the Johnson B, V and R filters. Absolute photometry, colour, and polarization were studied in a band 15 arc min wide, centred onto the coma and extending across the dust tail in the south-north direction to a few degrees. The colour index C(R/B), obtained by comparing the ratio of the blue and red images with that of the Sun, is found to be larger than one indicating a tail redder than the Sun. Such a reddening effect is quite common in comets, and was indeed observed in Comet West 1976 VI. In Comet Halley, the reddening is further found to increase with increasing distances from the nucleus,

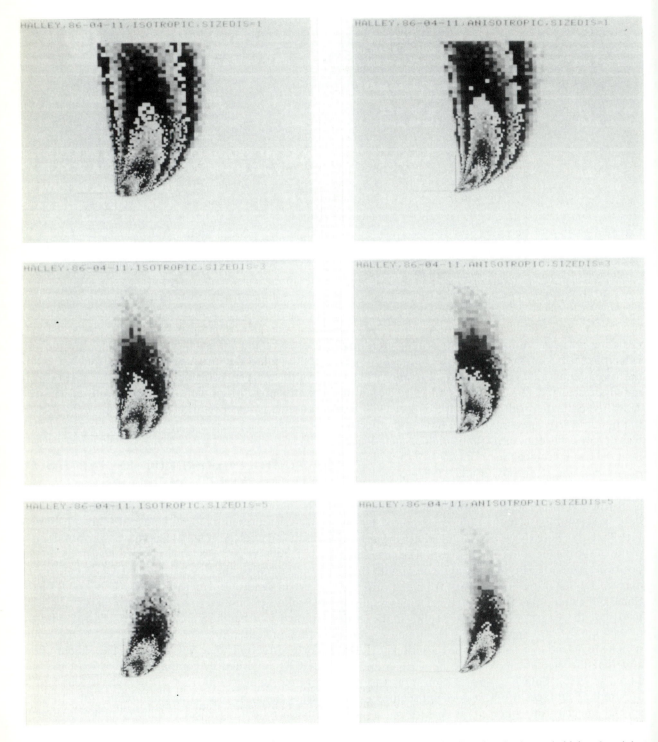

Fig. 16 — Numerical models of the dust tail of Comet Halley on 11 April 1986 for isotropic (left column) and anisotropic (right column) dust emissions and various size distributions denoted 1, 3 and 5 (prepared by K. Jockers, Max-Planck Institut für Aeronomie).

an effect which probably results from the spatial segregation of the grains as a function of their physical properties. The polarization in the R filter amounts to 0.04 at a phase angle of 31°, and to 0.075 at a phase angle of 45°.5. Future processing of other images will indicate if the decreasing branch of polarization with decreasing phase angle implies negative polarizations below 20° as found in the coma. The inferred presence of large rough grains there (Lamy *et al.* 1987a) would then be extended into the tail, a most interesting implication in terms of the physical properties of the grains, in particular their density. The polarization is further found to slightly increase with distance from the nucleus, but to be independent of wavelength in the B-V-R spectral domain.

The broadside view of the tail from February to April, never revealed during the 1910 apparition, offers a great potential for two-dimensional photometric modelling of its intensity distribution and variations with time. The presence of outbursts, of spikes and the marked anisotropic pattern of the dust emission from the nucleus (i.e., the sunward jets observed by the Vega and Giotto space probes) render the Finson & Probstein method inappropriate. The method developed by Kimura & Liu (1977) is therefore preferred; being purely numerical, it should handle the complex case of the dust tail of Comet Halley. Examples given for 11 April 1986 (Fig. 16) are intended to test the role of, first, the anisotropic versus the isotropic dust emission—the anisotropic ejection is restricted to a cone globally representing the sunward jets, and, second, the role of various size distributions. Both parameters appreciably influence the shape of the tail, but conclusions at this early stage are premature. Further work is required to fully understand this question and to derive the parameters characterizing the dust and its emission.

6 CONCLUSION: A FASCINATING TAIL INDEED!

The fine colour picture reproduced on the front cover was obtained by W. Liller on March 13.0, 1986 and is our final and spectacular illustration of the tails of Comet Halley: well separated from the blue (CO+ dominated), narrow plasma tail, the dust tail forms a broad, whitish fan.

Except for the striae, the dust tail of Comet Halley has displayed all possible structures observed so far in comets: streamers, spikes, and an anti-tail, offering an enormous potential for original and correlative studies of the dust, its physical properties, its size distribution, and the variation of its production rate. Indeed, the images obtained from February to May 1986 give a full display of the history of the dust production from 2 AU preperihelion to 1.3 AU postperihelion. The anti-tail and possibly the spikes give information on the large, submillimetric grains, while the edge-on view of the tail gives information on the terminal velocities ... . It remains to be seen if the information is compatible with that obtained by other means. Two outstanding issues clearly need further investigation: the relative importance of anisotropic (jets) versus isotropic dust emission in controlling the form of the tail, and the compatibility of the recurrent streamers with a rotation period.

ACKNOWLEDGEMENTS

I am much indebted to H. Pedersen (European Southern Observatory) for access not only to the wide-field CCD camera (WFCC) images but also to the instrument to carry out colour and polarization observations. My sincere thanks go to P. Malburet and Y. Valério (Laboratoire d'Astronomie Spatiale) for preparing the synchrone/syndyne graphs and the sky maps, to A. Llebaria (Laboratoire d'Astronomie Spatiale) for assistance in the image analysis, and to K. Jockers (Max Planck Institut für Aeronomie) for the photometric models.

REFERENCES

Bessel, W. (1836) *Astron. Nachr.* **13** 185

Burns, J.A., Lamy, P.L., & Soter, S. (1979) *Icarus,* **40** 1

Feldman, P.D., *et al.* (1987) *Astron. Astrophys.* **187** 325

Festou, M., Véron, P., & Ribes, J.C. (1985) *Les comètes, mythes et réalités,* Flammarion

Finson, M.L., & Probstein, R.F. (1966) AIAA Paper No. 66–32. (Presented at the *3rd Aerospace Sciences Meeting,* Jan. 24–26, 1966, New York)

Finson, M.L., & Probstein, R.F. (1968) *Astrophys. J.* **154** 327

Fulle, M. (1987) *Astron. Astrophys.* **181** L13

Jaegermann, R. (1903) *Prof. Dr. Th. Bredichins mechanische Untersuchungen über Komentenformen in systematischer Darnstellung,* St Petersburg

Kimura, H., & Liu, C.P. (1977) *Chin. Astr.* **1** 235

Koutchmy, S., & Lamy, P.L. (1978) *Nature,* **273** 522

Lamy, P.L. (1986) *Adv. Space Res.* **5** 317

Lamy, P.L., & Koutchmy, S. (1979) *Astron. Astrophys.* **72** 50

Lamy, P.L., Grün, E., & Perrin, J.M. (1987a) *Astron. Astrophys.* **187** 767

Lamy, P.L., Pedersen, H., & Vio, R. (1987b) *Astron. Astrophys.* **187,** 661

Moiseyev, N.D. (1925) *Astron. Zh.* **2** 73

Pansecchi, L., Fulle, M., & Sedmak, G. (1987) *Astron. Astrophys.* **176** 358

Sekanina, Z. (1986) In: *Proc. 20th ESLAB Symposium on the Exploration of Halley's Comet,* ESA SP-250, vol. II

Sekanina, Z., & Farrell, J.A. (1978), *Astron. J.* **83** 1675

Sekanina, Z., & Farrell, J.A. (1980), *Astron. J.* **85** 1538

Sekanina, Z., Larson, S.M., Emerson, G., Helin, E.F., & Schmidt, R.E. (1986) In: *Proc. 20th ESLAB Symposium on the Exploration of Halley's Comet,* ESA SP-250, vol. II 177

Sekanina, Z., & Schuster, H.E. (1978a) *Astron. Astrophys.* **65** 29

Sekanina, Z., & Schuster, H.E. (1978b) *Astron. Astrophys.* **68** 429

Tokunaga, A.T., Golish, W.F., Griep, D.M., Kaminski, C.D., & Hanner., M.S. (1986) *Astron. J.* **92** 1183

2

Comet Halley's dusty coma: *in situ* exploration with the dust impact detectors

J.A.M. McDonnell and G.S. Pankiewicz

1 INTRODUCTION

Halley's encounter by three separate space probes, each instrumented by a suite of dust detectors, has proven especially valuable in the light of the information gleaned. Though, now, the data sets are converging to a coherent image of the physics of the particulate components of the nucleus and coma, such coherence would not have resulted from a lesser onslaught from the armada of spacecraft (Fig. 1). The two Vega probes, encountering at some 8890 km and 8030 km miss distances, were not expected to reveal such high fluxes as Giotto's kamikaze encounter. Giotto's large sensor area (the dust shield) also led to an extension of the measured mass spectrum to much larger masses, where an intriguing change of the size distribution was discovered. Vega-1 and -2, however, encountered a much higher intrinsic nucleus emission than Giotto, a behaviour now attributed to a systematic postperihelion decline in Halley's output. Various instrumental intercalibration discrepancies and functional anomalies were also able to be sorted out by the redundancy afforded by multiple sensing; this overlapping of data has led to the reliable assessment of true flux variations as opposed to fluctuations in instrumental detection efficiency, perhaps caused by the heavy bombardment of the spacecraft. There was a time shortly after encounter when the comet enthusiast might be led to believe that three different comets had been encountered!

What, before encounter, might have been considered quite subtle instrumental differences between the three very similar mass spectrometer particulate composition analysers PIA, PUMA 1, and PUMA 2 on Giotto and Vega-1, -2 (Kissel 1986, Grard *et al.* 1986), did in fact lead to substantial advances in the understanding of the physics of impact mass spectrometry and particle compositions; the returned data from different target configurations, varying reflector voltages, and encounter scenarios each analysing the same type of particles was able to provide understanding of the impact process at mass—velocity regimes which could never be achieved in pre-flight laboratory calibration. The undetected, but now quite certain, failure of the flight cover to retract on one of Giotto DIDSY's subsystems led at first to indications of an absence of submicrometre particles on Giotto's approach to Halley. Comparison with the PIA data onboard the same spacecraft at first led to inexplicable differences, later confirmed by comparison with Vega's flux distribution (Moroz, personal communication). Though difficult to understand at first, the anomaly was soon revealed, and after DIDSY's cover was abraded by dust, the sensor continued to function until Giotto's departure from the coma. The three spacecraft data sets do now offer an unequalled opportunity for quantitative understanding of cometary physics, and perhaps almost as important, the 'calibration' of remote observational methods.

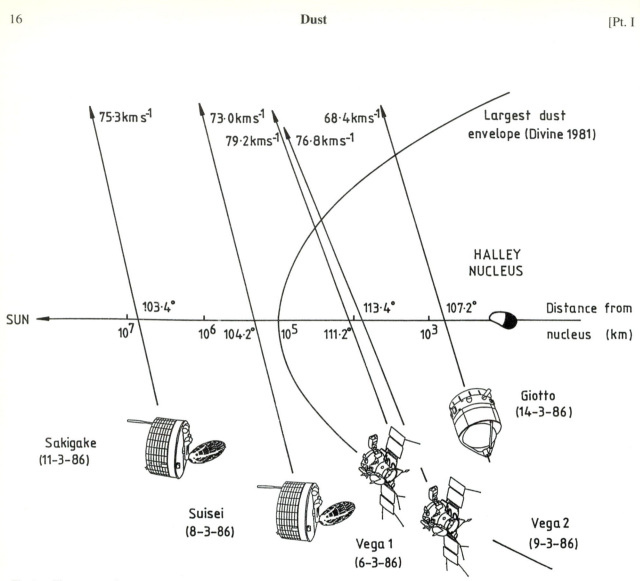

Fig. 1 — Five spacecraft encountered Halley during March 1986. Each trajectory is shown here on a logarithmically spaced distance scale with its relative velocity and geometry with respect to the Sun–comet line marked.

2 ENCOUNTER CONFIGURATIONS

2.1 Spacecraft encounter parameters in the coma

From the five spacecraft encounters, three have enabled detailed information to be obtained on the distribution of dust particles in the coma of Comet Halley. Weight constraints on the Japanese spacecraft, Sakigake and Suisei, did not permit the use of dust particle protection shields, and as earlier work showed that such particles were unlikely to be encountered some 200 000 km sunward of the comet nucleus (Divine 1981), the spacecraft were aimed to have closest approaches similar to this. However,

during the encounter of Suisei on 8 March, two relatively large dust particles struck the spacecraft, estimated to weigh several milligrams (Hirao & Itoh 1986).

As Comet Halley displays a retrograde orbit, inclined at 162° to the ecliptic, each spacecraft geometry can be imagined as very nearly being a head-on collision with the comet (remembering that the comet tail does not, in general, indicate the direction of motion of the comet). All the spacecraft passed sunward of the comet nucleus with relative velocities from 68.4 to 79.2 km s⁻¹, pressing the need for dust particle protection if a small encounter distance was desired.

The three spacecraft with dust protection were, therefore, able to fly by much closer. The Soviet Vega-1 flew by at a distance of 8890 km from the nucleus on the 6, and Vega-2 flew by at a distance of 8030 km on the 9 March. The European Space Agency spacecraft, Giotto, flew closer still on 14 March, the closest distance being a mere 600 km.

The trajectories of the spacecraft are shown in more detail in Fig. 1 on a logarithmically spaced diagram, to illustrate the large differences in the closest approach distances. It is because of these different distances that the integrated values of the fluxes (the fluence) are different for each spacecraft; there is also the time factor to be taken into account, between each flyby — the dust distribution in the coma is now seen to change greatly in a short period (less than 1 day), so that each flyby was observing a unique distribution in the coma. The important task is to relate these three encounters to a single model of the properties of the nucleus, set in the wider context of the more extended remote optical and thermal (infrared) measurements.

2.2 Sensors

Obtaining the *in situ* mass distribution of Halley required a variety of sensors, capable of measuring over about 15 decades of mass. Such a large mass range is required if we stop to consider the relative significance of each portion of the mass range existent in the comet coma. For example, it is at the high end of the mass spectrum (perhaps around 10^{-7} kg and above) where most of the mass is concentrated; in the region of 10^{-13} kg, we see the dominant visible grains, and below 10^{-15} kg the smallest, but the most abundant of particulates.

A large variety of detectors exist for detection of dust impacts (see McDonnell *et al.* 1978); the Halley encounters made especial use of three techniques:

(i) Penetration threshold sensing (DID-5 and DID-7 [McDonnell *et al.* 1986a], DUCMA [Simpson *et al.* 1986a]), where particulates above a given mass threshold penetrate a film, or, in the case of DID-5, the front (bumper) shield of the spacecraft.

(ii) Piezoelectric impact momentum sensing (SP-2 [Mazets *et al.* 1986], DID-2/3/4/5 and DID-1M [McDonnell *et al.* 1986a]). An impact on a target generates an acoustic wave, which is propagated as a bending wave, and may be detected by the compression of an ultrasonic resonant piezoelectric element, providing an output voltage proportional to the maximum deformation near the region of impact (McDonnell 1969). These sensors may therefore be deployed on surfaces of spacecraft, yielding large effective detection areas.

(iii) Impact charge sensing (SP-1 [Vaisberg *et al.* 1986], SP-2 [Mazets *et al.* 1986], DID-1P [McDonnell *et al.* 1986a]). A hypervelocity impact results in the release of a non-equilibrium plasma, whose ions and electrons may be separated under an applied electric field and collected on oppositely biased electrodes. The total charge Q measured is almost directly proportional to the projectile mass m, so that grain masses are estimated from preflight charge/mass calibrations of the form:

$$Q = km^a v^b \qquad (1)$$

where v is the projectile velocity, and k, a, and b are constants determined in the laboratory. The exponent a is generally unity, whilst b is typically 2.5–3.5 (McDonnell 1987).

Other sensing techniques used during the encounters included:

(iv) Optical detection (PIA and PUMA [Kissel 1986]), as a luminous flash produced as a result of the impact.

(v) Depolarization signals as a small volume of polarized material (PVDF) forming a film is displaced (DUCMA [Simpson *et al.* 1986a]).

(vi) Capacitor discharge sensing (DID-7 [McDonnell *et al.* 1986a]), where a 70 μm Mylar dielectric material, covered either side with a thin layer of aluminium, is perforated, thereby discharging through the impact-generated plasma.

(vii) Dust mass spectrometers (PIA and PUMA [Kissel 1986]), where dust particles impact on a silver target producing a light flash recorded by a photomultiplier, and projectile ions which enter the time of flight mass analyser.

All of the dust impact sensors used in the three spacecraft encounters are listed in Table 1, together with their sensing areas and mass ranges, which vary over a total range of 15 decades of mass.

Table 1.   Halley encounter dust impact sensors

System and brief description	Area (cm^2)	Mass range (kg)
VeGa-1, and -2		
DUCMA[a] impact depolarization detection	75	$1.5 \times 10^{-16} - 9 \times 10^{-14}$
PUMA[b] mass spectrometer	5	$3 \times 10^{-19} - 5 \times 10^{-13}$
SP-1[c] plasma	81	$10^{-19} - 10^{-13}$
SP-2[d] acoustic	500	$3 \times 10^{-16} - 2 \times 10^{-9}$
plasma	40	$10^{-19} - 10^{-14}$
Giotto		
DID[e] MSM/RSM acoustic	$\sim 2 \times 10^4$	$> 4 \times 10^{-12}$
CIS impact discharge	1000	$> 10^{-13}$
IPM acoustic	100	$4 \times 10^{-13} - 4 \times 10^{-11}$
plasma	100	$10^{-20} - 10^{-6}$
PIA[b] mass spectrometer	0.01 – 5	$3 \times 10^{-19} - 5 \times 10^{-13}$

[a] Simpson *et al.* 1986. [b] Kissel 1986. [c] Vaisberg *et al.* 1986. [d] Mazets *et al.* 1986. [e] McDonnell *et al.* 1986.

3 FLUX RATES MEASURED OVER 8 DAYS

Each flyby has given some indication of how the mass distribution rises toward closest approach, and then falls off afterwards according to the range of the mass spectrum each experiment measured. For a nucleus emitting dust into space isotropically, we would expect this overall flux variation to depend on the inverse square of the distance from the nucleus (R^{-2}), which has in general been observed from the measured flux rates (Mazets *et al.* 1986, McDonnell *et al.* 1986b, Simpson *et al.* 1986a); it is the departures from this trend seen by all experiments which provide exciting glimpses of the finer details of dust dynamics and emission such as jets, outbursts, and envelopes within the coma.

Comparison of flux rates by cumulative flux curves (Fig. 2) shows the changing dust distribution over the period of 8 days separating the Vega-1 and Giotto encounters. All three trajectories passed the nucleus along similar directions with respect to the Sun (which determines the region of highest dust emission), varying from an angle of 107.2° (between Giotto's trajectory and the Sun) to 113.4° for the same angle defined for Vega-2. The two curves in Fig. 2 have been obtained for the same distance from the nucleus (10 000 km), both before and after closest approach for all three spacecraft. (It should be noted that the varying uncertainty bars obtained by consideration of a Poisson number distribution on the Giotto data are due to the different areas for each of the DIDSY experiments.)

The fountain model of cometary dust describes dust travelling out from the nucleus, and then being 'reflected' away from the Sun by solar radiation pressure. This gives rise to parabolic trajectories (as in a fountain), and the overall dynamics of particles of a given size produce a boundary, or envelope, beyond which no particles are expected. This envelope is just the locus of points joining the extremities of each particle trajectory in the Sun's direction. Well defined envelope boundaries were observed by Simpson *et al.* (1986a), Vaisberg *et al.* (1986), and Mazets *et al.* (1986). Vaisberg *et al.* describe several dust envelopes with quite different density gradients varying from a dependence on $R^{-1.0}$ between about 20 000 and 60 000 km from the nucleus, to a dependence on $R^{-4.6}$ further away, possibly suggesting the existence of differing kinds of particle. Mazets *et al.* (1986) observe the spatial density of particles from 10^{-17} to 10^{-14} kg to rise at the expected envelope, but to drop quickly, giving rise to a flattened mass distribution far from the nucleus which would appear to show the strong dependence of radiation pressure efficiency to particle size.

Closer to the nucleus, Giotto DIDSY data (McDonnell *et al.* 1987) has shown pre-encounter fluxes lower than the R^{-2} dependence, associated with the approach from the dark side (see the flux-time profile in Fig. 3)—a dawn enhancement is then observed, although not as high as the transition expected from a nucleus source function which 'turns on' at the dawn terminator (Pankiewicz & McDonnell 1986). It is now

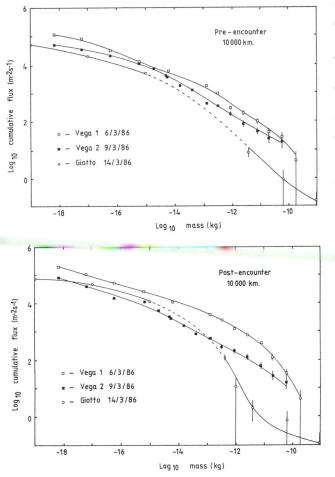

Fig. 2 — A comparison of cumulative fluxes over 10 mass ranges is made for each flyby, both pre- and post-encounter, at 10 000 km from the nucleus. Vega data are averaged over 10 seconds, whilst Giotto data are averaged over 20 seconds. Although near closest approach for the VeGa spacecraft, this distance is at a point along the trajectory of Giotto (146 seconds from encounter) where the actual number of particles is low, hence the large scatter.

certain that the DID-1 unit on board Giotto failed to have the 22 μm thick foil cover retract well before encounter (see Maas *et al.* 1986, McDonnell *et al.* 1987); however, this sensor sees a rapid increase in the flux rate over the dawn terminator, pointing to the total erosion of the cover so that smaller mass particles of higher spatial density were beginning to be seen here. The flux data first reported from PIA shown in Fig. 3 are very much 'flatter' here compared to an inverse square dependence, but are due to the non-inclusion of instrumental dead times; the data now

being analysed do demonstrate an enhanced gradient (McDonnell *et al.* 1989a).

The DUCMA experiment on VeGa-1 (Simpson *et al.* 1986b) shows a very rapid and perhaps anomalous increase in flux at closest approach for all ten mass channels, increasing by a factor of 45 in 5 seconds. Whilst such an increase may not be entirely dismissed, since other large flux increases are observed, it is difficult to account for this rise when compared to the SP-1 (Vaisberg *et al.* 1986) and SP-2 (Mazets *et al.* 1986) experiments on Vega-1, as well as the PUMA 1 flux which is now also yielding evidence against this behaviour. This rise occurs at a point in the trajectory with an angle of 22° from the Sun–comet line, and therefore cannot be accounted for in terms of a dawn terminator enhancement, which has an angle of 90° from the Sun–comet line, and would also give rise to a lower density at this distance than that observed by Giotto.

Traversal of jets is identified in the inner coma by all experiments, where fluxes are much higher than the R^{-2} dependence exhibited elsewhere. The distinct change in the slope of the mass distribution seen in the Giotto post-encounter data of Fig. 2 is probably due to passage through such a jet, giving rise to a large number of smaller particles, and a relative depletion of particles in the range 10^{-13} to 10^{-10} kg.

The high flux of small particles (which have a higher relative velocity away from the nucleus) may be due to Giotto seeing these particles impinging on the spacecraft shield from a relatively new jet—if Giotto had passed this point several hours later, an enhancement of the more slowly moving larger particles might have been seen instead. In fact, when some of the more diffuse jets observed by Vaisberg *et al.* (1986) and Mazets *et al.* (1986) are looked at carefully, it is found that the jets peak initially in the higher mass channels and later on in the lower (less massive) mass channels, indicating a velocity distribution of particles emitted from a nucleus rotating in a prograde sense relative to the spacecraft trajectory.

Apart from these local effects seen by each spacecraft, the encounters reveal a gradually decreasing overall flux of particles in the 8-day period, as can be seen from the normalized fluence curves of Fig. 4 (explained in more detail in the next section). This postperihelion decrease is expected as a result of the

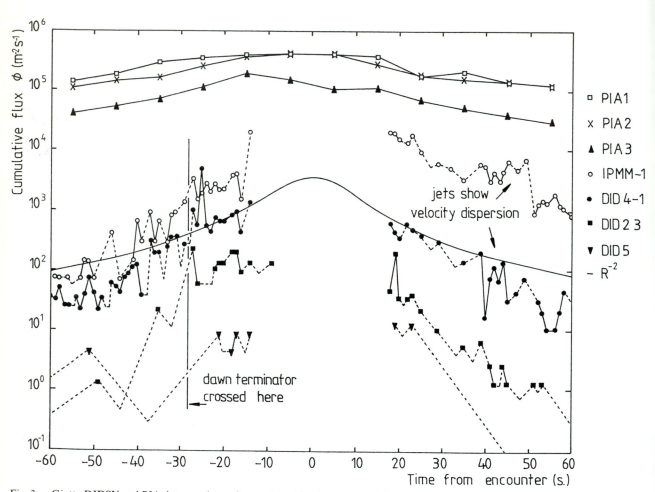

Fig. 3 — Giotto DIDSY and PIA data are shown from −1 to +1 minute around closest approach. The general trend is an inverse square dependence, but significant deviations in this time period include the crossing of the dawn terminator and the traversal of jets (for example at about 45 seconds post-encounter). The 'flat' distribution of PIA data close to encounter has been attributed to saturation. Dead-time corrections have been applied to produce revised fluxes

comet moving further away from the Sun (perihelion was on 9 February 1986), thereby having less solar heating on the nucleus surface during the course of the spacecraft encounters.

4 DERIVED PARAMETERS OF THE *IN SITU* DUST DISTRIBUTION

Measuring the mass distribution in the coma can give us information on several parameters of astrophysical significance. By considering the total fluence of the *in situ* measurements, we have the best method of obtaining the overall distribution of particles in the coma. This is usually interpreted through the behav-

iour of the cumulative (logarithmic) spectral mass index, α.

Consider an infinitesimal distance along a spacecraft trajectory, $dx = v_s.dt$, where v_s is the relative coma–spacecraft velocity and dt is the time it takes to fly along dx. Then the cumulative flux $\phi \,(> m, x)$, $(m^{-2} \ s^{-1})$ is the number of particles with masses greater than mass m that the spacecraft measures at a distance x from the closest approach point. One might expect this to increase with decreasing mass in the following power law form:

$$\phi \,(> m, x_1) = K_1 \, m^{-\alpha_1} \qquad (2)$$

where K_1 is a constant at distance x_1, and α_1 is the

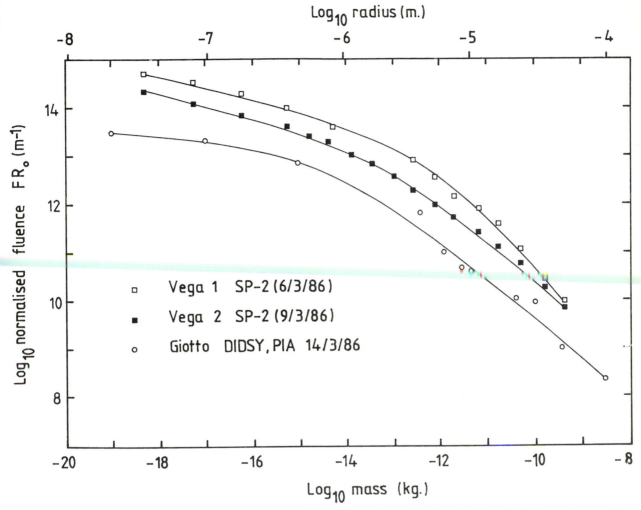

Fig. 4 — Fluences normalized to take account of closest approach distances are shown with data taken from the SP-2 experiments on Vega (Mazets *et al.* 1986) and the DIDSY and PIA experiments of Giotto (McDonnell *et al.* 1987). Grain radius is also shown as a comparison with mass—these are computed from the grain model employed by Divine *et al.* (1986).

cumulative spectral mass index at this time. For the spherically symmetric coma model mentioned earlier, and in the light of the fact that values of α vary along the flyby trajectories (see, for example, Zarnecki *et al.* 1986 or Mazets *et al.* 1986), the cumulative flux can be written more generally as

$$\phi(>m, x) = (k \cdot R_0^2 / R(x)^2) \cdot m^{-\alpha(x)}, \quad (3)$$

k being a constant throughout the coma, $R(x)$ the distance to the nucleus at point x along the trajectory, and R_0 the closest approach distance. The fluence, F,

which is the cumulative flux integral along the trajectory, would therefore be

$$F(>m) = \int_{x_1}^{x_2} \phi(>m, x) \, \mathrm{d}x$$

$$= km^{-\langle\alpha\rangle} \cdot \int_{-x}^{+x} (R_0^2 / R(x)^2) \, \mathrm{d}x$$

$$= km^{-\langle\alpha\rangle} \cdot \pi R_0 \quad (4)$$

where a time averaged cumulative spectral mass index $<\alpha>$ has been used, and x_1 and x_2 are the entry and exit distances into the coma, which are taken as $-\infty$ and $+\infty$ here, as the function asymptotically goes to zero at these limits. It can be seen from (4) that the fluence for a given flyby is clearly dependent on the closest approach distance, R_0.

If we assume the peak flux during encounter (which would occur at closest approach in this model) has a cumulative spectral mass index $\alpha_0 = <\alpha>$, then the fluence can also be written as

$$F\,(>m) = \phi\,(>m,0).\pi R_0. \qquad (5)$$

The peak flux, however, is a function of the closest approach distance for a given flyby, so that

$$\phi\,(>m,0).R_0{}^2 = \phi\,(> m, x).R(x)^2 = C, \qquad (6)$$

where C is a constant through the coma, but which will vary along the comet orbit. Therefore, a normalized fluence, F_N, has been chosen to enable a comparison of each flyby:

$$\begin{aligned} F_N\,(>m) &= F(>m).R_0 \\ &= \pi C \end{aligned} \qquad (7)$$

using equations (5) and (6). This is shown in Fig. 4 for the three closest Halley encounters between the 6 and 14 March 1986, showing the general decrease in activity over that period.

Fig. 4 also shows the averaged cumulative spectral mass index $<\alpha>$ to vary with mass; for example the value of $<\alpha>$ measured on Giotto varies from 0.14 at 10^{-17} kg to 0.85 between 4×10^{-13} kg and 10^{-9} kg. However, the 'discrete' data from the Giotto DIDSY experiments (McDonnell et al. 1987) have revealed a much lower value of $<\alpha>$ at masses greater than about 10^{-9} kg than that seen in Fig. 4, resulting in an 'abundance' of larger particles and a relative depletion at around 10^{-8} kg.

To analyse the multiple detection discrete events of Giotto DIDSY, which data give rise to this lower slope, the momentum transferred to the front shield must first be considered. This is greater than the momentum carried by the particle itself, as shield material is vapourised and ejected (McDonnell et al.

1984). The momentum p_1 of particles that strike the front shield, but do not penetrate, is given an enhancement factor of 11 for the relative Giotto–coma velocity, so that

$$p = 11.mv \qquad (8)$$

for a particle of mass m at relative velocity v. If the particle penetrates the front shield, this enhancement must be derated, thereby allowing for ejecta which continue through the shield. The momentum p_2 of these particles is

$$P_2 = 11(m_{PEN}/m)^{\gamma}.mv \qquad (9)$$

where m_{PEN} is the penetrating mass threshold ($= 3 \times 10^{-9}$ kg for the front shield), and γ is a momentum derating exponent (Wallis 1986). For these particles of mass $m > m_{PEN}$, a momentum p_3 is transferred to the rear shield:

$$P_3 = [1 - (m_{PEN}/m)^{\gamma}].mv. \qquad (10)$$

In this way, discrete data events below the penetrating mass limit can be analysed by using equation (8) to determine the value of $<\alpha>$ here. For discrete data above the penetrating mass, a suitable value of γ in equations (9) and (10) is chosen which will enable a continuity of the flux slope between these penetrating particles and the lower mass non-penetrating events. This results in a derating factor of $\gamma = 0.55$, and hence a value of $<\alpha> = 0.41$, significantly lower than that of the smaller particles.

The total impacting mass of particles can be obtained independently from the Giotto Radio Science experiment (Edenhofer et al. 1986), and the deceleration recorded by the ESA tracking centre ESOC (Morley & Fertig 1986). By extrapolating the DIDSY 'binned' data (count rates) to the largest single particle that would have hit the spacecraft, an average large particle index $<\alpha>$ of 0.54 ± 0.02 is obtained that satisfies the total effective mass that is required from the spacecraft deceleration.

Alternatively, one could use the best fit value of the large particle index, $<\alpha> = 0.41$, joining this to the binned data fluence, and then let the index resume its value of 0.85 for even larger masses, again satisfying

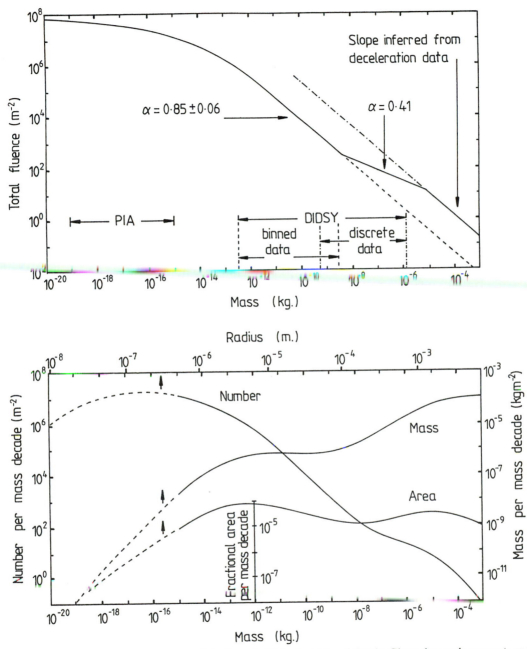

Fig. 5 — A fluence plot is shown in the upper part of the figure (solid line) which satisfies the Giotto data and represents an intermediate stage between an outburst (chain line) and an inactive period some time later (dotted line). Distribution functions per mass decade from the fluence above are shown in the lower half of the figure. The lower value of the cumulative mass index applicable to particles with 10^{-8} kg $<m<10^{-5}$ kg implies an increase of the relative contributions to both the mass and geometric area functions.

the deceleration data. This results in the slope relaxing back to $<\alpha> = 0.85$ at above masses of about 10^{-5} kg; this is shown for the complete Giotto fluence in Fig. 5.

Using this total mass distribution from the fluence, it is then possible to find the cross-sectional area of the particles as a fractional area per mass decade for the complete distribution; this is also shown in Fig. 5. This therefore represents the area that would give rise to light scattering for a sample of particles in a given mass range, and so gives an idea of the scattered light and thermal emission that would have been seen from the Earth when folded with scattering efficiencies.

Observations taken some 5 hours before the Giotto encounter (Hanner *et al.* 1987) indicate that a size distribution for particles greater than about 10^{-19} kg, with a value of $\alpha \sim 0.85$ (similar to that actually measured for smaller masses), would have resulted in a satisfactory fit to the thermal emission spectrum— but they would not be able to account for an excess of large grains (with $\alpha \sim 0.54$) seen by Giotto.

However, these observations can be compatible if we consider the sudden outbursts so commonly associated with Halley. Imagine material from an outburst leaving the nuclear surface at about 1800 UT on 13 March. Small grains with velocities in the region of 600 m s^{-1} (Gombosi 1986) would have travelled some 2160 km in the first hour. This is far enough away from the nucleus to give rise to an infrared spectrum showing a silicate emission feature at around 10 μm, typically produced by a large abundance of these smaller and hotter grains. This was, in fact, seen by groundbased observers (e.g. Hanner *et al.* 1987) at around this time.

Meanwhile, the larger grains would have travelled only about 360 km if their terminal velocities were 100 m s^{-1} (Gombosi 1986). As groundbased observations are generally averaged over apertures ranging from 3000 to 20 000 km (transformed to the comet coma) these larger grains would occupy too small a region to show their effects, so that at this time, apertures would have been dominated by the small grains. By 0000 UT on 14 March, six hours have elapsed since the initial outburst, and the small grains have travelled about 13 000 km, but the larger grains have travelled only 2160 km; they are now the dominant grains seen in the apertures, and suppress the 10 μm feature which may be emitted only from the (now fewer) grains of <30 μm. The larger grains are also close enough to the sunward side of the nucleus, and within distances <4000 km from the nucleus—where most impacts occurred on the Giotto encounter. Thus, although the groundbased observations appear to be incompatible

with the Giotto data, a nucleus model which incorporates dust emission that is able to vary by factors of up to 30 in surface rates can explain the mass distributions obtained from the two techniques (Perry *et al.* 1988). This is apparent in Fig. 5, where the earlier high fluence reduces in time to that seen by Giotto for particles <10^{-9} kg.

By looking at the temporal variation of α, we can find how the distribution changes across the coma, and perhaps obtain some information on the age of any jets a spacecraft passes through. For example, a high value of α indicates an excess of smaller grains, and therefore a relatively young jet is being observed—the spacecraft is flying through a place where the small grains in the 'front' of the jet have already reached.

Additionally, we note that by integrating the differential flux, $\phi(m,t)$, and assuming it to be emitted from an area S of the nucleus surface, we find the total emission rate of dust from the surface,

$$M = S. \int_0^{m_1} \varphi(m, t) \, dm \qquad (11)$$

where m_1 is the largest 'grain' size observed. Integrating below a value for m_1 of 10^{-8} kg, we find from the Giotto encounter that the total emission rate per unit area transformed to the nucleus is 6.7×10^{-6} kg m^{-2} s^{-1}; if the active surface area is $\pi \times (4 \text{ km})^2$, we obtain $M \sim 340$ kg s^{-1}. Further emission rates that have been derived from the total Giotto fluence transformed to the nucleus are given in Table 2. These values are obtained by assuming that the emission rate at the nucleus was the same during the time that all the grains that were detected were ejected from the nucleus, and that the active region on the nucleus surface is homogeneous. This is of course at one point in the comet orbit, close after perihelion, and we expect the mass loss to decrease as the comet moves further away from the Sun.

5 DYNAMICS OF THE DUST SCENE

We have already seen how isotropic emission from the nucleus leads to the inverse square law dependence on the distance from the nucleus at close approaches. Eventually, most grains will be deflected away from

Table 2

Mass indices: 10^{-13} kg $< m < 10^{-8}$ kg, $\alpha = 1.02$ (typical)
at nucleus) 10^{-8} kg $< m$ $\alpha = 0.71$ (effective at encounter time)

Averaged surface emission rates per square metre:

$< 10^{-8}$ kg $- 6.7 \times 10^{-6}$ kg m^{-2} s^{-1}

$< 10^{-3}$ kg $- 5.7 \times 10^{-5}$ kg m^{-2} s^{-1}

< 1 kg $- 4.2 \times 10^{-4}$ kg m^{-2} s^{-1}

Upper mass limit (kg)	Grain diameter	Total surface emission rate to upper mass limit (kg s^{-1} for 4 km nuclear radius)
1.58×10^{-18}	0.1 μm	7.5×10^{-4}
1.60×10^{-17}	1.0 μm	0.78
6.31×10^{-13}	10.0 μm	37
10^{-8}	0.62 mm	340
10^{-3}	1.34 cm	2 900
10^{-2}	2.88 cm	5 500
10^{-1}	6.20 cm	11 000
1	13.4 cm	21 000

the Sun by solar radiation pressure, giving rise to the typical comet dust tail seen spreading up to 10^7 km. However, close to the nucleus, the important dynamical effect is that of an expansion velocity resulting from the sublimed gas from the nuclear surface (Divine *et al.* 1986).

Fig. 6 illustrates a sample of Giotto and Vega data taken for particles of masses greater than about 6×10^{-15} kg, and shows the flux value measured as a function of distance from the nucleus. The dotted line shows the inverse square dependence one would expect from such a simple body; the figure shows various distinctive and significant peaks and troughs, indicating more dense or less dense regions respectively.

It is believed now that Comet Halley's nucleus contains discrete active regions on the surface (Sekanina & Larson 1986), which, when subjected to solar heating, release large quantities of gas and dust. The dust is then accelerated in the outflowing gas, finally decoupling to leave the dust grains with a characteristic terminal velocity. The effects of these are seen by groundbased observers as jets of dust, often appearing to curve around the nucleus in a

direction controlled by the rotational nature of the nucleus. Fig. 7 shows a small region of a CCD image taken at 1839 UT on 13 March 1986, just over 5 hours before the Giotto encounter. A projection of the trajectory of Giotto is also shown, together with times relative to encounter. One jet can be seen as a distortion in the isophotes, arcing over from the top left of the nucleus toward the left of the image. If we assume the jet to be emitting from the nucleus in the sunward direction at this time, then for grains of 10^{-15} kg mass with a typical terminal velocity of 500–600 m s^{-1} (Gombosi 1986), they would have travelled approximately 10 000 km in this direction during the course of about 5 hours, perhaps resulting in the jet seen by Giotto at this distance. This might explain the predominance of small particles in the mass distribution (Fig. 2) at this distance, but may not explain the depletion of the larger particles.

It may be postulated that this depletion could be caused by fragmentation. We could suppose a stream of particles emitted from the nucleus with a 'normal' coma distribution, perhaps like that seen by Vega -2 in Fig. 8. If, for example, particles originally 'glued' together as a conglomerate fragmented in flight because of sublimation of water ice holding the dust together (Simpson 1986c), then the mass distribution would rise in the smaller mass region, and fall in the high mass region (perhaps like the data of Giotto in Fig. 8), simply owing to increasing numbers of smaller particles. The explanation of the distribution in Fig. 2, post-encounter, might therefore arise from a combination of jetting and fragmentation. However, owing to the form of the mass distribution, the depletion of larger particles could be more evident by the excess of small particles. The loss of 1 particle of, say, mass 10^{-8} kg could create up to 10^{11} particles of mass 10^{-19} kg! The actual fluence ratio measured for Giotto at these masses is nearer to about 10^0.

Considering the dynamical evolution of a jet, its spatial concentration will be dispersed because of variations in the terminal velocity distribution, and differing optical properties affecting the solar radiation pressure on the grains; eventually the effect of many jets is to coalesce into the less dense, diffuse background of dust in the coma, seen by remote observations.

More probably, we may find a more satisfactory

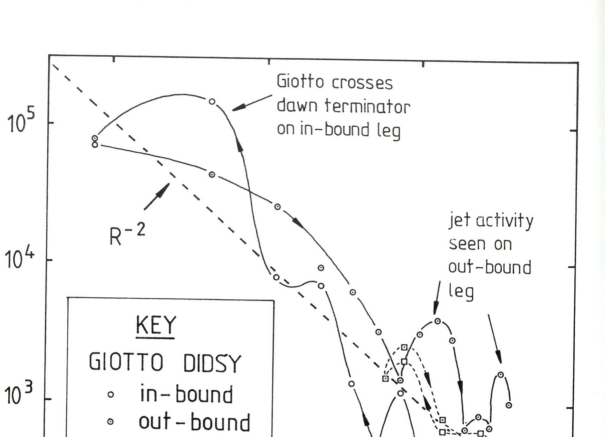

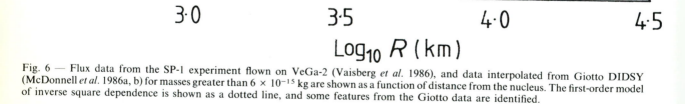

Fig. 6 — Flux data from the SP-1 experiment flown on VeGa-2 (Vaisberg *et al.* 1986), and data interpolated from Giotto DIDSY (McDonnell *et al.* 1986a, b) for masses greater than 6×10^{-15} kg are shown as a function of distance from the nucleus. The first-order model of inverse square dependence is shown as a dotted line, and some features from the Giotto data are identified.

explanation of the variations of size distribution from examination of more elegant modelling of the dynamics. Larger particles have been found from the *in situ* measurements to differ substantially in their dynamics in the coma of Comet Halley, from the fact that

their individual orbits are dependent to a much lesser extent on solar radiation pressure. Consider the dimensionless quantity β, which is the ratio of solar radiation acceleration, a_{RAD}, to solar gravitational acceleration, a_{GRAV}, for a particle of radius s, and

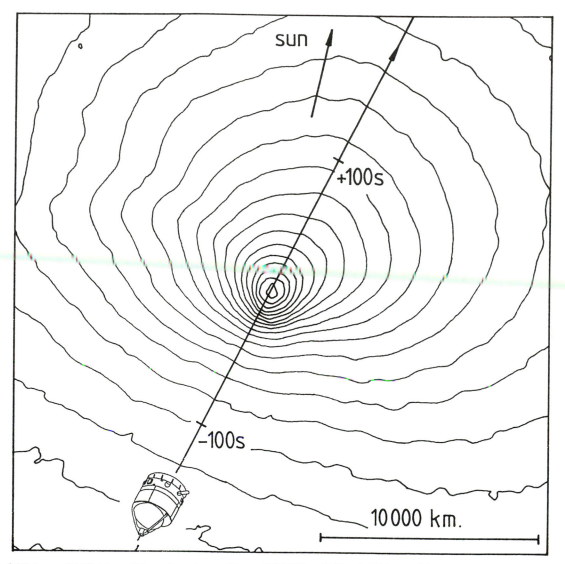

Fig. 7 — CCD image 13/463 taken with a red continuum filter at 1839 UT on 13 March 1986, over 5 hours before the Giotto encounter. Isophotes are at intervals of 20%. The projected path of Giotto is shown as a solid line with times from closest approach.

having radiation pressure efficiency Q_{PR} (see, for example, Divine *et al.* 1986). If the particle has density ρ, then we can write

$$\beta = \frac{a_{RAD}}{a_{GRAV}} = K \cdot \frac{Q_{PR}}{\rho s} \qquad (12)$$

with $K = 5.74 \times 10^{-4}$ kg m^{-2}.

Particle dynamics are therefore largely governed by their size and type, different materials (notably conducting and non-conducting) having different

radiation pressure efficiencies, as shown in Fig. 9. Using data from Divine *et al.* (1986) for radiation pressure efficiencies, we find that, typically, particles of 1 μm radius would have $0.3 \leq \beta \leq 0.4$, so that they are accelerated relatively quickly into the tail (<1 day), whereas particles of 0.1 mm and upwards (corresponding to masses above about 10^{-9} kg) have $\beta \leq 0.01$, regardless of type. These will be more highly influenced by terminal velocities (though small) resulting in highly asymmetrical cometocentric trajec-

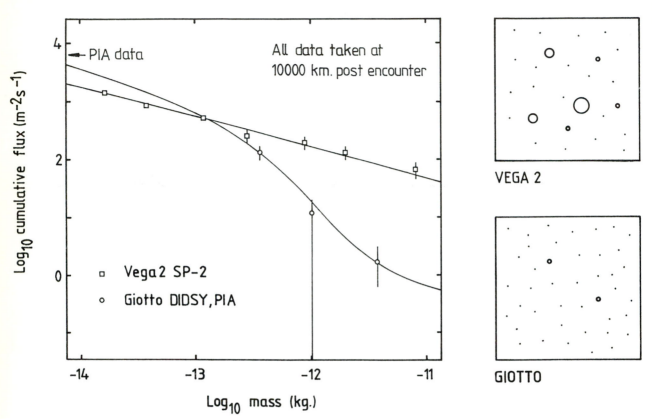

Fig. 8 — Cumulative fluxes taken from Fig. 2 are shown here for Vega-2 and Giotto. The data were taken at 10 000 km from the nucleus post-encounter over a mass range 10^{-14} to 10^{-11} kg, where the two distributions cross. The boxes show the number of particles of radius 1 μm, 3.6 μm, 6.4 μm, and 11 μm that would be detected from the distributions given, over an area of 100 cm^2 during 1 second.

tories (Fertig & Schwehm 1984), many from the sunlit side of the nucleus being displaced forward of the comet's orbital motion, so that the spacecraft encounters should have seen more large ($> 10^{-9}$ kg) particles pre-encounter. This effect is not due to any asymmetry in the particulate emission velocity from the comet surface, but a to combined effect of solar radiation pressure and the resultant heliocentric velocity vectors. Giotto DIDSY (McDonnell *et al.* 1987) did in fact see an asymmetrical distribution of particles of mass $\sim 10^{-9}$ kg and greater, as shown in Fig. 10. Data between -1 and $+1$ minute around closest approach have been excluded, as numbers detected here are very high and symmetrical about the Sun—comet line, as expected from both the fountain model (see Divine 1981) and the modelling of Fertig & Schwehm. This enhances the asymmetry of the large particles so that 71% of these are seen before 1

minute to closest approach, the first impacting at -43 minutes; compared to the other 29% seen after 1 minute from closest approach, the last of these particles occurring at $+7$ minutes.

6 DUST EMISSION AT THE NUCLEUS

Each spacecraft flyby records the rate of impacts of particles at a given point in the coma. Since each particle was originally emitted from the nucleus, we can make an estimate of the locations on the nucleus that these particles would have been ejected from if we know the following:

(i) the dynamical forces on the particles following emission from the nucleus, including the velocity and mass distributions, and radiation pressure effects;

(ii) the rotational properties of the nucleus, to determine the **emission** location in a comet-centred

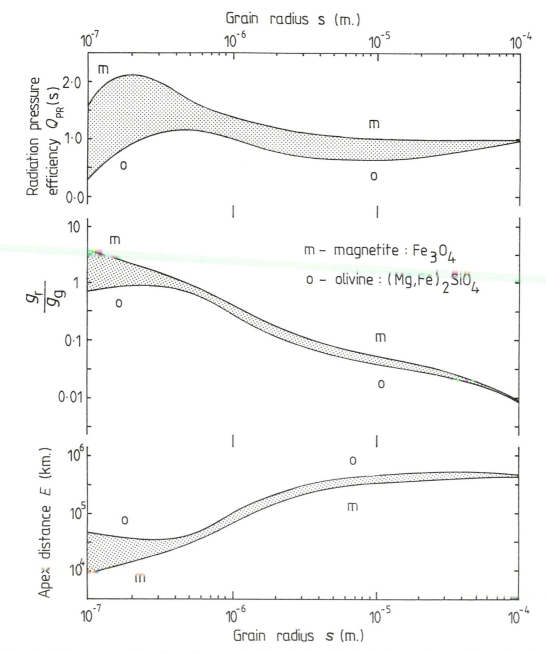

Fig. 9 — Radiation pressure efficiencies are shown as a function of grain radius, for both magnetite and olivine spheres (representing dark absorbing and dielectric grains respectively) with data taken from Divine *et al.* (1986). The dimensionless parameter β which describes the effective radiation pressure strength at the heliocentric distance of the Giotto encounter at 0.89 AU can then be derived. Using the terminal velocity distribution of Gombosi (1986), the apex distances from the fountain model are then obtained. In all cases, the shaded areas represent particles with properties between the extremes of magnetite and olivine.

system of co-ordinates (e.g. longitude and latitude on the surface).

Overall, it can be assumed that most of the particles smaller than about 10^{-6} kg observed less than 10^4 km

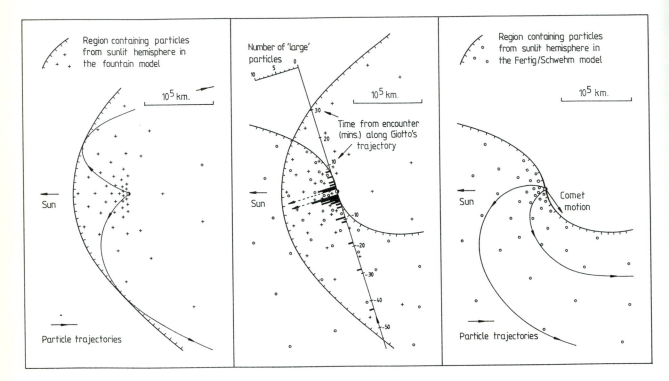

Fig. 10 — Two dust dynamic models are shown for particles of mass $\sim 10^{-9}$ kg. On the left is the simple fountain model showing symmetrical behaviour about the nucleus on the Sun–comet line, where modelling from Divine (1981) is used to calculate the apex distance. This is smaller by a factor of 5 than the apex in Fig. 9, because of the lower terminal velocity used at this mass. Crosses schematically show the spatial density distribution according to this model. The right-hand side of the figure shows Fertig & Schwehm modelling (adapted from Fertig & Schwehm 1984), resulting in highly asymmetrical trajectories about the Sun–comet line, with circles to represent the density distribution here. Results of Giotto DIDSY for large particle events of $\sim 10^{-9}$ kg and greater (McDonnell *et al.* 1987) are shown in the centre. The numbers of these particles, taking account of telemetry loss post-encounter, are shown along the trajectory every 30 seconds up to -1 minute, and from $+1$ minute, where event rates are much higher and more symmetrical. Note that the first of these particles occurs at -43 minutes, and the last at $+7$ minutes.

away from the nucleus have arrived on a radial trajectory. Small particles are 'reflected' back from the Sun by the solar radiation pressure, their full trajectories being described by parabolae. As their apex distances (the furthest distance they travel before turning back) have been found to be of the order of 10^5 km (McDonnell *et al.* 1986b, Simpson *et al.* 1986a), we may assume that their early flights ($< 10^4$ km) are in straight lines. This will also mean that although the majority of particles detected are direct from the nucleus, a small amount due to the reflected component will also be measured. In fact, if E is the apex distance of the particles concerned, then to a good approximation near the nucleus, the distance they will have travelled is $2E$, so that their flux would be

$$\phi_{\text{REFLECTED}} \sim a \cdot \frac{1}{(2E)^2}. \qquad (13)$$

The direct flux, as mentioned earlier, is just

$$\phi_{\text{DIRECT}} \sim b \cdot \frac{1}{R^2} \qquad (14)$$

if R is the cometocentric distance, and hence the total flux measured at the spacecraft will be

$$\phi \sim \frac{a}{4E^2} + \frac{b}{R^2}. \qquad (15)$$

If the emission rate is assumed constant, then we can write $a = b$, and it is clear that at 10^4 km, the ratio of reflected to direct is only 1:400; and at distances less than this, the largest component is from direct emission.

If we trace back the trajectory from the impact point to the nucleus with the help of the above dynamics, and use a typical velocity distribution (for example those of Divine 1981 or Gombosi 1986), we end up at a point on the nucleus surface that is different from the sub-spacecraft point at the time of impact, because the nucleus will have rotated in some sense between the time of emission and the time of impact. Pre-encounter data, including work from the 1910 observations of the comet, arrived at a simple spin period of around 2.2 days. However, observations from this apparition by Millis & Schleicher (1986) have revealed a second nuclear period of 7.4 days. This has been interpreted by Sekanina (1987) as an ellipsoidally shaped body precessing with a 2.2 day period having a precessional angle of 77°, and with a true rotation of 7.4 days, producing a much more complex rotation.

Thus, to obtain a true picture of the dust emission at the nucleus from flyby measurements will require a full understanding of the particle dynamics and rotational state of the nucleus, as well as looking for some commonality between the three encounters in terms of the nuclear surface.

The full DIDSY dataset has recently been produced, based on a complete calibration of sensor response times from in-flight data and evaluation of multiply detected discrete events from impacts of large particles. In addition, the PIA data have been revised to take account of the saturation effects near closest approach (McDonnell *et al.* 1989a).

The discrete data confirm the presence of a large particle excess pre-encounter with a smaller excess post-encounter (as predicted by Fertig and Schwehm-type models). The presence of this excess over a large spatial distance implies a true excess of large mass grains on the cometary surface. Fluxes, dust production rates and dust-to-gas ratio as a function of mass are described in McDonnell *et al.* (1989b, 1989c).

ACKNOWLEDGEMENTS

We acknowledge valuable formative discussions with colleagues, especially C.H. Perry and S.F. Green. The Giotto DIDSY project was funded within the UK by the Science and Engineering Research Council.

REFERENCES

Divine, N. (1981) Numerical models for Halley dust environments, *Proc. Joint NASA/ESA Working Group Meeting*, ESA SP-174, 25–30.

Divine, N., Fechtig, H., Gombosi, T.I., Hanner, M.S., Keller, H.U., Larson, S.M., Mendis, D.A., Newburn, R.L. Jr., Reinhard, R., Sekanina, Z., & Yeomans, D.K. (1986) The Comet Halley dust and gas environment, *Space Science Review* **43** 1–104.

Edenhofer, P., Buschert, H., Porsche, H., Bird, M.K., Volland, H., Brenkle, J.P., Kursinsky, E.R., Mottinger, N.A., & Stelzried, C.T. (1986) Dust distribution of Comet Halley from the Giotto Radio Science Experiment, In: *20th ESLAB Symposium on the Exploration of Halley's Comet*, ESA SP-250, Vol. II, 215–218.

Fertig, J., & Schwehm, G.H. (1984) Dust environment models for Comet P/Halley: Support for targeting of the Giotto S/C, *Adv. Space Res.* Vol. 4, 9, 213–216.

Gombosi, T.I. (1986) A heuristic model of the Comet Halley dust size distribution, In: *20th ESLAB Symposium on the Exploration of Halley's Comet*, ESA SP-250, Vol. II, 167–171.

Grard, R., Gombosi, T.I., & Sagdeev, R.Z. (1986) The VeGa missions, in *Space missions to Halley's Comet*, ESA SP-1066, 49–70.

Hanner, M.S., Tokunaga, A.T., Golisch, W.F., Griep, D.M., & Kaminski, C.D. (1987) Infrared emission from Halley's dust coma during March 1986, *Astronomy and Astrophysics*, **187** 653–660.

Hirao, K., & Itoh, T. (1986) The Planet–A Halley encounters, *Nature* **321** 294–297.

Kissel, J. (1986) The Giotto Particulate Impact Analyser, In: *The Giotto mission—its scientific investigations*, ESA SP-1077, 67–83.

Maas, D., Göller, J.R., Grün, E., Lange, G., Aigner, S., Igenbergs, E., Rott, M., & Weishaupt, U. (1986) Impact simulation experiments with nano- to microgram particles at the Munich Plasma Drag Accelerator, In: *20th ESLAB Symposium on the Exploration of Halley's Comet*, ESA SP-250, Vol. II 337–340.

Mazets, E.P., Sagdeev, R.Z., Aptekar, R.L., Golenetskii, S.V., Guryan, Yu.A., Dyachkov, A.V., Ilyinskii, V.N., Panov, V.N., Petrov, G.G., Savvin, A.V., Sokolov, I.A., Frederiks, D.D., Khavenson, N.G., Shapiro, V.D., & Shevchenko, V.I. (1986) Dust in comet Halley from VeGa observations, In: *20th ESLAB Symposium on the Exploration of Halley's Comet*, ESA SP-250, Vol. II, 3–10.

McDonnell, J.A.M. (1969) Calibration studies on a piezoelectric sensing diaphragm for the detection of micrometeorites in space, *J. Sc. Instruments (J. Phys. E.)* Series 2, **2** 1026.

McDonnell, J.A.M. (1978) Microparticle studies by space instrumentation, In: *Cosmic dust* ed. McDonnell, J.A.M. (Chichester and New York: Wiley), 337–426.

McDonnell, J.A.M. (1987) The Giotto Dust Impact Detection System, *J. Phys. E: Sci. Instrum.* **20** 741–758.

McDonnell, J.A.M., Alexander, M., Lyons, D., Tanner, W., Anz, P., Hyde, T., Chen, A.-L., Stevenson, T.J., & Evans, S.T. (1984) The impact of dust grains on fast fly-by spacecraft: momentum multiplication, measurements and theory, *Adv. Space Res.*, Vol. 4, 9, 297–301.

McDonnell, J.A.M., Alexander, W.M., Burton, W.M., Bussoletti, E., Clark, D.H., Evans, G.C., Evans, S.T., Firth, J.G., Grard, R.J.L., Grün, E., Hanner, M.S., Hughes, D.W., Igenbergs, E., Kuczera, H., Lindblad, B.A., Mandeville, J.-C., Minafra, A., Reading, D., Ridgeley, A., Schwehm, G.H., Stevenson, T.J., Sekanina, Z., Turner, R.F., Wallis, M.K., & Zarnecki, J.G. (1986a) The Giotto Dust Impact Detection System, In: *The Giotto mission—its scientific investigations* ESA SP-250, 85–107.

McDonnell, J.A.M., Kissel, J., Grün, E., Grard, R.J.L., Langevin, Y., Olearczyk, R.E., Perry, C.H., & Zarnecki, J.C. (1986b) Giotto's Dust Impact Detection System DIDSY and Particulate Impact Analyser PIA: Interim assessment of the dust distribution and properties within the coma, In: *20th ESLAB Symposium on the Exploration of Halley's Comet*, ESA SP-250, Vol. II 25–38.

McDonnell, J.A.M., Alexander, W.M., Burton, W.M., Bussoletti, E., Evans, G.C., Evans, S.T., Firth, J.G., Grard, R.J.L., Green, S.F., Grün, E., Hanner, M.S., Hughes, D.W., Igenbergs, E., Kissel, J., Kuczera, H., Lindblad, B.A., Langevin, Y., Mandeville, J.-C., Nappo, S., Pankiewicz, G.S., Perry, C.H., Schwehm, G.H., Sekanina, Z., Stevenson, T.J., Turner, R.F., Weishaupt, U., Wallis, M.K., & Zarnecki, J.C. (1987) The dust distribution within the inner coma of Comet P/Halley 1982i: encounter by Giotto's impact detectors, *Astronomy and Astrophysics*, **187** 719–741.

McDonnell, J.A.M., Green, S.F., Grün, E., Kissel, J., Nappo, S., Pankiewicz, G.S., & Perry, C.H. (1989a) In situ exploration of the dusty coma of comet Halley at Giotto's encounter: flux rates and time profits from 10^{-19} kg to 10^{-5} kg, *Adv. Space Res.* Vol.-9, **3** 277–280

McDonnell, J.A.M., Pankiewicz, G.S., Birchley, P.N.W., Green, S.F., & Perry, C.H. (1989b) The comet nucleus: Ice and dust morphological balances in a production surface of comet P/Halley, *Proc. XXth Lunar & Planet. Sci. Conf.*, in press.

McDonnell, J.A.M., Lammy, P.L., & Pankiewicz, G.S. (1989c) Physical properties of cometary dust, In: *Comets in the Post Halley Era*, Kluwer, in press.

Millis, R.L., & Schleicher, D.G. (1986) Rotational period of Comet Halley, *Nature* **325** 326–328.

Morley, T.A., & Fertig, J. (1986) Giotto's encounter with Comet Halley, a braking and shaking experience, *ESA Bulletin* **46** 71–73.

Pankiewicz, G.S., & McDonnell, J.A.M. (1986) Dust emission studies in Halley's inner coma: an approach to modelling the nucleus emission characteristics, In: *20th ESLAB Symposium on the Exploration of Halley's Comet*, ESA SP-250, Vol. II, 201–206.

Perry, C.H., Green, S.F., & McDonnell, J.A.M. (1988) A possible explanation for the inconsistency between the Giotto grain mass distribution and groundbased observations, In: *Infrared observations of Comets Halley and Wilson and properties of the grains*, workshop held at Cornell University 10–12 July 1987, NASA conference publication **3004** 178–180.

Sekanina, Z. (1987) Nucleus of Halley's Comet as a torque-free rigid rotator, *Nature* **325** 326–328.

Sekanina, Z., & Larson, S.M. (1986) Dust jets in Comet Halley observed by Giotto and from the ground, *Nature* **321** 357–361.

Simpson, J.A., Sagdeev, R.Z., Tuzzolino, A.J., Perkins, M.A., Ksanfomality, L.V., Rabinowitz, D., Lentz, G.A., Afonin, V.V., Erö, J., Keppler, E., Kosorokov, J., Petrova, E., Szabó, & Umlauft, G. (1986a) Dust Counter and Mass Analyser (DUCMA) measurements of Comet Halley's coma from VeGa spacecraft, *Nature* **321** 278–280.

Simpson, J.A., Rabinowitz, D., Tuzzolino, A.J., Ksanfomality, L.V., & Sagdeev, R.Z. (1986b): Halley's Comet coma dust particle mass spectra, flux distributions and jet structures derived from measurements on the VeGa-1 and VeGa-2 spacecraft, In: *20th ESLAB Symposium on the Exploration of Halley's Comet* Vol. II, 11–16.

Simpson, J.A. (1986c) Encounters of the Soviet VeGa spacecraft with Halley's Comet: A preliminary report on the scientific observations, In: *Papers read at a joint meeting of The Royal Society and The American Philosophical Society* **2** 109–136.

Vaisberg, O., Smirnov, V., & Omelchenko, A. (1986) Spatial distribution of low-mass particles ($m \leq 10^{-10}$ g.) in Comet Halley coma, In: *20th ESLAB Symposium on the Exploration of Halley's Comet*, Vol. II, 17–23.

Wallis, M.K. (1986) Hypervelocity dust impulses on the Comet Halley probes, *Planet. Space Sci.* **34** 1087–1089.

Zarnecki, J.C., Alexander, W.M., Burton, W.M., & Hanner, M.S. (1986) Mass distribution of particulates measured by Giotto's Dust Impact Detection System (DIDSY) in the close encounter period, In: *20th ESLAB Symposium on the Exploration of Halley's Comet*, Vol. II 185–190.

3

The dust coma structure of Comet Halley

O.L. Vaisberg

1 INTRODUCTION

The structure of the cometary dust coma is an important source of scientific information relevant to the study of the cometary nucleus and to the properties of the dust itself. The following factors control the motion of dust particles and the structure of the dust coma:

- source function or dust emission pattern of the nucleus, including temporal variations;
- acceleration of dust particles by expanding gas in close proximity to the nucleus region;
- solar gravitation;
- radiation pressure of solar electromagnetic emission on cometary particle;
- rotation of nucleus;
- electromagnetic forces;
- sublimation and splitting of particles.

Our degree of understanding of these factors is not uniform. Progress had been made by analysis of astronomical observations and by theoretical calculations before the first spacecraft encounter with Comet Halley in 1986. We will briefly discuss the main features of Comet Halley's dust coma: large-scale structure with shells and/or jets and central condensation, sunward cone of strong dust emission with jet-like structures, and small-scale inhomogeneities. Analysis of these structures as observed by spacecraft even at this preliminary stage contributes significantly to what was known from astronomical observa-tions. Subsequent discussion of observed structural features and some processes controlling dust behaviour is mostly based on personal experience with data from the SP-1 plasma impact detector on Vega spacecraft.

2 STRUCTURE OF COMA

Fig. 1, taken from Larson et al. (1986) clearly shows the large-scale structure of the dust coma: shells and jets and central condensation. Jets, fans, arcs, and halos are dominant features of Comet Halley's coma. They show permanent emission of dust from discrete sources on the sunlit side of the rotating nucleus (Sekanina & Larson 1984). Nearly circular halos approximately centred on the nucleus are permanent features of Comet Halley. They strongly differ from predicted paraboloid envelopes (Fig. 2) that have a 2:1 ratio of semilatus rectum to the vortex distance. (Divine et al. 1986).

It is possible to evaluate from Fig. 1 the linear scale of the main features. Distinct outer shells are seen at $30–50 \times 10^3$ kilometres on the sunward side and at $50–70 \times 10^3$ km on the flank. With contemporary electronic detectors (CCD—charge coupled device) the traces of dust are seen at 2×10^5 km from the nucleus (Green & Hughes 1986). The central dust condensation has a radius of $\sim 10 \times 10^3$ km (Fig. 1).

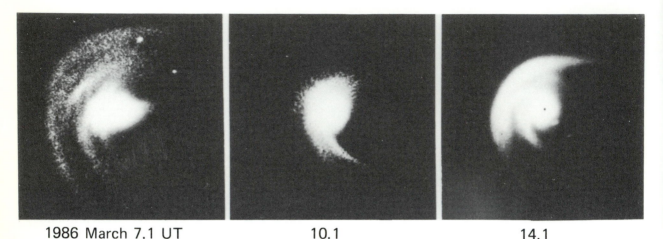

1986 March 7.1 UT   10.1   14.1

Fig. 1 — Enhanced broadband CCD images of Halley's Comet taken shortly after Vega-1 (7 March 1986 left), and the Vega-2 (10 March, middle) encounters, and at the time of the Giotto encounter (14 March, right) (Larson *et al.* 1986). The Sun is to the left.

3 MOTION OF DUST PARTICLES

The theory of dust motion was developed by several authors (Probstein 1969, Divine & Newburn 1984, Gombosi *et al.* 1982). The outflowing gas accelerates dust particles close into the nucleus region (~ 250 km), and the dust grains then decouple from the gas. The terminal velocity of the dust grains is given by

$$V_{\text{term}} \sim s^{-0.5} \tag{1}$$

where a is the radius of a (spherical) particle. The smallest particles attain a velocity close to the gas velocity. In the region not far from the nucleus the differential number density of grains is

$$n_{\text{k}} = \frac{q_{\text{k}} \cdot f}{V_{\text{term}} R^2}, \tag{2}$$

where q_{k} is the differential number flux of particles of kind k emitted in unit time, f the normalized source function, and R the cometocentric distance. Owing to (antisunward) pressure of solar radiation, the initially straight trajectory of the dust particle becomes curved. Particles of the same size and the same initial velocity move (in the cometocentric coordinate system) along parabolas within the dust paraboloid (Fig. 2). The distance from the cometary nucleus to the subsolar point of this paraboloid, or apex, is given by

$$E = \frac{V_{\text{term}}^2}{2g_{\text{r}}}, \tag{3}$$

where g_{r} is the acceleration due to radiation pressure, which is usually considered as normalized by solar gravitational acceleration g_{g} in the form

$$\beta = \frac{g_{\text{r}}}{g_{\text{g}}} \tag{4}$$

The value of β is strongly dependent on the radius and on the kind of particle. It reaches a maximum ~ 0.5 for nonconducting spheres of ~ 0.2 μm radius, and ~ 2 for conducting spheres with a ~ 0.1 μm radius (Burns *et al.* 1979).

The above approximation is valid for fairly short time intervals after the dust particle's release from

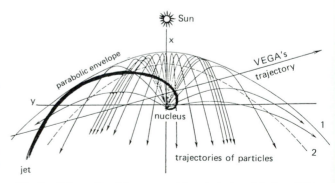

Fig. 2 — Schematics of dust coma with trajectories of dust particles, jet, and parabolic envelope (1). Envelope (2) is for particles with higher absorption efficiency of solar light or with lower terminal velocities, or both. x axis is toward the Sun, y axis is perpendicular to Sun–comet line.

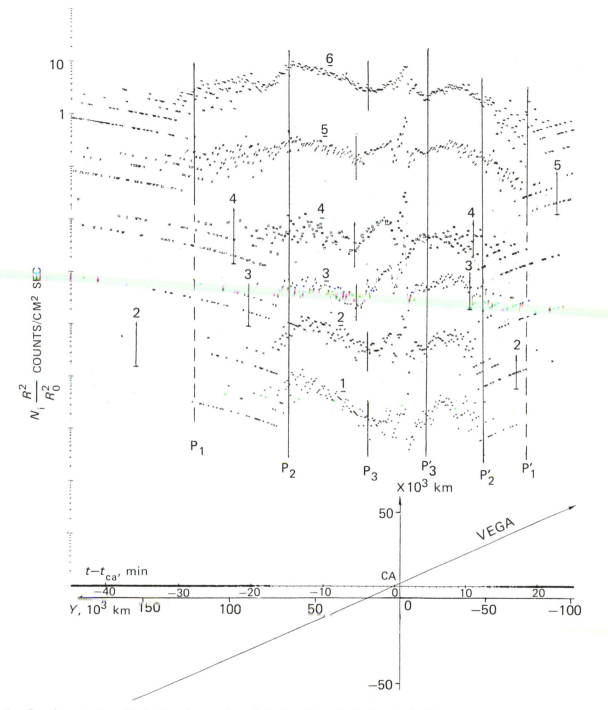

Fig. 3 — Counting rates N_i profiles in 6 decade mass channels (indicated by underlined number) of SP-1 detector on Vega-2, normalized by squared cometocentric distance R, $N_i R^2/R_o^2$ (R_o—distance at closest approach, CA). Transformation of channel numbers to mass ranges is given by $m_i = (10^{-N_i-9} \div 10^{-N_i-10})$ g. Scale is for channel 6; each successive profile is shifted down by one order of magnitude to avoid overlapping. Sensitive area of detector is 81 cm². Horizontal scales are time relative to closest approach and the distance along y-axis. Vega's trajectory is shown. Boundaries P_1 and P_2 were determined by sharper decreases of counting rates; boundary P_3 was determined by change of counting rate slope. Expected locations of envelope for different mass ranges are shown by vertical numbered lines (those for 1st and 6th channels and one for 5th channel are beyond the frame).

nucleus—less than about 2 days (Probstein 1969). For greater time intervals, the approximation breaks down because of the gradient of the solar gravitation force.

4 DISTRIBUTION OF PARTICLES

Fig. 3 shows the distribution of dust particles in the mass ranges from $\sim 10^{-16}$ g to $\sim 10^{-10}$ g along the path of Vega-2 on 9 March 1986 (Vaisberg et al. 1986a). The first particle was recorded at $\sim 320\,000$ km on the inbound trajectory. The counting rates in 6 decade mass ranges were multiplied by R^2/R^2_o, where R_o is the radial distance at closest approach (CA), to emphasize the large-scale inhomogeneities by smoothing out the increase of phase space with cometocentric distance (see equation (2)). Several boundaries P_i are easily identified by steeper gradients of dust number densities on the inbound trajectory profiles, but are less marked on the outbound trajectory profiles. The significance of these boundaries is emphasized by observed changes in the mass distribution of dust at the boundaries (Vaisberg et al. 1986b), accompanied by changes of the penetration abilities of dust grains at the same boundaries (Smirnov et al. 1986). The P_1 boundary is distinct in the 10^{-14}–10^{-16} g mass range, and is suggested in the 10^{-13}–10^{-14} g mass range profile. Boundary P_2 is easily identified in all mass ranges, occupying 6 orders of magnitude in particle mass. The P_3 boundaries are located at different distances (if they indeed represent the same feature).

Table 1. Apexes of dust shells

Spacecraft	Leg	Apex, 10^3 km		
		P_1	P_2	P_3
Vega-1	In	41.4	21.3	10.8
Vega-2	In	42.7	24.5	11.2
Vega-2	Out	45.3	30.6	16.5
Giotto (McDonnell et al. 1986)	In + Out		$\sim 120\,000$	

Note: These figures were obtained (Vaisberg et al. 1986b) with the use of slightly different criteria from those used in the present paper.

It is not easy to identify specific boundaries with the main paraboloid or with jets or shells without detailed modelling and comparison with ground-based obser-

vations. Projected along the paraboloid the boundaries P_1 and P_2 give values of apexes $(44.0 \pm 1.3) \times 10^3$ km and $(27.5 \pm 2.0) \times 10^3$ km, respectively (Table 1). Vaisberg et al. (1986a) suggested that boundary P_1 was the main paraboloid. Mazets et al. (1986) have suggested that the strong gradient on the inbound profile at $\sim 70\,000$ km is the boundary of the paraboloid. Boundary P_2, when extrapolated to the subsolar point, is in good agreement with apex $E = 26\,000$ km determined from ground-based observations for particles having a calculated $V_{term} \approx 440$ m s^{-1} and $\beta \approx 0.5$ (Sekanina 1981).

The existence of common boundaries in the outer coma for particles covering six orders of magnitude by mass (or two orders of magnitude by radius if all these particles are spheres with the same density) does not seem compatible with existing data on radiation pressure efficiency that show a change by a factor of ~ 10, while the radius of particles changes in one direction from the value corresponding to the maximum of radiation pressure (Burns et al. 1979).

Contradictions between observed boundaries and

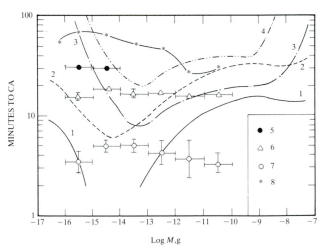

Fig. 4 — Comparison of boundaries in dust envelope observed with SP-1 (Vega-2) with calculations of their locations along Vega-1 inbound trajectory (Massonne & Grün 1986). Calculations were made for: magnetite, low velocities (1), magnetite, high velocities (2), olivine, low velocities (3), olivine, high velocities (4). Scale of time to closest approach was transformed to logarithmic with allowance for differences between Vega-1 and Vega-2 velocities and Sun–nucleus–spacecraft angles. Locations of observed boundaries P_1 (5), P_2 (6), and P_3 (7) shown with respective uncertainties indicated whenever they surpass the dimension of the respective mark. Times when first particles were registered in 7 mass channels (including mass range $3 \times 10^{-17} - 1 \times 10^{-16}$ g) are shown by asterisks (8).

those determined by radiation pressure are demonstrated by Fig. 4. Neither the locations of boundaries P_1 and P_2 nor the first recorded particles in different mass ranges obey the predictions of radiation pressure-influenced boundaries. It is possible that when following the 'same paraboloid' through different mass ranges we have to 'jump' from one boundary to another, but we found no direct evidence in favour of this.

Yet it seems that P_3 boundaries tell us something about radiation pressure influence (Figs 3, 4). Indeed, strong increase of particle number density in the mass range 10^{-13}–10^{-14} g with decreasing distance at $\sim 20\,000$ km on the inbound trajectory is clear evidence of the pile-up of these particles near this

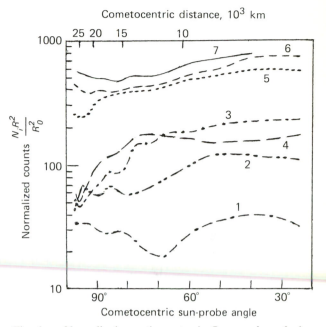

Fig. 6 — Normalized counting rates in 7 mass channels (see captions to Figs 3, 4) for small cometocentric distances ($\lesssim 30\,000$ km) versus Sun–comet–spacecraft angles.

range of distances. The location of this boundary is not far from that expected for dielectric particles near the maximum radiation pressure efficiency (Fig. 4), that is, $a \sim 0.25\ \mu m$, $\rho \sim 1\,\mathrm{g\ cm^{-3}}$, $m \sim 2 \times 10^{-14}$ g. It is possible that corresponding boundaries in other mass ranges are masked by other features.

The operation of the radiation pressure mechanism is seen on mass spectra of dust grains (Fig. 5). Depression in mass range 10^{-13}–10^{-14} g is already seen at $\sim 10\,000$ km from the nucleus, in accordance with model calculations (Bertaux & Cot 1982). This depression is more evident in mass spectra observed at greater cometocentric distances. But unlike the model calculations, this depression is not so strong, and it shows much weaker development with distance, compared to theoretical models.

Particles suffering strongest radiation pressure influence are most effective in light scattering. The location of a boundary at $\sim 20\,000$ km for 10^{-13}–10^{-14} g particles compares well with the dimension of central condensation (Fig. 1). This suggests that the dust particle counter sees this central condensation as the increase of number density of 10^{-13}–10^{-14} g particles.

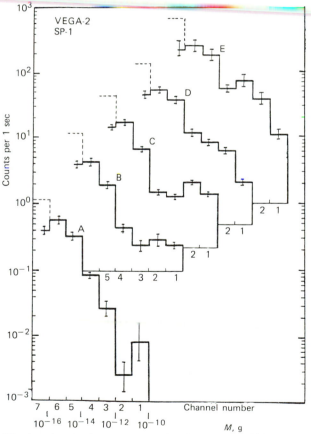

Fig. 5 — Averaged mass spectra of grains observed with SP-1 on inbound trajectory of Vega-2 at distances $\gtrsim 120\,000$ km (A), $\sim 90\,000$ km (B), $\sim 40\,000$ km (C), $\sim 20\,000$ km (D), and $\sim 10\,000$ km (E). Radial intervals are also indicated on Fig. 11b. Error bars indicate the error of the average obtained from a set of 10-sec. averaged spectra for specific intervals of distances.

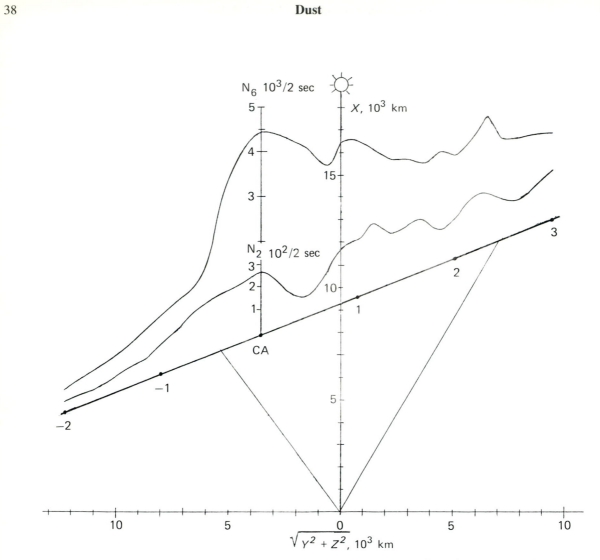

Fig. 7 — Counting rates profiles along Vega-1 trajectory in two mass ranges of SP-1: 3×10^{-17}–10^{-13} g (N_6) and 10^{-12}–10^{-11} g (N_2) within sunward–centred cone of dust activity. Trajectory of spacecraft is shown in solar-oriented cylindric coordinate system. Time is indicated in minutes relative to closest approach (CA). Note the different scales of counting rates.

5 DUST COMA IN SOLAR DIRECTION

Readers familiar with comets will understand the astonishment of participants in the study of Halley's Comet in March 1986 when they suddenly saw on the TV monitor a picture taken from several tens of thousand km which showed a strong Sun-directed luminous cone. With a vertex angle of 70°–80° this cone nearly coincided with the part of the Vega trajectory where the dust counter recorded strong jet-like inhomogeneities (Vaisberg *et al.* 1986a). This kind of fan-shaped coma, where most of the dust (and gas) release takes place, had been found earlier from astronomical observations (Sekanina 1979).

Background dust shows a moderate increase of number density with a decrease of the probe–nucleus–Sun angle (Fig. 6). Except for strong gradient along the trajectory in mass ranges 10^{-13}–10^{-14} g (and 10^{-12}–10^{-13} g) that we attribute to radiation pressure influence, the overall increase of number density from terminator to Sun-directed emission cone is by a factor of about 2. So the most important contribution to dust emission in the subsolar region should come from jets, in agreement with conclusion of Sekanina (1986) that discrete jets are the main form of nucleus activity.

Fig. 7 shows the variations of counting rates in two

mass ranges of dust for the Vega-1 crossing of the enhanced sunward emission cone at distances of $\sim (9–12) \times 10^3$ km. It could be seen that variations lie, on the average, within a factor of about 2, suggesting jet activity. It could be seen that although small dust grains $(3 \times 10^{-17}–10^{-13}$ g) show, in many cases, similarly to $10^{-12}–10^{-11}$ g particles, the profile of small grains is smoother than that of larger ones. A similar range of variations was observed during VeGa-2 pass through the sunward hemisphere at the distances $\sim (8–13) \times 10^3$ km (Vaisberg et al. 1986, Mazets et al. 1986).

One strongly different counting rate peak was found on Vega-2. Unlike all other peaks that did not show noticeable spatial displacement of different masses, this one had a strong spatial dispersion of masses (Fig. 8a) (Vaisberg et al. 1986a). The mass spectrum of dust within this feature was different from that of surrounding dust (Fig. 8b). It was interpreted as a narrow jet emitted from a rotating nucleus with

$$\varphi_i - \varphi_o = \frac{T}{2\pi} \cdot \frac{R_i}{V_i} \qquad (5)$$

where φ_i and R_i are the azimuth and cometocentric distance where ith masses were observed, V_i the respective terminal velocity, T the period of rotation of the nucleus, and φ_o the azimuth of the jet's source on the nucleus.

A self-consistent solution of (5) can be obtained by using a spatial dispersion curve $[t(R,\varphi), m_i]$ (Fig. 8c), where $t(R,\varphi)$ is the spacecraft time and m_i the effective mass of ith mass range. Equation (5) can be satisfied with $T \approx 52$ h and with prograde direction of rotation (Sekanina & Larson 1986, Sagdeev et al. 1986), with $\varphi_o = 35°$ west of noon meridian (in agreement with observations of one of the jets by a TV-camera on Vega-2 (K. Szegö and B. Smith, personal communications 1986), and with a velocity dispersion curve not markedly different from those obtained from solutions of gas lift equations (Divine & Newburn 1984, Gombosi et al. 1982). The result is shown on Fig. 9 in comparison with other observational evidence and with theoretical curves (Divine & Newburn 1984, Gombosi et al. 1982). Agreement between our data and theoretical curves is not surprising, as we selected one of several possible solutions for different φ_o comparing with theoretical calculations. Values of V_{term} obtained by Simpson et al. (1986) from interpretation of jets are significantly lower. The possibility of explaining this difference lies in the fact that dust measurements analysed (Simpson et al. 1986) give significantly lower counting rates in similar mass channels compared to SP-1 and SP-2 data (Vaisberg et al. 1986, Mazets et al. 1986), suggesting that Simpson et al.'s measurements in fact correspond to significantly larger masses.

The values of velocity from A'Hearn et al. were obtained from interpretation of CN jets. It was argued by Sekanina (1986) that the source of CN is ~ 0.1 μm grains, so we attributed the value of velocity from A'Hearn et al. to this mass range.

A value ≈ 440 m s^{-1} was obtained from the analysis of the width of the Comet Halley dust tail as observed

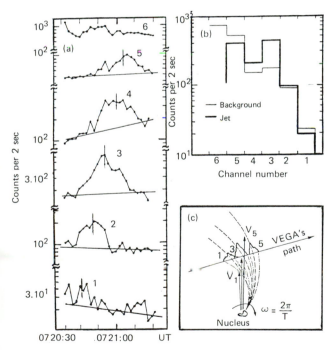

Fig. 8 — Dispersed inhomogeneity as observed with SP-1 on Vega-2 near Sun–comet line. (a) Counting rate profiles in 6 mass channels (see Fig. 3 caption for explanation). Note different counting rate scales. Trends in background counting rates are shown by straight lines. Vertical bars show the locations of 'centres of gravity' of excess counting rates above the background in each mass channel. (b) Mass spectra in inhomogeneity and in the background. (c) Interpretation of observed spatial mass-dispersion as a velocity-dispersed narrow jet from rotating with period T nucleus.

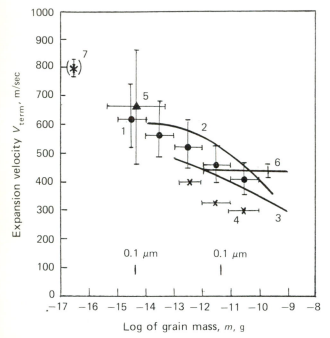

Fig. 9 — Terminal velocity dispersion of grains obtained from interpretation of dispersive jet (1, dots •) (Vaisberg *et al*. 1986a). Velocity error bars correspond to ± 5° error in location of jet's source on nucleus. Theoretical curves are: (2) from Divine and Newburn (1989), (3) from Gombosi *et al*. (1982). Crosses (X) (4) are from Simpson *et al*. (1986). Triangle (▲) (5) is an interpretation of CN jets from A'Hearn *et al*. (1986) that may be associated with ∼0.1 μm grains (Sekanina 1986). Horizontal bar (6) is located at the velocity value determined from the width of the tail during the 1910 apparition (Sekanina 1981), and may correspond to particles with $m \gtrsim 10^{-12}$ g providing largest scattering area. Asterisk * (7) is the expansion velocity of gas (Lämmerzahl *et al*. 1986). Masses of grains with radii 0.1 μm and 1.0 μm with densities 1 g cm^{-3} are indicated.

in the 1910 apparition (Sekanina 1981). This value of velocity and calculated $\beta \approx 0.5$ correspond to apex E

$\approx 26\,000$ km that is in reasonable agreement with the dust coma dimension in the solar direction (Sekanina 1981). Measured mass spectra of dust (Vaisberg *at al*. 1986, Mazets *et al*. 1986) show that the scattering properties of coma are mainly determined by particles with $m \gtrsim 10^{-12}$ g. So the value of dust velocity from (Sekanina 1981) is not in disagreement with our determination.

Clear mass dispersion of the discussed jet, unlike other observed inhomogeneities and differences of mass spectrum of this jet compared to surrounding dust (Fig. 8), suggest that the source of this jet is different from the source of the background dust. It was suggested that discrete jets originate from vents in cometary crust (Sekanina 1986). It is possible that the mass dispersive jet observed on Vega-2 supports this supposition.

We can speculate on the expected positions of the paraboloid by using our velocity dispersion curve (if it can be applied not only to jet but also to other cometary dust), and with the β value (equation (4)) for basalt (Burns *et al*. 1979). Calculated values are listed in Table 2, and are marked on Fig. 3. It does not seem that the observed behaviour in different mass ranges conforms to this simple consideration. Additional analysis and modelling are needed to clarify the situation.

6 DUST PROFILES

It appears that even small-scale fluctuations of the spatial distribution of particles may be meaningful. Deviations of counting rates of dust counters from Poisson statistics have been found (Simpson *et al*. 1986, E. Mazets personal communication 1986).

Table 2.   Calculated locations of paraboloids for different mass ranges

Quantity	Mass (g)				
	3×10^{-11}	3×10^{-12}	3×10^{-13}	3×10^{-14}	3×10^{-15}
V_{term}, m s^{-1}	406	457	523	575	620
β for basalt (Burns *et al*. 1979)	0.095	0.235	0.51	0.58	0.33
$E = \dfrac{V_{\text{term}}^2}{2 g_g \beta}$, km	1.04×10^5	5.33×10^4	3.22×10^4	3.41×10^4	6.98×10^4
R inbound, km	3.24×10^5	1.66×10^5	1.00×10^5	1.06×10^5	2.18×10^5
R outbound, km	1.53×10^5	7.85×10^4	4.74×10^4	5.02×10^4	1.03×10^5

$g_g = 0.8334$ cm sec^{-2} for Vega-2

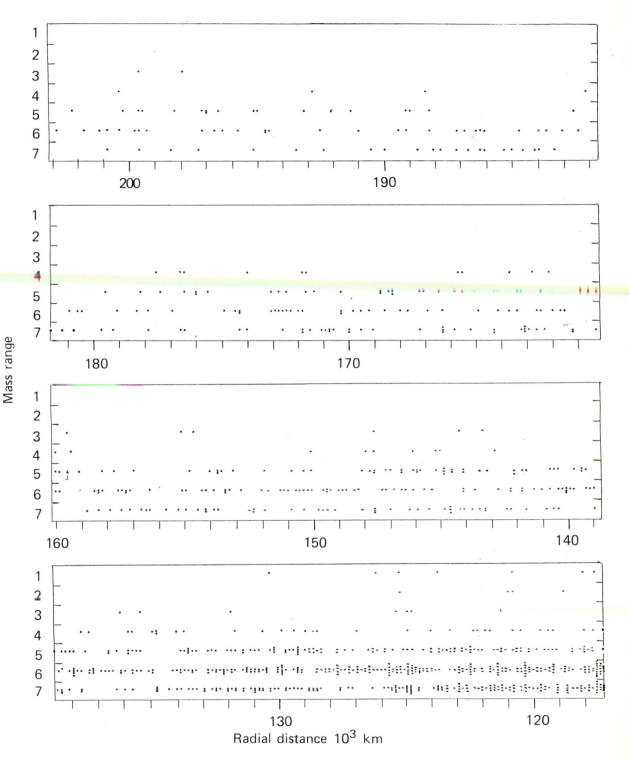

Fig. 10 — Dust particles recorded with SP-1 beyond P_1 envelope on inbound trajectory of Vega-2.

Examples of clustering of small particles ($m \lesssim 10^{-15}$ g) as observed with SP-1 on Vega-2 at distance $\sim 250\,000$ km are shown in Fig. 10.

The 3-day interval between the Vega-1 and Vega-2 encounters with Comet Halley allow one to look for temporal variations. Fig. 11 shows that on 6 March 1986 (Vega-1 flyby) the dust coma was about 2.5 times denser than during Vega-2 flyby on 9 March 1986. It seems that not only the Vega-2 inbound profile and outbound profile, but also the Vega-1 inbound profile show similar structural features.

They demonstrate that respective boundaries were

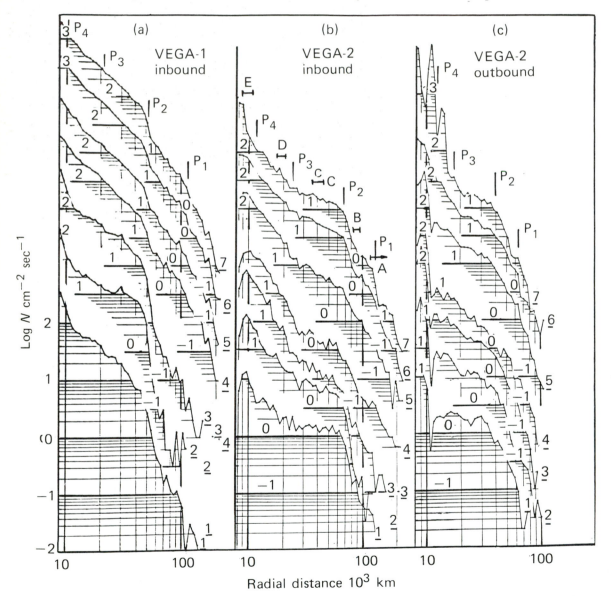

Fig. 11 — Radial dust profiles from SP-1 on Vega-1 inbound (a) and Vega-2 inbound (b) and outbound (c). Vega-1 profiles are (underlined numbers): *1*, channels $1+3$; *2*, channel 2; 3—channels $1+2+3$, 4—channels $4+6$, 5—channels $5+7$; 6, channels $4+5+6+7$; 7, channels $1+2+3+4+5+6+7$. Underlined numbers on Vega-2 profiles correspond to channels numbers. Conversion of channel numbers in mass ranges is given in Fig. 3 caption. Numbers near profiles are counting rates (logarithms). Boundaries P_i are indicated. Bars A to E on Vega-2 inbound profiles are for radial intervals where mass spectra on Fig. 5 were obtained.

about 1.2 times further from the nucleus for Vega-2 compared to Vega-1. The larger spatial extent of Vega-2 dust envelope is confirmed by the comparative distances at which the first particles were recorded: ~260 000 km and ~320 000 km for Vega-1 and Vega-2 respectively. The cometocentric distance of the dust boundary is $\sim g_r$ (see equation (3)), and for Vega-1 g_r was about 1.13 times larger because of the smaller heliocentric distance of the comet. So the observed differences in linear scales can be explained to a large extent by a decrease of radiation pressure, although other factors may be important, including a change of gas expansion velocity or the effect of mass loading of gas by dust.

If correspondence between inbound boundaries and outbound ones on Vega-2 profiles was properly established, it means that outbound boundaries are located at larger cometocentric distances in respect to a tentative envelope. This suggests a higher dust velocity in the 'evening' side of the comet, and influence of the thermal inertia of the surface of the nucleus.

It has been suggested (Boehnhardt 1986, Sekanina 1986) that electromagnetic forces may play a significant role on the motion of small grains. It was argued that these forces are not important, as the distribution of small ($\lesssim 0.1$ μm) grains does not show asymmetry (Sekanina et al. 1986).

It is interesting to consider whether the observed structures are merely associated with jet activity, or there are really several paraboloids inserted one in another. Nearly abrupt changes of dust mass spectra at boundaries observed in Comet Halley coma and apparent similarity between structures that existed during Vega-1 and Vega-2 encounters (Vaisberg et al. 1986b, Smirnov et al. 1986) suggest that dust particles in different structures or paraboloids have different properties. So in situ measurements confirm the conclusion obtained from astronomical data that two different components exist in the Comet Halley dust coma (Sekanina 1981). These may be two populations, suggested by measurements with thin foils: one dominating the mass range $m \lesssim 10^{-13}$ g with $\rho \gtrsim 1$ g cm^{-3} and one dominating the mass range $m \gtrsim 10^{-13}$ g with low density and/or fluffy structure (Lämmerzahl et al. 1986).

Some evidence of dust activity was found. CN jets were observed, strongly suggesting the outgassing of small grains (A'Hearn et al.).

There is some evidence of the secondary origin of small grains: existence of a common boundary of grains differing by 6 orders of magnitude by mass, weak cometocentric dependence of depletion in mass range where radiation pressure influence should be strongest, absence of smallest particles in dispersive jet, similar behaviour of grains with masses 3×10^{-17}–10^{-16} g and 10^{-16}–10^{-15} g throughout the coma, and their existence well beyond the main envelope (Vaisberg et al. 1986b), existence of clusters (Simpson et al. 1986, Mazets personal communication 1986, Vaisberg et al. 1986b), depletion of masses with $m \sim 10^{-12}$ g at $R < 4000$ km (McDonnell et al. 1986). This preliminary conclusion comfortably links with evidence of different structures of larger and smaller grains (Lämmerzahl et al. 1986).

7 CONCLUSIONS

In situ measurement of the dust coma structure of Comet Halley have strengthened the main conclusions obtained from astronomical observations, have shown the existence of different populations of grains, and have given some evidence of the activity and evolution of grains with time.

ACKNOWLEDGEMENTS

The author is indebted to R.Z. Sagdeev for help, to L.S. Gorn and M.V. Iovlev for development of the SP-1 detector, and to V.N. Smirnov, E. Grün, J.A.M. McDonnell, E.P. Mazets, V.D. Shapiro, and V.I. Tarnopolsky for useful discussions.

REFERENCES

A'Hearn, M.F., Hoban, S., Birch, P.V., Bowers, C., Martin, R., & Klinglesmith, D.A. III (1986) CN jets in Comet P/Halley, Nature 324 N 6098, 649–651
Bertaux, J.-L., & Cot, C. (1982) Segregation of cometary dust grains by differential radiation pressure, In: Cometary exploration, ed. Gombosi, T. vol. II, 113–120
Boehnhardt, H. (1986) The charge of fluffy dust grains of silicate and carbon near P/Halley and P/Giacobini–Zinner, In: Exploration of Halley's Comet, ed. Battrick, B., Rolfe, E.J., & Reinhard, R., ESA SP-250, 207–213
Burns, J.A., Lamy, Ph.L., & Soter, S. (1979) Radiation forces on small particles in the solar system, Icarus 40 N 1, 1–48
Gombosi, T.I., Szëgo, K., Gribov, B.E., Sagdeev, R.Z., Shapiro, V.D., Shevchenko, V.I., & Cravens, T.E. (1982) Gas dynamic

calculations of dust terminal velocities with realistic dust size distributions, In: *Cometary explorations*, ed. Gombosi, T.I., vol. II, 99–112

Divine, N., & Newburn, R.L., Jr. (1984) Numerical models for cometary dust environments, In: *Cometary exploration*, ed. Gombosi, T.I., vol. II, 81–98

Divine, N., Fechtig, H., Gombosi, T.I., Hanner, M.H., Keller, H.U., Larson, S.M., Mendis, D.A., Newbern, Ray L., Jr., Reinhard, R., Sekanina, Z., & Yeomanis, D.K. (1986) The Comet Halley dust and gas environment, *Space Sci. Rev.* **43** 1–104

Green, S.F., & Hughes, D.W. (1986) Near nucleus CCD imaging of Comet Halley at the time of the spacecraft encounters, In: *Exploration of Halley's Comet*, ed. Battrick, B., Rolfe, E.J., Reinhard, R., ESA SP-250 vol. II, 157–162

Lämmerzahl, P., Krankowsky, D., Hodges, R.R., Stubbeman, U., Woweries, J., Herrwerth, I., Berthelier, J.J., Illiano, J.M., Eberhardt, P., Dolder, U., Schulte, W., & Hoffman, J.H. (1986) Expansion velocity and temperature of gas and ions measured in the coma of Comet Halley, In: *Exploration of Halley's Comet*, eds. Battrick, B., Rolfe, E.J., & Reinhard, R., ESA SP-250, 179–182

Larson, S., Sekanina, Z., Levy, D., Tapia, S., Senay, M. (1986) Comet Halley near-nucleus phenomena in 1986. In: *Exploration of Halley's Comet*, eds. Battrick, B., Rolfe, E.J., Reinhard, R., ESA SP-250, vol. II, 145–150

Massonne, L., & Grün, E. (1986) Structures in Halley's dust coma, In: *Exploration of Halley's Comet* eds. Battrick, B., Rolfe, E.J., Reinhard, R., ESA SP-250, vol. III, 319–322

Mazets, E.P., Sagdeev, R.Z., Aptekar, P.L., Golenetskii, S.V., Guryan, Yu.A., Dyachkov, A.V., Ilyinskii, V.N., Panov, V.N., Petrov, G.G., Savvin, A.V., Sokolov, I.A., Frederiks, D.D., Khavenson, N.G., Shapiro, V.D., & Shevchenko, V.I. (1986) Dust in Comet Halley from Vega observations. In: *Exploration of Halley's Comet*, eds. Battrick, B., Rolfe, E.J., Reinhard, R., ESA SP-250, vol. II, 3–10

McDonnell, J.A.M., Kissel, J., Grün, E., Grard, R.J.L., Langevin, Y., Olearczyk, R.E., Perry, C.H., & Zarnecki, J.C. (1986) Giotto's dust impact detection system DIDSY and particulate impact analyzer PIA: interim assessment of the dust distribution and properties within the coma, In: *Exploration of Halley's Comet*, eds. Battrick, B., Rolfe, E.J., Reinhard, R., ESA SP-250, vol. II, 25–38

Probstein, R.F. (1969) The dusty gas dynamics of comet heads, In: *Problems of hydrodynamics and continuum mechanics*, ed. Bishop *et al.*, Society of Industrial and Applied Mathematics, Philadelphia 568–583

Sagdeev, R.Z., Blamont, J., Galeev, A.A., Moroz, V.I., Shapiro, V.D., Shevchenko, V.I., & Szegö, K. (1986) Vega spacecraft encounters with Comet Halley, *Nature* **321** N 6067, 259–268

Sekanina, Z. (1979) Fan-shaped coma, orientation of rotation axis, and surface structure of a cometary nucleus. 1. Test of a model on four comets. *Icarus* **37** N 2, 420–442

Sekanina, Z. (1981) Properties of dust particles in Comet Halley from observations made in 1910 during its encounter with the Earth. In: *The Comet Halley dust and gas environment*, ESA SP-174, 55–65

Sekanina, Z. (1986) Dust environment of Comet Halley, In: *Exploration of Halley's Comet*, eds. Battrick, B., Rolfe, E.J., & Reinhard, R., ESA SP-250, vol. II, 131–143

Sekanina, Z., & Larson, S.M. (1984) Coma morphology and dust-emission pattern of periodic Comet Halley. II. Nucleus spin vector and modelling of major dust features in 1910. *Astron, J.*, **89** 1408–1425, 1446–1447

Sekanina, Z., & Larson, S.M. (1986) Dust jets in Comet Halley observed by Giotto and from the ground, *Nature* **321** N 6067, 361–363

Sekanina, Z., Larson, S.M., Emerson, G., Helin, E.F., & Schmidt, R.E. (1986) Sunward spike and equatorial plane of Halley's Comet, In: *Exploration of Halley's Comet*, eds. Battrick, B., Rolfe, E.J., & Reinhard, R., ESA SP-250, vol. II, 177–181

Simpson, J.A., Rabinowitz, D., Tuzzolino, A.J., Ksanfomality, L.V., & Sagdeev, R.Z. (1986) Halley's Comet dust particle mass spectra, flux distributions and jet structures derived from measurements on the Vega-1 and Vega-2 spacecraft, In: *Exploration of Halley's Comet*, eds. Battrick, B., Rolfe, E.J., & Reinhard, R., ESA SP-250, vol. II, 11–16

Smirnov, V., Vaisberg, O., & Anisimov, S. (1986) An attempt to evaluate the structure of cometary dust particles, In: *Exploration of Halley's Comet*, eds. Battrick, B., Rolfe, E.J., Reinhard, R., ESA SP-250, vol. II, 195–199

Vaisberg, O.L., Smirnov, V.N., Gorn, L.S., Iovlev, M.V., Balikchin, M.A., Klimov, S.I., Savin, S.P., Shapiro, V.D., & Shevchenko, V.I. (1986a) Dust coma structure of Comet Halley from SP-1 detector measurements, *Nature* **321** N 6067, 274–276

Vaisberg, O.L., Smirnov, V.N., & Omelchenko, A. (1986b) Spatial distribution of low-mass dust particles ($m \leq 10^{-10}$ g) in Comet Halley coma, In: *Exploration of Halley's Comet*, eds. Battrick, B., Rolfe, E.J., & Reinhard, R., ESA SP-250, vol. II 17–23

Wallis, M.K., & Hassan, M.H.A. (1982) Electrodynamics of cometary dust particles, In: *Cometary exploration*, vol. II, ed. Gombosi, T., 57–63

4

Dust jet morphology in 1986 ground-based images of Comet Halley

Stephen Larson

1 INTRODUCTION

The cometary coma is a temporary atmosphere of gas and dust whose distribution is governed by its source at the nucleus and its subsequent interaction with the surrounding environment. An inhomogeneous nucleus with discrete sources of dust emission will produce a coma pattern which changes as the nucleus rotates and/or the source production rate varies. Dusty comets, such as Comet Bennett (1970 II), may have dust jets with spectacular pinwheel patterns due to nucleus rotation (Larson & Minton 1971).

The dust jets seen in Comet Halley in 1910 and 1985–86 have been studied as tracers of the motion of the jet sources on the nucleus (Sekanina & Larson 1984, 1986a). Dust jets, observed visually in 1682 by Hevelius (1685) and in 1835–36 by Maclear (1838), Herschel (1847), Struve (1839), Schwabe (1835) and Bessel (1836), are a well established feature of Comet Halley. The jets were also seen in 1910 and recorded in photographs taken from several observatories (Comstock 1915, Bobrovnikoff 1931, Perrine *et al.* 1934). Qualitative studies described the daily changes in the form of the coma features, but the usually short nightly observing window at a single observatory prevented unambiguous identification of the same feature on other nights.

The organization of the International Halley Watch presented the opportunity to obtain data with sufficient temporal and spatial resolution to study the evolution of the dust jets in detail. The Near-Nucleus Studies Net, established to obtain high-resolution ground-based images from several longitudes, generated an unprecedented amount of images during Comet Halley's most active phases (November 1985–June 1986). Over 3500 images are contained in the Halley Archive; those shown here represent only one data set. Thanks to the cooperation of the net members, the archive contains enough data to keep many scientists busy for years studying the detailed evolution of jet sources.

2 IMAGE PROCESSING

A major problem in studying dust jets in ground-based images is that they are of low contrast with respect to the general coma background which has a steep intensity gradient from the nucleus. Visual drawings, while usually suppressing the radial gradient and emphasizing the subtle, low-contrast features, have variable and unknown astrometric precision from which quantitative information might be derived. Photographs, on the other hand, contain positional information, but extracting features with contrasts of a few percent would require techniques not available to the astronomers of 1910.

Modern digital image processing was applied to many of the 1910 images only recently (Larson &

Sekanina 1984, 1985), and showed for the first time the evolution of features over as much as three days. After digitizing the original plates on a scanning microdensitometer, a radial and rotational shift-difference algorithm was applied to provide a map of the edges of sharp intensity discontinuities such as jets (Larson & Sekanina 1984). Other algorithms were tried with varying degrees of success. Azimuthal averaging did not work because the strong sunward asymmetry in coma brightness produced an antisunward 'hole' and not enough compensation in the sunward side. Common unsharp masking required 'sharp' masks that produced distracting edge effects. For the uncalibrated 1910 photographs in density space, the central brightness peak is compressed relative to the faint outer coma, so the combination radial/rotational shift differencing gave the most useful and unambiguous result (Fig. 1). The modern solid state devices (charge-coupled devices, or CCDs) used in 1985–86 exhibit good linearity over ranges of 4 or 5 decades, so the same algorithm does not work so well in intensity space. In the latter case, a double rotational shift enhances the jets down to the central condensation. Such shift-differenced images produce maps of intensity discontinuities, so the jets are seen by their edges.

3 1910 COMA FEATURES

The coma morphology seen in 1910 included spirals, arcs, and ragged-edged envelopes. Most of the dust was emitted in the sunward direction, and the spiral jets were seen to evolve into quasi-parabolic envelopes (Fig. 2). The behaviour of the jet edges was consistent with Keplerian motion of dust particles emitted continuously from the sunlit hemisphere of a rotating nucleus and deflected tailward by solar radiation pressure. The maximum sunward apex of dust particles was 6–7×10^4 km from the nucleus when 0.7–1.1 AU from the Sun. Crude photometry from the uncalibrated plates indicated a continuous range of local particle density enhancements of up to 30 times in strong, narrow jets. The conceptual basis for modelling the jets is described by Sekanina & Larson (1984). By varying the particle velocity, radiation pressure, and location of the source on the nucleus, it was possible to match the model exactly with the

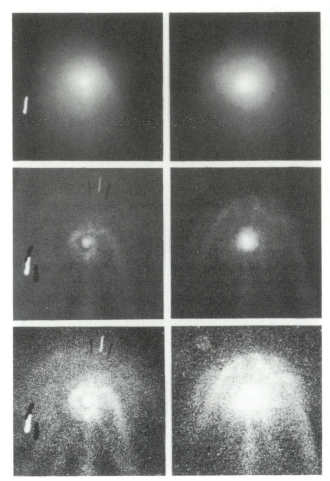

Fig. 1 — Coma feature enhancement by rotational/radial shift-differencing about the brightness peak for 2 images on June 4.737 (left) and June 5.718 (right), 1910. The upper row is digitized, but unprocessed, and the lower two rows are two contrasts of the enhanced version. The originals were taken by G.W. Ritchey with the Mt Wilson 1.5 m telescope. The Sun is up, and the edge of each frame is 180 000 km at the comet.

measured feature positions over several days. Adjusting the model parameters for many jets seen over a large range of viewing geometry, it was possible to set constraints on the nucleus spin vector. The observed jet spirals in early May indicated a clockwise nucleus rotation, while those in late May (after solar conjunction) indicated counterclockwise rotation. The change in sense of the spirals occurred on 22 May when the Earth crossed the nucleus's equator. Other constraints were the measured projected outflow velocities of 0.2–0.4 km s^{-1} and the frequency of distinctive anti-

Fig. 2 — Progression of spiral jet on May 7.493, 8.489, and 9.488 (left to right), 1910. Same scale, orientation and origin as in Fig. 1. Note how the expanding jet evolves into the outer envelope. The sense of the jet indicates clockwise rotation of the nucleus.

sunward jets. A self-consistent solution for all of the measured jets indicated a rotation period near 2.2 days and a 30° obliquity. The large number of variables makes it difficult to establish uniqueness, but at the time, it provided the best estimate of the spin vector (see Sekanina's Chapter 15 in this volume on the rotation problem). The limited number of images and already large number of variables precluded consideration of multiple spin modes, diurnal variation in ejection velocity, particle fragmentation and erratic vent migration.

4 1985–86 COMA FEATURES

Variable coma activity was seen when Comet Halley was 6 AU inbound (Spinrad *et al.* 1984) and perhaps earlier if the brightness variations commonly observed in 1983 were not modulated by nucleus rotation alone. Regular appearance of discrete jets seems to have commenced in November 1985 when the comet was 1.6 AU from the Sun. The activity level that produced sustained jet activity coincided with the appearance of the ion tail. Brightness 'outbursts' in late November and in December were related to the onset of bright jets. By early January 1986, the anti-sunward jet was seen at least twice 2 days apart, and dust jets were visible every day.

The postperihelion appearance in February–April 1986 was marked by much higher dust production and more spectacular jet activity. The roughly two-day appearance of new jets gave the impression that an 'active' hemisphere was rotating into sunlight, while

the sense and general shape of the jets were also consistent with those features recorded in the 1910 images. Qualitatively, the jet morphology in 1910 and 1986 was very similar (Larson *et al.* 1987; see Fig. 3). As indicated by the 1910 images, the March 1986 jets indicated clockwise rotation of the nucleus, and in mid-April, the jets appeared straight because the Earth was near the equatorial plane, and close to the Sun–comet line.

Since brightness outbursts coincided with the production of strong jets, it was not surprising that the periodic variations in CN, C_2, and continuum production rates identified first by Millis & Schleicher (1986) correlated with dust jet activity. When daily red-band (mostly continuum) images were enhanced in the same way and compared for the March–May 1986 period, the correlation is quite apparent (Larson & Sekanina 1987). The signal counts within constant linear distances of the photocentre of the CCD images followed the Millis–Schleicher lightcurve on the same days, and showed a decrease in amplitude as jet activity decreased in May. When images taken in March and early April (the phase angle was constant to within 18 degrees) are phased according to the 7.36 day period, a smooth progression of jet development can be seen (Fig. 4 and its accompanying Table 1). This is particularly impressive because of the repetition of complex jet configurations over 5 rotations in the sample. This smooth progression is not apparent when the images are phased to either a 2.2 day period, or $\frac{1}{2}$, $\frac{1}{3}$, or $\frac{1}{4}$ of the 7.4 day period.

The measured outflow velocities and jet curvature

1910

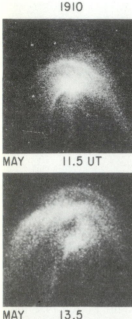

MAY   11.5 UT

MAY   13.5

MAY   23.2

JUNE   5.2

1986

JAN   2.1 UT

MARCH   18.1

JAN   4.1

MARCH   19.1

Fig. 3 — Comparison of 1910 (left column) and 1986 (right column) dust morphology. The Sun is up, and each frame is 75 000 km across at the comet. May 11.5 and June 5.2 were taken by Ritchey at Mt Wilson, May 13.5 by Curtis at Lick, and May 23.2 by Shaw at Helwan. Jan. 2 was taken by Larson and Levy at Catalina, and the other 1986 images were taken by Tapia and Senay at Boyden. The 1986 images are all red-band CCD images. All images have been enhanced in the same way.

give another story. On images taken a few hours apart (Fig. 5), the projected outflow velocities range from 0.2 to 0.5 km s^{-1}. Possible foreshortening makes these the lower limits on the actual outflow velocity. The curvature of the jets for this range of velocities requires nucleus rotation of 1.8–2.5 days. Detailed modelling of a few jets, assuming a pole with no obliquity, reinforces this conclusion (Larson & Sekanina 1987). Although the uncertainty in the apparently complex nucleus spin vector prevents extensive modelling at this time, the ground-based images will provide the key to determining the spin vector and lifetimes of the dust jet sources.

Table 1.   Dates of images phased to 7.36 day period (perihelion = 0)

First row (l–r)	Rotation	
March 24.030	5	0.784
March 17.102	4	0.843
March 31.992	6	0.866
March 10.953	3	0.891
March 25.034	5	0.921
March 3.128	2	0.944
March 18.077	4	0.975
Second row (l–r)		
April 2.054	7	0.010
March 26.025	6	0.055
March 19.116	5	0.116
April 3.019	7	0.141
March 12.096	4	0.163
March 27.026	6	0.191
March 20.101	5	0.250
Third row (l–r)		
April 4.055	7	0.282
March 13.087	4	0.297
March 28.042	6	0.329
March 21.037	5	0.377
Feb. 27.112	2	0.399
April 5.014	7	0.412
March 14.101	4	0.435
Fourth row (l–r)		
March 7.129	3	0.488
March 22.041	5	0.514
Feb. 28.127	0.536	
March 15.106	4	0.571
March 30.051.	6	0.602
March 23.068	0.653	
March 31.078	6	0.742

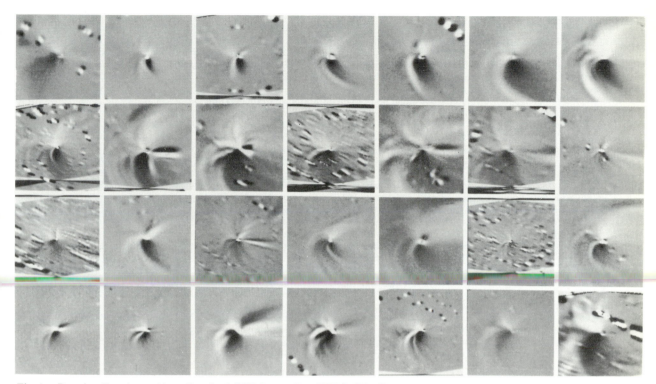

Fig. 4 — Rotationally enhanced broadband red CCD images from 27 Feb–5 April 1986 phased to a 7.36 day period. The montage covers 5 rotations and starts at a minimum in the light curve (see Table 1). All images are oriented north up and east to the left and are 75 000 km each side at the comet. The phase angle varied only 18 degrees. These were taken at the Boyden Observatory by Tapia, Senay, and Larson.

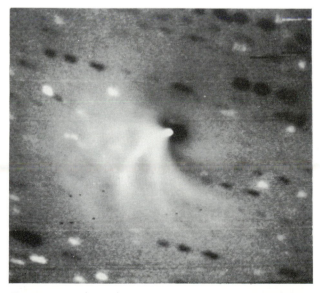

Fig. 5 — Differenced red-band CCD images from Boyden Observatory taken 2.6 hours apart on 6 April 1987. Any feature is seen because of radial outflow from the nucleus.

Dust features are sometimes related to gas features, such as in one documented sudden outburst which produced gas and dust shells in 1910 (Larson & Sekanina 1985). Discrete CN jets were first reported by A'Hearn *et al.* (1986), but they did not correlate very well with simultaneously observed dust jets (Larson *et al.* 1987). The number and morphology of the CN jets followed the 7.4 day period (A'Hearn *et al.* 1987). Because of the diffusion rate and scale length of CN, a population of small, invisible particles has been proposed as a source of CN far from the nucleus (A'Hearn *et al.* 1986). How this specialized population may relate to the visible particles is currently unknown. Much work remains to derive a consistent and coherent picture of dust/gas interaction.

REFERENCES

A'Hearn, M.F., Hoban, S., Birch, P.V., Bowers, C., Martin, R., & Klinglesmith, D.A. (1986) *Nature* **324** 649

A'Hearn, M.F., Hoban, S., Samarasinha, N., & Klinglesmith, D.A. (1987) Periodicities in the morphology of gaseous jets in Comet Halley. *Bull. Amer. Astron. Soc.* **19** 866

Bessel, F.W. (1836) Beobachtungen uber die physische Beschaffenheit des Halleyschen Kometen und dadurch veranlasste Bemerkungen. *Astron. Nachr.* **13** 185–232

Bobrovnikoff, N.T. (1931) Halley's Comet in its apparition of 1909–1911. *Publ. Lick Obs.* **17** 305–482

Comstock, G.C. (1915) *Report of Comet Committee 1909–1913*, Amer. Astron. Soc. 179–218

Herschel, J.F.W. (1847) *Results of astronomical observations at the Cape of Good Hope*. Smith & Elder, London. Chapter 5, 393–413

Hevelius, J. (1685) *Annus Climactericus* (Rhetius, Danzig)

Larson, S.M., & Minton, R.B. (1972) Photographic Observations of Comet Bennett, 1970 II. In: *Comets: scientific data and missions*, ed. Kuiper, G.P., & Roemer, E. (University of Arizona, Tucson), 183–208

Larson, S.M., & Sekanina, Z. (1984) Coma morphology and dust emission pattern of Periodic Comet Halley. I. High-resolution images taken at Mt. Wilson in 1910. *Astron. J.* **89** 571–578, 600–606

Larson, S.M., & Sekanina, Z. (1985) Coma morphology and dust emission pattern of Periodic Comet Halley. III. Additional high-resolution images taken in 1910. *Astron. J.* **90** 823–826, 917–923

Larson, S.M., & Sekanina, Z. (1987) Dust jet morphology and the light curve of Comet Halley. *Bull. Amer. Astron. Soc.* **19** 866

Larson, S.M., Sekanina, Z., Levy, D., Tapia, S., & Senay, M. (1987) Comet Halley near-nucleus phenomena in 1986. *Proceedings of the 20th ESLAB Symposium on the Exploration of Halley's Comet*. ESA SP-250 ed. Battrick, B., Rolfe, E.J., & Reinhard, R., (ESTEC, Noordwijk, The Netherlands). vol. II, 145–150

Maclear, T. (1838) Observations of Halley's Comet, Made at the Royal Observatory, Cape of Good Hope, in the years 1835 and 1836. *Mem. Roy. Astron Soc.* **10** 91–155

Millis, R.L., & Schleicher, D.G. (1986) Rotational period of Comet Halley. *Nature* **324** 646–649

Perrine, C.D., Winter, R., Symonds, F., & Glancy, A.E. (1934) Observaciones del Cometa Halley durante su aparicion en 1910. *Publicado por el Observatorio Cordoba* **25** 1–87

Schwabe, H. (1836) Der Halley'sche Komet. *Astron Nachr.* **13** 145–152

Sekanina, Z., & Larson, S.M. (1984) Coma morphology and dust emission pattern of Periodic Comet Halley. II. Nucleus spin vector and modelling of major dust features in 1910. *Astron. J.* **89** 1408–1425, 1446–1447

Sekanina, Z., & Larson, S.M. (1986a) Coma morphology and dust emission pattern of Periodic Comet Halley. IV. Spin vector refinement and map of discrete dust sources for 1910. *Astron. J* **92** 462–482

Sekanina, Z., & Larson, S.M. (1986b). Dust jets in Comet Halley observed by Giotto and from the ground. *Nature* **321** 357–361

Spinrad, H., Djorgovski, S., & Belton, M.S. (1984) Periodic Comet Halley (1982i), *IAU Circular* 3996

Struve, F.G.W. (1839) *Beobachtungen des Halleyschen Cometen auf der Dorpater Sternwarte*. Kaiserl. Akad. Wiss., St. Petersburg, 132 pp.

5

Infrared studies of Comet Halley during the 1985–1986 apparition

J.C. Zarnecki

1 INTRODUCTION

Until the arrival of the first spacecraft in the region of a cometary coma in 1986, our knowledge of cometary dust was derived predominantly from three types of observation, namely: (i) the study of scattered sunlight and thermal emission from the dust grains; (ii) observations of the structure and dynamics of cometary dust tails; and (iii) collection and analysis of micrometeoroids in the Earth's stratosphere. It is in the first category that infrared observations have been making an increasingly significant contribution, and, prior to the spacecraft encounters, much of the understanding of the properties of cometary dust arose through such observations. This chapter reviews such observations carried out at the recent Halley apparition, predominantly from ground-based telescopes but including also airborne and spacecraft-based measurements. For a description of the other techniques described above, the reader is referred to the reviews by Sekanina (1980), Brownlee (1985), and McDonnell (1986).

The infrared region of the electromagnetic spectrum is particularly rich for the study of the dust since it encompasses not only scattered solar radiation, but also thermal emission from the grains and, by employing higher resolution observations, spectral features potentially able to identify the composition of the dust. Furthermore, in this spectral region, the solid nucleus makes only a minor contribution to the total

flux (except for close flyby measurements), while additionally the contribution from the gases is expected to be generally small (though with some notable exceptions). This spectral region thus appears nearly ideal for studying the cometary dust. Fig. 1 (from Ney 1982) shows infrared spectra (relating to Comet West) typical of those measured prior to the Halley apparition. Several interesting features are apparent. First, at near infrared wavelengths, a slightly reddened solar spectrum is observed. Above ~ 3 μm, thermal emission from the grains is dominant, with an increasing grain temperature as the comet approaches the Sun (see Fig. 1). The derived temperature is generally found to be higher than the equilibrium blackbody temperature at the appropriate heliocentric distance, implying that the grains responsible for the emission are hot, absorbing grains of size ~ 10 μm (Hanner 1983). Furthermore, many comets at small heliocentric distances show an emission feature near 10 μm, evident also in Fig. 1, indicative of the presence of small silicate grains, although the appearance and behaviour of this feature has not been totally consistent from comet to comet. However, the broad features of the infrared spectrum have seemed to show remarkable consistency over the range of comets observed, from the first such observation of comet Ikeya–Seki by Becklin & Westphal (1966) through to the recent Halley apparition.

INFRARED OBSERVATIONS OF COMETS

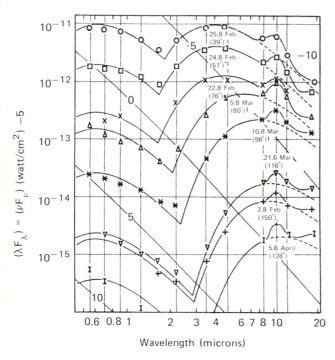

Fig. 1 — The behaviour of Comet West with scattering angles of 39°
to 150°. Dates and scattering angles are indicated (Ney 1982).

Using the approach followed by the Infrared
Spectroscopy and Radiometry Network of the International
Halley Watch, we can divide the infrared
Halley observations into five separate categories,
namely: (i) photometry, (ii) intermediate and (iii) high
resolution spectroscopy; (iv) imaging; and (v) polarization
observations. Some of these types of observation—e.g.
high resolution spectroscopy or polarization—will
be possible only at a relatively small
number of observatories where the specialized instruments
are available to observers. In addition, the
brightness of the comet will obviously dictate what
types of observation are possible at any time, with, for
example, high resolution spectroscopy possible perhaps
for only some months around perihelion, while at
the other extreme, photometry is possible over a time
span of perhaps as long as three years. It is still too
early to determine exactly how many infrared observations
have been carried out in each category, partly
because much of the data has yet to be fully analysed
and published, but also because, at the time of writing,

observations are still being carried out, though
admittedly only with the larger telescopes as the comet
fades. However, it is instructive to use the information
on infrared observations that have been reported to
the IHW. These represent observations both planned
and carried out, so we can expect them to be neither
totally accurate nor complete, but they should give a
reasonable indication of the relative importance of the
various types of observation. Fig. 2 shows the monthly
log of the infrared observations reported in this way,
divided into the five categories of observation, from
the first in December 1984 through to July 1986.
Summing all the reported observations, we find that
the total infrared observing time is divided up
approximately as follows: photometry 45%; intermediate
resolution spectroscopy 26%; high resolution
spectroscopy 3%; imaging 19%; and polarization 7%.
As one might expect, this simple analysis shows that
nearly half the infrared observations were photometric,
with the majority of the remaining observations
being either in the intermediate resolution spectroscopic
or imaging categories. High resolution spectroscopy
and polarization observations require both
specialized instrumentation and the comet to be
relatively bright, resulting in these categories accounting
for about 10% of all infrared observations.

Since infrared astronomy is a relatively new branch
of astronomy (dictated by the development of appropriate
instrumentation), no infrared observations
were carried out at the 1910 apparition. The very first
infrared observation of Comet Halley was therefore
performed at this apparition, in fact on 20 December
1984 by Birkett *et al.* (1985), who detected the

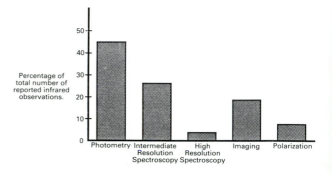

Fig. 2 — Division of infrared observations by category as reported
to the IHW for the period December 1984 to July 1986.

scattered solar radiation at J (1.25 μm) after a 30 min. integration with the United Kingdom Infrared Telescope (UKIRT). This observation was repeated by Hanner & Tokunaga (1985) who, some four weeks later, also detected the comet at J at the NASA Infrared Telescope Facility (IRTF). At this time the comet was still further than 5 AU from the Sun. The first detection of thermal radiation (at N, 10 μm) was by Zarnecki and McDonnell at UKIRT on 24 August 1985 (Green *et al.* 1986).

a small aperture will lead to difficulties in centring the aperture on the coma. Tracking may also prove to be a problem at some telescopes, particularly when the comet is faint. These and other problems associated with comet photometry have been described by Green *et al.* (1986). An excellent detailed description of the procedures to be followed for the highest accuracy photometry has been provided by Tokunaga (1986). These procedures have been followed by several of the Halley infrared observers in order to achieve precision results and to allow reliable inter-comparison of measurements by different observers.

2 PHOTOMETRY

Photometry both in the scattered solar continuum (predominantly J, H, and K) and at thermal infrared wavelengths has been reported by a large number of observers, and, as already indicated, judged by the number of observations, this type of observation has constituted the major activity of the IR Net of the IHW. For photometry of the highest accuracy, comets present particular problems which may not be present for stellar photometry. Since the coma represents a region of extended emission, great care is needed in the choice of the amplitude and direction of the beam chop. Additionally, at wavelengths > 3 μm, noise increases with aperture size, yet the resultant choice of

2.1 Scattered light colours

The colour of the radiation scattered by grains of size comparable to the wavelength of the radiation is dependent on grain size, composition, and surface roughness, as well as the scattering angle, although Campins & Hanner (1982) have suggested that JHK cometary colours, by themselves, are not good indicators of composition or size of the dust particles. Although JHK observations have been reported by many observers, for the greatest consistency and precision of JHK colours, the relatively large datasets accumulated at the IRTF, UKIRT, and ESO only are considered here. Table 1 shows the J–H and H–K colours derived from the observations.

Table 1.   Mean near infrared colours derived from observations at the IRTF, UKIRT, and ESO

Observatory and reference	J–H	H–K	No. of nights/observations and span of observations
IRTF Tokunaga *et al.* (1986b)	0.48 ± 0.01	0.17 ± 0.01	10/10 23 Aug 85–8 Jan 86
UKIRT Green *et al.* (1986)	0.45 ± 0.01	0.15 ± 0.01	11/19 10 Apr 85–3 May 86
ESO Bouchet *et al.* (1987)	0.40 ± 0.01	0.21 ± 0.01	15/42 26 Sept 85–24 Jan 86

The errors quoted have all been derived by the author from the errors in the individual measurements published by the appropriate observers, and are *standard errors* (i.e. the standard deviation in the appropriate mean). This has been done as there is some confusion in the published errors. For example, for the IRTF data (Tokunaga *et al.* 1986b), two different errors have been published by the authors, namely, for the H–K colour, ±0.003 in section IV(b) of the text, and ±0.01 in table V for the same parameter. The discrepancy most likely represents the fact that in one case it is the sample standard deviation that is quoted, while in the other case, it is the standard error. To eliminate possible confusion, the author has therefore derived weighted means and errors for all these measurements, using a common technique.

The agreement between the IRTF and UKIRT colours is close, and within the errors. The ESO measurements are perhaps slightly discordant with the others, J–H being somewhat low by comparison and H–K high. This may be significant, and it is interesting to note that the J–K colours from the three sets of observations are in closer agreement (i.e. 0.65, 0.60, and 0.61). It should also be noted that the internal scatter of the ESO dataset is considerably greater than for the IRTF or UKIRT data (i.e. for the J–H colour, the standard deviation is 0.09, 0.05, and 0.03 for the ESO, UKIRT, and IRTF data respectively).

These average colours seem to be wholly consistent with values measured from other comets (e.g. Campins & Hanner 1982). Additionally, both Tokunaga *et al.* (1986b) and Bouchet *et al.* (1987) have indicated that their measured colours are essentially unchanging with respect to heliocentric distance or scattering angle. To confirm this assertion, the author has taken all the J–H colours determined from the IRTF, UKIRT, and ESO under consideration here and considered their variation with phase angle, b (180°—scattering angle). For a range of 11° to 55° for phase angle (or 125° to 169° for scattering angle), the J–H colour is essentially constant, although statistically the best fit is -0.0024 ± 0.0007 mag. per degree (phase angle). Campins & Hanner (1982) show a similar behaviour for J–H over this range of angles based on observations of comets P/Tuttle, P/Meier, P/Stephan-Oterma, and Bowell.

In common with observations at other wavelengths—i.e. visible (Le Fevre *et al.* 1984, West *et al.* 1986), and ultraviolet (Festou *et al.* 1986)—short-term variability in the infrared of up to 2 magnitudes has been noted by many observers (Russell *et al.* 1986, Suto *et al.* 1987, Lynch *et al.* 1986, Bouchet *et al.* 1987) associated with either rotation effects or outbursts. Tokunaga *et al.* (1986b) claim possible indications of a change of optical properties of grains associated with an outburst, when on 12 December 1985, their J–H and H–K colours were bluer than the average and the comet was significantly brighter than on the following day. However, more work is needed before it can be claimed with certainty that a change of grain type has been observed associated with outbursts of the nucleus.

2.2 Thermal emission and grain temperature

The temperature of the emitting dust grains is normally characterized by fitting a blackbody spectrum to the thermal infrared observations ($\lambda > 3 \ \mu m$). This approach does have limitations. For example, if particles of a wide range of sizes are responsible for the thermal emission, then temperatures derived at different infrared wavelengths will yield different temperatures—i.e. temperatures obtained by fitting to shorter wavelength data will be higher than those derived from fits at longer wavelengths. Thus, such modelling of the thermal emission can potentially yield information on the size of particle predominantly responsible for the particular thermal emission. For example, Harvey *et al.* (1985) claim that the shape of the spectrum in the 40–160 μm region, obtained from airborne observations, implies the existence of a significant population of grains with sizes of order a few tens of micrometres.

A search of the literature has yielded some 55 reports of dust temperature derived by fitting blackbody curves to spectrophotometric data both from groundbased and airborne observations (out to ~ 160 μm). These observations cover a range of heliocentric distances from 2.83 AU preperihelion. These data have been combined to investigate the variation of dust grain temperature as a function of heliocentric distance, and are shown in Fig. 3. For clarity, no error bars are shown on these values. Where quoted, the errors vary considerably, generally being large when the emission is weak at large heliocentric distances (up to ± 25 K) and reasonably small (at best ± 5 K) at closer distances. Previously, Divine (1981) and Eaton (1984) have studied the similar variation of dust temperature with distance for a variety of comets. Divine (1981), based on a sample of 6 comets, showed a dependence of temperature, T, on heliocentric distance, r (in AU), of the form $T = 310 \ r^{-0.58}$. More recently, Eaton (1984) derived $T = 329 \ r^{-0.53}$ based on a larger sample of 12 comets. For comparison, the equilibrium temperature of a rapidly rotating blackbody is given by $T = 277 \ r^{-0.5}$. The data shown in Fig. 3 have been investigated for a similar power law dependence. Note that the discrepant point at $T = 730$ K and $r = 0.79$ AU from Phillips *et al.* (1986) has been omitted from the fit. Investigation of this result shows that the temperature is derived from observations at J, H, K, and L (i.e. 1.25–3.45 μm), at the extreme short wavelength end of the thermal spectrum, making a temperature derivation very uncertain owing to the difficulty of subtracting the scattered continuum contribution. Giving all data points equal weighting in the fit, a relation of the form $T = 336\,r^{-0.52}$ is derived, with an uncertainty of $\pm\ 0.04$ in the index. This is very close to $T = 329r^{-0.53}$ derived by Eaton (1984), and shows that in this admittedly

rather crude approach to dust characterization, Halley does not differ significantly from the behaviour of other comets studied in this manner.

2.3 Dust production rate

Measurement of the thermal emission allows the dust production rate to be calculated. This can be done either by fitting a theoretical spectrum for an appropriate size distribution to the observed data (Tokunaga *et al.* 1986b) or more simply by obtaining

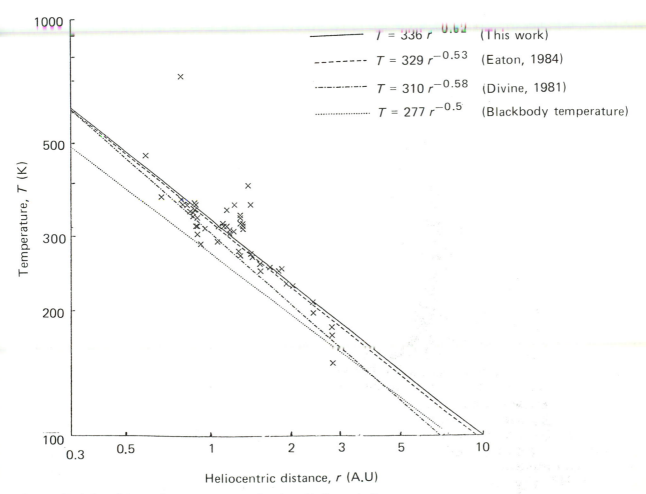

Fig. 3 — Variation of dust grain temperature as a function of heliocentric distance.

the value of the peak flux density as described by Ney (1982). However, relatively few of the Halley thermal measurements have yet been analysed in this way to derive dust production rates. The only extensive set of calculations appears to be those of Tokunaga *et al.* (1986b) who have calculated preperihelion dust production rates varying from 10^5 g s^{-1} at 2.81 AU to 9×10^6 g s^{-1} at 0.90 AU. Green *et al.* (1986) report that their measurements are in broad agreement with these, but otherwise few such dust production rates calculated from infrared measurements have been presented to date.

3 POLARIZATION

Prior to the current Halley apparition, there had been few measurements of the infrared polarization of comets, although some measurements have been reported for Comet West (Oishi *et al.* 1978). For Comet Halley, the only extensive measurements reported to date have been those by Brooke *et al.* (1986). These observers measured the linear polarization at J, H, and K over a time span of 8 months at both CTIO (Cerro Tololo Inter-American Observatory) and the IRTF, covering a range of phase angles from 9°.4 to 65°.3 . The major observational results are shown in Fig. 4 where the percentage linear polarization is plotted as a function of scattering angle (180° − phase angle) for the three filters. Also shown are results for two other comets (Hartley–Good and Thiele) observed over the same period. For Halley, the JHK polarizations vary approximately linearly with phase angle over the range measured here, with an overall shape similar to that measured in the visible for the dust continuum (e.g. Bastien *et al.* 1986).

A new result claimed by these authors is that there is a change in the wavelength dependence of the polarization as a function of phase angle. For small phase angle β, the dependence of polarization with wavelength is flat, while at higher values, polarization tends to increase with wavelength, although it should be cautioned that the uncertainty in some of the data points is high. They also point out that this trend observed in the infrared is strongly supported by the

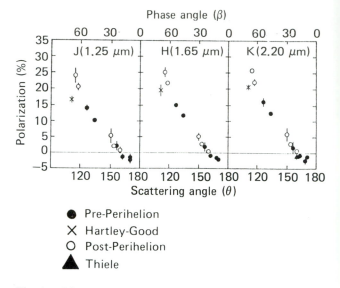

Fig. 4 — Linear polarization of Comet Halley and other comets (Brooke *et al.* 1986)

observations of Mukai *et al.* (1986) which show a similar trend in the visible.

Brooke *et al.* (1986) use Mie theory calculations in an attempt to fit the polarization measurements described here to various grain size distributions. Using the size distributions measured by the Giotto DIDSY (McDonnell *et al.* 1986) and VeGa SP-2 (Mazets *et al.* 1986), they show that no single component size model is able to fit the observed infrared polarization vs phase angle curve. The authors conclude that it is necessary to invoke a two-component grain model, involving a moderately absorbing material and a 'dirty' silicate. Scattering by the silicates gives the negative branch at large scattering angles, while the absorbing material dominates the polarization at smaller scattering angles. However, the limitations of the application of Mie theory in this context should be noted—namely, it is strictly applicable only to the case of homogeneous spherical grains. The true picture may, of course, be very different from this idealized picture (cf. Brownlee particles). However, these observations should prove invaluable in providing a constraint for dust models for Comet Halley, as any realistic model must be able to show consistency with these observations. To complicate matters further, however, Mukai *et al.*

(1986), who performed visible polarization measurements mentioned previously, report that their data may be fitted by a single-component model, although they do not rule out multi-component models.

4 SPECTROSCOPY

The infrared spectral region is potentially well suited for spectroscopic studies of cometary dust as various frost species (e.g. H_2O, CO_2, CH_4, NH_3, H_2S, and NH_4SH) show spectral features here (Combes 1982) as well as the well known spectral signatures of silicates in the 10 μm and 20 μm regions. Before Halley, spectroscopic studies of comets had yielded few positive unambiguous results, apart from observations of the broad silicate features, mentioned above, from several comets. The only other spectral feature of note was probably the marginal detection of an ice feature at 3 μm from Comet Cernis claimed by Hanner (1984).

However, it may transpire in the long term that of all infrared observations of Halley, spectroscopic observations will have proved to be the most fruitful. Several spectroscopic features have been observed, most of the information being quite novel in the context of cometary studies. At this early stage of interpreting the data from this apparition, the complete picture is very far from clear and unambiguous, and identifications are still rare. However, it seems likely that these observations may well be the most profound in determining the nature of the cometary dust. They are potentially able to provide the tightest constraints to cometary dust models.

4.1 Gaseous H_2O

Chronologically, probably the first positive spectral observations were those by Knacke et al. (1986b) in late 1985. Normally, water band observations are not possible from the ground because of interference by terrestrial water vapour. However, these workers took advantage of the fact that if the geocentric velocity of

a comet is sufficiently high, certain lines can be Doppler shifted to wavelengths at which adequate atmospheric transmission is achievable. In particular they identified the intercombination vibrational bands of H_2O near 1.4 μm and 1.9 μm as suitable. Using a circular variable filter (CVF) spectrometer, the band 1.4–2.5 μm was observed from UKIRT and the IRTF in late October and early November 1985 with a resolution of 130 and 30 respectively. Both observations show features at around 1.4 and 1.9 μm but at low statistical significance. It should be appreciated that observations near the terrestrial water bands are notoriously difficult, and that the results should be interpreted accordingly. The derived H_2O production rate is in reasonable agreement with the independently measured OH production rate, supporting the suggestion that OH is the daughter molecule derived from the observed H_2O.

Support for the previous observation came with the VeGa encounters in March 1986. Both spacecraft carried a three-channel spectrometer (TKS) including an infrared instrument covering the band 0.95–1.97 μm with a resolution of 70. This instrument failed on VeGa-1, but VeGa-2 returned infrared spectra (Krasnopolsky et al. 1985, 1986). An important feature of the spectrum is an emission band at 1.38 μm, identified with H_2O—presumably the same feature as observed earlier by Knacke et al. (1986b). Also detected in this observation were the (0,0) band of CN at 1.1 μm, and the OH vibrational-rotational bands in the long wavelength part of the observed spectrum.

To reduce (though not completely eliminate) the problem of atmospheric water, observations can be made from aircraft altitudes, and in fact such observations were carried out on the NASA-Kuiper Airborne Observatory, using the University of Arizona high resolution Fourier transform spectrometer (FTS) with a resolution of about 10^5 (Mumma et al. 1986). Observations on two separate days in December 1985 of the 2.61 to 2.71 μm band yielded detections of nine Doppler-shifted lines in the ν_3 band of H_2O. These are mostly detected with high signal to noise, and these observations can therefore properly be regarded as the first definite detection of H_2O in a comet. From the measured intensities of the observed lines, the authors are able to derive the H_2O production rate, the ortho-para water ratio, the nuclear-spin temperature, and the populations of various rotational states in the ground vibrational states. This observation, in particular, illustrates the pivotal role that high-resolution infrared spectroscopy is going to play in the future development of infrared spectroscopy.

Further observations with the same instrument were carried out postperihelion on four separate dates in March 1986 (Weaver *et al.* 1986) but covering a slightly wider spectral band than previously (2.5 to 2.7 μm). Fifteen spectral lines of H_2O in the v_3 band were detected, including all of the lines detected preperihelion. Of the six newly discovered lines, most (though not all) were not observed preperihelion purely because, according to these observations, the comet had significantly stronger emission post rather than preperihelion. Another particularly interesting feature of these data is the variability of the H_2O data over the span of six days of observation, reflecting presumably the variability of nucleus activity. As mentioned earlier, also of significance is the asymmetry in water production rate pre- and postperihelion at a heliocentric distance of about 1 AU where these observations were performed. The derived production rates postperihelion are typically an order of magnitude higher than those preperihelion; this contrasts with the conclusions from IUE data at similar heliocentric distances (Feldman *et al.* 1986). The reasons for this discrepancy are still not entirely clear.

Fourier transform spectroscopy was also carried out by Maillard *et al.* (1986) at the Canada–France–Hawaii (CFH) telescope. Amongst various features observed is one at 2.44 μm which is tentatively identified with the $(v_1 + v_3 - 2v_2)$ band of H_2O. Seven other lines are detected in this observation, four being identified with CN emission.

The same H_2O band as observed by the KAO FTS was covered by the onboard IKS infrared spectrometer on VeGa-1 (Combes *et al.* 1986a). This instrument employed a CVF, and therefore had only modest resolution (~ 40) compared with the high-resolution Fourier transform spectrometer. Therefore the many components of the v_3 band are unresolved by this instrument, which records a single emission feature at around 2.7 μm. According to Encrenaz (1988), the distribution of H_2O, derived from these data, as a function of nuclear distance, R, follows the expected

$1/R^2$ distribution for parent molecules. A short review of infrared observations of H_2O in Comet Halley has been presented by Larson *et al.* (1986).

4.2 3 μm ice band

Direct observational evidence for the existence of water ice is weak, even though models for the composition of cometary nuclei predict that water ice should be the most abundant icy component. For Comet Halley, Bregman *et al.* (1985) reported a detection of water ice, but at a low signal to noise ratio. Tokunaga *et al.* (1986a) observed the band 2.85–3.37 μm with a 32-channel spectrometer when the comet was at a heliocentric distance of 1.19 AU No ice reflection or absorption feature was detected at a strength of 15% of the continuum or greater. These observers have indicated that this negative observation may be consistent with the possibly positive observations of Bregman *et al.* (1985) for Halley, of Campins *et al.* (1983) for Bowell, and of Hanner (1984) for Cernis. These observations were carried out at heliocentric distances of 1.9 AU, 3.4–3.5 AU, and 3.3 AU respectively. This is significant since, as indicated by Hanner (1981), the lifetime of dirty ice against destruction by sublimation is a strong function of heliocentric distance, being perhaps as short as 1000 s or less at 1 AU, and correspondingly longer at greater distances. If correct, this would make detection of water ice from Earth extremely difficult at smaller heliocentric distances as the apertures employed would correspond to distances at the comet much greater than occupied by the 'icy grain halo'.

4.3 3–4 μm band

Arguably, the 3–4 μm band has been the most productive region for groundbased spectroscopy of Comet Halley, although the first definite detection in this band was probably made by the IKS infrared spectrometer on-board VeGa-1 on 6 March 1986 (Combes *et al.* 1986a, c). This represented the only *in situ* spacecraft measurement in this band, as the similar instrument on VeGa-2 produced no data, owing to a failure of the cryogenic system, while Giotto, Sakigake, and Suisei carried no such instruments. With a resolution of 40, IKS observed, in this region, a broad emission feature at 3.2–3.4 μm, while other features are apparent in the published spectrum. Over the following ten weeks, at least five groups observed features in this spectral region (Knacke *et al.* 1986a,c, Wickramasinghe & Allen 1986, Baas *et al.* 1986, Danks *et al.* 1986, 1987, Tokunaga *et al.* 1987). These observations were performed at CTIO, IRTF, AAO (Anglo-Australian Observatory), UKIRT, and ESO with spectral resolutions varying from 67 at CTIO to ~480 at UKIRT. These separate observations generally agree in their broad conclusions, namely (i) the existence of emission features at 3.30, 3.38, and 3.51 μm (with uncertainties of perhaps ± 0.02 μm) and (ii) that the line to thermal continuum ratio is variable, implying the existence of different populations for the continuum and emission features. Some other spectral features are present in some of the data sets, and in particular Danks *et al.* (1986, 1987) suggest the existence of perhaps ten distinct features over the range 2.8–4.2 μm. Several of the authors have compared these features with those seen in other astrophysical environments (for a brief review, see Encrenaz 1988). Some similarities exist with features at around 3.30 and 3.38 μm observed in various objects (e.g. 3.30 μm in emission near UV sources and 3.4 μm in absorption in dense clouds), but nowhere are these two features seen together in emission. Clearly, the spectrum of Halley in this region does not bear a strong resemblance to that for any other astronomical object.

The consensus opinion at present seems to indicate that the 3.30 and 3.38 μm features are associated with C-H stretching modes of both saturated and unsaturated hydrocarbons. The hydrocarbon molecules are solar ultraviolet pumped to produce infrared resonance fluorescence or scattering. Baas *et al.* (1986) suggest that an abundance of ~0.15% relative to H_2O is required of the hydrocarbon molecule(s) to produce the observed flux. Wickramasinghe & Allen (1986) suggest that the data are indicative of an origin in the solid phase, although a contribution from molecules of relatively high molecular weight cannot be excluded. The present data do not allow a definite identification of the suggested organic material, but should certainly allow valuable constraints on future dust grain models.

4.4 6–20 μm band

The original analysis of the VeGa-1 IKS data (Combes *et al.* 1986a) showed, apart from the first measurement of the 3–4 μm features, a broad emission feature centred around 7.5 μm. However airborne observations in December 1985 and April 1986 by the NASA Ames Faint Object Grating (AFOG) spectrometer onboard the KAO, with a resolution of ~60 at these wavelengths, showed no comparable feature (Campins *et al.* 1986c). These authors suggest that this discrepancy may indicate that this emission was prominent only in the region of the nucleus, and would therefore have been diluted in observations from Earth. However, it has been reported recently (Encrenaz 1988) that the original IKS analysis was probably in error, so that the corrected spectrum is now in full agreement with the KAO data.

The most prominent features in this spectral region are probably the silicate features at around 10 and 20 μm—these features have been observed in many comets (Ney 1982) at varying strengths relative to the continuum. This is often interpreted as being indicative of a two-component dust model, made up of hot absorbing grains and silicate grains (Campins & Hanner 1982). The silicate features have also been seen in Comet Halley by several observers at this apparition (Gehrz & Ney 1986, Combes *et al.* 1986a, Tokunaga *et al.* 1986b, Herter *et al.* 1986, Campins *et al.* 1986a, c). Several points of note are apparent in these observations. For example, Tokunaga *et al.* (1986b) claim that the 10 μm feature, when first observed by them on 13 December 1985, was much broader than is usual. Gehrz & Ney (1986) point out that the silicate features were generally absent from Halley at heliocentric distances greater than 1 AU, while at small distances, the signature was found to vary significantly. This the authors take to be an indication of variation in grain composition, with a reduction in the silicate feature being associated with a reduction in grain albedo. This implies that darker carbonaceous grains dominate when the silicates are missing. Campins *et al.* (1986c) find significant short-term variability, with, for example, the 10 μm feature being very much stronger on 10 April 1986 than two days earlier, while Campins *et al.* (1986a) studied the spatial distribution of thermal emission at three wavelengths and noted that the 10 μm feature is strongest at the nuclear condensation and decreased in all directions, but most slowly in the sunward direction. This is interpreted as showing that the sunward fan has a higher relative abundance of small grains than the rest of the coma.

Undoubtedly there is a good deal of further information tied up in the data on the silicate features. Much modelling and interpretation still needs to be done, and this will certainly be one of the main areas of activity in interpreting the Halley infrared data.

4.5 $\lambda > 20$ μm

At least four infrared observations at $\lambda > 20$ mm have been reported from airborne observatories. Using the KAO, Herter *et al.* (1986) observed the 16–30 μm band. Apart from possibly detecting the 20 μm silicate feature, a possible feature at 28 μm was reported along with weaker features at 23.8 and 26.7 μm, although the possibility that they are due to incorrect atmospheric correction cannot be absolutely ruled out. Glaccum *et al.* (1986) observed the 20–65 μm band, again from the KAO, one of their observations being three days after those of Herter *et al.* (1986) mentioned above. There was no 28 μm feature apparent at this time, suggesting that, if real, it must have weakened by more than a factor of 3 with respect to the continuum. In common with other observers, significant short term variation is observed. Campins *et al.* (1986b) and Harvey *et al.* (1985) observed the band 40–160 μm from the KAO with a 6 channel far infrared photometer. It is claimed that the spectrum flattened over a period of two days. In addition, the 50 μm/100 μm flux ratio is observed to decrease at angular distances 30″ and 60″ in the tailward direction. This they interpret as indicating evidence for a lower dust colour temperature or a larger average dust particle size away from the nuclear region. Finally, Stacey *et al.* (1986) report the detection of a rotational transition of OH at 119.4 μm, the first such detection in a comet, using a Fabry–Perot spectrometer. This observation enabled the OH production rate to be determined.

5 IMAGING

It is only very recently that infrared imaging instruments have become available for astronomical observations—i.e. the NASA/MSFC (Marshall Space Flight Center) 5×4 bolometer array (Telesco *et al.* 1986) or the UKIRT IRCAM (infrared camera) array—and it is only with instruments such as these that true imaging can be performed. Very few such images of Halley have been reported. To date, the only such true imaging has been with the NASA/MSFC 20 channel bolometer used to observe Halley at 10.8 μm in November 1985 at the Wyoming Infrared Observatory (Hammel *et al.* 1986) and in March 1986 at the IRTF (Campins *et al.* 1986a). In the November observations, simultaneous visible observations were taken with a CCD camera on the University of Hawaii 2.24 m telescope, allowing the construction of a map of the spatial variation of the geometric albedo as well as the distribution of infrared emission. Compared with the albedo for Comet P/Giacobini–Zinner, reported by the same observers (Hammel *et al.* 1986), that for Halley is claimed to be consistently higher (0.25 to 0.45) and shows less variation, though as with Giacobini–Zinner the region of lowest albedo is in the anti-Sun direction and there is a trend for albedo to increase radially from the nucleus, except along the tail. The postperihelion observations with the same instrument were carried out on three consecutive days in March 1986. The spatial distribution of the thermal emission shows two main features, namely a sunward fan and the absence of a well defined tail. This latter observation is interpreted by the observers as being indicative of a relative lack of large particles (radius > 100 μm), at least when compared with Comet Giacobini–Zinner. Further information is derived by comparing the maps made at 12.8 and 19.2 μm to yield a map of colour temperature. This shows a maximum at the nuclear condensation. Clearly, these data have only so far undergone preliminary analysis, and can still yield a good deal more information.

Other techniques of imaging without using an imaging array include mapping the coma with a single aperture, or scanning across the coma. A good example of the former technique is the observation by Hayward *et al.* (1988) at 10.3 μm on 13 March 1986, within hours of the Giotto encounter. This image shows a sunward fan, composed of several distinct jets. This is qualitatively similar to what was observed by Giotto (Keller *et al.* 1986, Edenhofer *et al.* 1986, Zarnecki *et al.* 1986) but on a scale of thousands of km. Subsequent analysis of this data (Hayward *et al.* 1988) has yielded an albedo map with values predominantly in the range 0.04–0.08, much lower (and generally consistent with values measured for other comets) than the values described earlier. However, this difference can be explained by the fact that the measurements of Hammel *et al.* (1986) were taken when Halley was close to opposition, possibly resulting in enhanced backscattering, while these results were taken at a phase angle of 64°.

Multi-aperture photometry, whereby a set of measurements using different sized apertures, all centred on the nucleus, are employed, does not constitute imaging of the infrared emission, but it does give some information on the dust distribution. For example, a constant radial outflow of dust at a constant velocity would yield a dependence of observed flux F_λ at wavelength λ from an aperture of diameter d at the coma of $F_\lambda \propto d^\alpha$ with $\alpha = 1$. Stanga *et al.* (1986) present such data for several dates between October 1985 and March 1986, generally indicating $\alpha = 1$, but on two dates, a considerably steeper dependence ($\alpha \sim 2$) is measured. At one of these times, increased infrared flux was observed at all measured wavelengths, suggesting that an increase in α is related to increased cometary activity. Other possible explanations include the limited lifetime against evaporation of icy grains and the varying efficiency of radiation

pressure on different-sized grains. Both of these factors would produce deviations from the simple dependency. Bouchet *et al.* (1987) also performed multi-aperture photometry on at least six nights, and found the situation to change considerably. For example, on 28 December 1985, the simple dependence is well fitted out to a distance of 22 000 km, while on 25 December there is a departure from the linear dependence for a diameter as small as 10 000 km. Green *et al.* (1986) performed detailed measurements at J, H, K, L, and L′ on 4 December 1985. All filters showed a similar dependence with aperture, yielding $\alpha = 1.36$.

ACKNOWLEDGEMENTS

The author acknowledges those Halley infrared observers who communicated their results ahead of publication. Assistance by, and discussions with, Stefano Nappo, Simon Green, and Tony McDonnell of the Unit for Space Sciences, University of Kent, are gratefully acknowledged. Peter Hingley (Royal Astronomical Society) assisted with literature searches, and Alison Rook (University of Kent) typed the manuscript. Finally, it must be acknowledged that much of the success of the infrared observations of Comet Halley was due to the organization and encouragement of the Infrared Network of the International Halley Watch, under the leadership of the discipline specialists, Roger Knacke and Thérèse Encrenaz.

REFERENCES

Baas, F., Geballe, T.R., & Walther, D.M. (1986) Spectroscopy of the 3.4 micron emission feature in Comet Halley *Astrophys. J.* **311** L97

Bastien, P., Ménard, F., & Nadeau, R. (1986) Linear polarization observations of P/Halley, *Mon. Not. R. astr. Soc.* **223** 827

Becklin, E.E., & Westphal, J.A. (1966) Infrared observations of Comet 1965f *Astrophys. J.* **145** 445

Birkett, C.M., Green, S.F., Longmore, A.J., & Zarnecki, J.C. (1985) *IAU Circular no.* 4025

Bouchet, P., Chalabaev, A., Danks, A., Encrenaz, T., Epchtein, N., & Le Bertre, T. (1987) Infrared photometry of Comet P/Halley before perihelion, *Astron. Astrophys.* **174** 288

Bregman, J.D., Witteborn, F.C., Rank, D.M., & Wooden, D. (1985) *IAU Circular No.* 4149

Brooke, T.Y., Knacke, R.F., & Joyce, R.R. (1986) Near-infrared studies of Comet Halley: polarization and color, *ESA SP-250* **II** 87

Brownlee, D.E. (1985) Cosmic dust: Collection and research, *Ann. Rev. Earth Planet. Sci* **13** 147

Campins, H., & Hanner, M.S. (1982) Interpreting the thermal properties of cometary dust, In: *Comets*, Wilkening, L.L. ed. University of Arizona Press, Tucson, 341–356

Campins, H., Rieke, G.H., & Lebofsky, M.J. (1983) Ice in Comet Bowell, *Nature* **301** 405

Campins, H., Telesco, C.M., Decher, R., & Ramsey, B.D. (1986a) Thermal infrared imaging of Comet Halley, *ESA SP-250* **II** 91

Campins, H., Joy, M., Harvey, P.M., Lester, D.F., & Ellis, H.B. Jr. (1986b) Airborne photometry of Comet Halley from 40 to 160 microns, *ESA SP-250* **II** 107

Campins, H., Bregman, J.D., Witteborn, F.C., Wooden, D.H., Rank, D.M., Allamandola, L.J., Cohen, M., & Tielens, A.G.G.M. (1986c) Airborne spectrophotometry of Comet Halley from 5 to 9 microns, *ESA SP-250* **II** 121

Combes, M. (1982) Infrared observations of Comet Halley, *Proceedings of the ESO Workshop on the need for coordinated ground-based observations of Halley's Comet*, Véron, P., Festou, M., & Kjär, K. ed. ESO, Munich, 83–91

Combes, M., Moroz, V.I., Crifo, J.F., Lamarre, J.M., Charra, J., Sanko, N.F., Soufflot, A., Bibring, J.P., Cazes, S., Coron, N., Crovisier, J., Emerich, C., Encrenaz, T., Gispert, R., Grigoryev, A.V., Guyot, G., Krasnopolsky, V.A., Nikolsky, Yu.V., & Rocard, F. (1986a) Infrared sounding of Comet Halley from VeGa-1, *Nature* **321** 266

Combes, M., Moroz, V., Crifo, J.F., Bibring, J.F., Coron, N., Crovisier, J., Encrenaz, T., Sanko, N., Grigoriev, A., Bockelee-Morvan, D., Gispert, R., Emerich, E., Lamarre, J.M., Rocard, F., Krasnopolsky, V., & Owen, T. (1986b) The 2.5 to 5 microns spectrum of Comet Halley from the IKS instrument of VeGa, *Adv. Space Res.* **5** 127

Combes, M., Moroz, V., Crifo, J.F., Bibring, J.P., Coron, N., Crovisier, J., Encrenaz, T., Sanko, N., Grigoriev, A., Bocklee-Morvan, D., Gispert, R., Emerich, C., Lamarre, J.M., Rocard, F., Krasnopolsky, V., & Owen, T. (1986c) Detection of parent molecules in Comet Halley from the IKS-VeGa experiment, *ESA SP-250* **I** 353.

Danks, A., Encrenaz, T., Bouchet, P., Le Bertre, T., Chalabaev, A., & Epchtein, N. (1986) Observation of an emission feature at 3.4 microns in the spectrum of Comet Halley, *ESA SP-250* **III** 103

Danks, A.C., Encrenaz, T., Bouchet, P., Le Bertre, T., &

Chalabaev, A. (1987) The spectrum of Comet P/Halley from 3.0 to 4.0 μm, *Astron. Astrophys.* **184** 329

Divine, N. (1981) A simple radiation model of cometary dust for P/Halley, *ESA SP-174* 47

Eaton, N. (1984) Comet dust-applications of Mie scattering, *Vistas in Astronomy* **27** 111

Edenhofer, P., Buschert, H., Porsche, H., Bird, M.K., Volland, H., Brenkle, J.P., Kursinsky, E.R., Mottinger, N.A., & Stelzried, C.T. (1986) Dust distribution of Comet Halley from the Giotto radio science experiment, *ESA SP-250* **II** 215

Encrenaz, T. (1988) The infrared spectrum of Comet P/Halley, In: *Comets to Cosmology. Lecture Notes in Physics No. 297*, ed. Lawrence, A. Springer Verlag, Berlin, 48–59.

Feldman, P.D., Festou, M.C., A'Hearn, M.F., Arpigny, C., Butterworth, P.S., Cosmovici, C.B., Danks, A.C., Gilmozzi, R., Jackson, W.M., McFadden, L.A., Patriarchi, P., Schleicher, D.G., Tozzi, G.P., Wallis, M.K., Weaver, H.A., & Woods, T.N. (1986) IUE observations of Comet Halley: evolution of the UV spectrum between September 1985 and July 1986, *ESA SP-250* **I** 325

Festou, M.C., Feldman, P.D., A'Hearn, M.F., Arpigny, C., Cosmovici, C.B., Danks, A.C., McFadden, L.A., Filmozzi, R., Patriarchi, P., Tozzi, G.P., Wallis, M.K., & Weaver, M.A. (1986) IUE observations of Comet Halley during the VeGa and Giotto encounters, *Nature* **321** 361

Gehrz, R.D., & Ney, E.P. (1986) Infrared temporal development of P/Halley, *ESA SP-250* **II** 101

Glaccum, W., Moseley, S.H., Campins, H., & Loewenstein, R.F. (1986) Airborne spectrophotometry of P/Halley from 20 to 65 microns, *ESA SP-250* **II** 111

Green, S.F., McDonnell, J.A.M., Pankiewicz, G.S.A., & Zarnecki, J.C (1986) The UKIRT observational programme, *ESA SP-250* **II** 81

Hammel, H.B., Storrs, A.D., Cruikshank, D.P., Telesco, C.M., Decher, R.M., & Campins, H. (1986) Albedo maps of comets P/Giacobini–Zinner and P/Halley, *ESA SP-250* **II** 73

Hanner, M.S. (1981) On the detectability of icy grains in the comae of comets, *Icarus* **47** 342

Hanner, M.S. (1983) The nature of cometary dust from remote sensing, *Cometary Exploration II*, ed. Gombosi, T., Hungarian Academy of Sciences, Budapest, 1

Hanner, M.S. (1984) Comet Cernis: icy grains at last? *Ap. J.* **277** L75

Hanner, M.S., & Tokunaga, A.T. (1985) *IAU Circular no.* 4034

Harvey, P.M., Campins, H., Ellis, H.B., Joy, M., & Lester, D.F. (1985) Far infrared photometry of Comet Halley from NASA's Kuiper Airborne Observatory, *Adv. Space Res.* **5** 335

Hayward, T.L., Gehrz, R.D., & Grasdalen, G.L. (1987) Ground-based infrared observations of Comet Halley, *Nature* **326** 55

Hayward, T.L., Grasdalen, G.L., & Green, S.F. (1988) An Albedo Map of P/Halley on 13 March 1986, *Infrared Observations of Comets Halley and Wilson and Properties of the Grains*, NASA Conference Publication, 151.

Herter, T., Gull, G.E., & Campins, H. (1986) Airborne spectrophotometry of P/Halley from 16 to 30 microns, *ESA SP-250* **II** 117

Keller, H.U., Arpigny, C, Barbieri, C., Bonnet, R.M., Cazes, S., Coradini, M., Cosmovici, C.B., Delamere, W.A., Huebner, W.F., Hughes, D.W., Jamar, C., Malaise, D., Reitsema, H.J., Schmidt, H.U., Schmidt, W.J.K.H., Seige, P., Whipple, F.L., & Wilhelm, K. (1986) First Halley multicolour camera imaging results from Giotto, *Nature* **321** 320

Knacke, R.F., Brooke, T.Y., & Joyce, R.R. (1986a) The 3.2 to 3.6 micron emission features in Comet Halley—comparison with interstellar and laboratory spectra, *ESA SP-250* **II** 95

Knacke, R.F., Noll, K.S., Geballe, T.R., Tokunaga, A.T., & Brooke, T.Y. (1986b) Ground-based detection of water in Comet Halley, *ESA SP-250* **III** 99

Knacke, R.F., Brooke, T.Y., & Joyce, R.R. (1986c) Observations of 3.2–3.6 micron emission features in Comet Halley, *Astrophys. J* **310** L49

Krasnopolsky, V.A., Gogoshev, M., Moreels, G., Moroz, V.I., Krysko, A.A., Barke, V.V., Jegulev, V.S., Parshev, V.A., Sanko, N.F., Tomashova, G.V., Tkachuk, A.Yu., Troshin, V.S., Novikov, Perminov, V.G., Sulakov, I.I., Fedorov, O.S., Gogosheva, Ts., Sargoichev, S., Palazov, K., Georgiev, A., Nedkov, I., Kanev, K., Clairmidi, J.K., Vincent, M., Mougin, B., Parisot, J.P., Zucconi, J.M., Lepage, J.P., Runavot, J., Bertaux, J.L., Blamont, J.E., Festou, M., Herse, M., Valnicek, B., & Vanysek, V. (1985) Near infrared observations of Comet Halley from VeGa-2, *Adv. Space Res.* **5** 12, 143

Krasnopolsky, V.A., Gogoshev, M., Moreels, G., Moroz, V.I., Krysko, A.A., Gogoshev, Ts., Palazov, K., Sargoichev, S., Clairmidi, J., Vincent, M., Bertaux, J.L., Blamont, J.E., Troshin, V.S., & Valnicek, B. (1986) Spectroscopic study of Comet Halley by the VeGa-2 three-channel spectrometer, *Nature* **321** 269

Larson, H.P., Mumma, M.J., Weaver, H.A., & Davis, D.S. (1986) Spectroscopy of comets; an infrared perspective, *ESA SP-250* **III** 223

Le Fevre, O., Lecacheux, J., Mathez, G., Lelièvre, G., Baudrand, J., & Lemonnier, J.P. (1984) Rotation of Comet P/Halley: recurrent brightening observed at the heliocentric distance of 8 AU, *Astron. Astrophys.* **138** L1

Lynch, D.K., Russell, R.W., Retig, D.A., Rice, C.J., & Young, R.M. (1986) Preperihelion observations of Comet Halley at 2–13 microns *ESA SP-250* **III** 479

Maillard, J.P., Crovisier, J., Encrenaz, T., & Combes, M. (1986) The spectrum of P/Halley between 0.9 and 2.5 microns, *ESA SP-250* **I** 359

McDonnell, J.A.M. (1986) Extraterrestrial material analysis: achievements and future opportunities for laboratory analysis in NASA and ESA planetary programmes, *Adv. Space Res.* **6** 7, 21

McDonnell, J.A.M., Kissel, J., Grün, E., Grard, R.J.L., Langevin, Y., Olearczyk, R.E., Perry, C.H., & Zarnecki, J.C. (1986) Giotto's dust impact detection system DIDSY and particulate impact analyser PIA: interim assessment of the dust distribution and properties within the coma, *ESA SP-250* **II** 25

Mazets, E.P., Sagdeev, R.Z., Aptekar, R.L., Golenetskii, S.V., Guryan, Yu. A., Dyatchkov, A.V., Illyinskii, V.N., Panov, V.N., Petrov, G.G., Sarvin, A.V., Sokolov, I.A., Frederiks, D.D., Khavenson, N.G., Shapiro, V.D., & Shevenenko, V.I. (1986) Dust in Comet Halley from VeGa observations, *ESA SP-250* **II** 3

Mukai, T., Mukai, S., & Kikuchi (1986) Role of small grains in the visible polarization of Comet Halley, *ESA SP-250* **II** 59

Mumma, M.J., Weaver, H.A., Larson, H.P., Scott Davis, D., & Williams, M. (1986) Detection of water vapor in Halley's Comet, *Science* **232** 1523

Ney, E.P. (1982) Infrared observations of comets In: *Comets*, ed. Wilkening, L.L., University of Arizona Press, Tucson, 323–340

Oishi, M., Kawara, K., Kobayashi, Y., Maihara, T., Noguchi, K., Okuda, H., Sato, S., Ilijima, T., & Ono, T. (1978) *Publ. Astr. Soc. Japan* **30** 149

Phillips, J.P., Mampaso, A., & Gorzon, F. (1986) Near-infrared profiles of Comet Halley, *Astron. Astrophys.* **161** L17

Russell, R.W., Lynch, D.K., Rudy, R.J., Rossano, G.S., Hackwell, J.A., & Campins, H.C. (1986) Multiple aperture airborne infrared measurements of Comet Halley, *ESA SP-250* **II** 125

Sekanina, Z. (1980) Physical characteristics of cometary dust from dynamical studies In: *Solid particles in the solar system* IAU Symp. No. 90, ed. Halliday, I., & McIntosh, B.A., Reidel, Boston, 237–250

Stacey, G.J., Lugten, J.B., & Genzel, R. (1986) Detection of OH from Comet Halley in the far-infrared, *ESA SP-250* **I** 369

Stanga, R., Munieri, A., Lorenzetti, D., Saraceno, P., & Strafella, F. (1986) Infrared monitoring of Halley's Comet *ESA SP-250* **III** 365

Suto, H., Maihara, T., Mizutani, K., Yamamoto, T., & Thomas, J.A. (1987) Near infrared spectrophotometric observations of Periodic Comet Halley (1982i), Publ. Astr. Soc. Japan, **39**, 925

Telesco, C.M., Decher, R., Baugher, C., Campins, H., Mozurkewich, D., Thronson, H.A., Cruikshank, D.P., Hammel, H.B., Larson, S., & Sekanina, Z. (1986) Thermal-infrared and visual imaging of Comet Giacobini–Zinner *Ap. J.* **310** L61

Tokunaga, A.T. (1986) *IRTF photometry manual*, University of Hawaii, Institute of Astronomy, Honolulu

Tokunaga, A.T., Smith, R.G., Nagata, T., DePoy, D.L., & Sellgren, K. (1986a) 3 micron spectroscopy of Comet Halley (1982i) *Astrophys. J.* **310** L45

Tokunaga, A.T., Golisch, W.F., Griep, D.M., Kaminski, C.D., & Hanner, M.S. (1986b) The NASA infrared telescope facility Comet Halley monitoring program. I. Preperihelion Results, *Astron. J.* **92** 1183

Tokunaga, A.T., Nagata, T., & Smith, R.G. (1987) Detection of a new emission band at 2.8 μm in comet P/Halley, *Astron. Astrophys.* **187**, 519

Weaver, H.A., Mumma, M.J., Larson, H.P., & Davis, D.S. (1986) Postperihelion observations of water in Comet Halley, *Nature* **324** 441

West, R.M., Pedersen, H., Monderen, P., Vio, R., & Grosbol, P. (1986) Postperihelion imaging of Comet Halley at ESO, *Nature* **321** 363

Wickramasinghe, D.T., & Allen, D.A. (1986) Discovery of organic grains in Comet Halley, *Nature* **323** 44

Zarnecki, J.C., McDonnell, J.A.M., Alexander, W.M., Burton, W.M., & Hanner, S. (1986) Mass distribution of particulates measured by Giotto's dust impact detection system (DIDSY) in the close encounter period, *ESA SP-250* **II** 185

6

The modelling of cometary dust features

L. Massonne

1 INTRODUCTION

The principal tool for modelling cometary dust features is a coma model, which is a method of computing the spatial density of the cometary dust at all locations in the coma. In scientific practice the coma model would normally have the form of a program computing the dust density at a range of locations, but pen and paper are generally sufficient to model a comet.

The results are determined by a set of model parameters describing the cometary nucleus and the dust particles. A comparison of the coma model results with *in situ* measurements and astronomical observations can help to improve the model parameters.

A coma model is of interest to a variety of users. In the preparation of a spacecraft mission to a comet some prediction of the dust densities is needed for the design of the probe and the scientific instruments. As in the Halley missions, the spacecraft operators may use a coma model to predict the risk of a fast flyby at different distances. In the evaluation of the measurements the model will be used to explain certain features in the dust flux and to deduce nuclear parameters such as rotation period and jet activity. Astronomers may use the coma model to simulate the overall structure of the comet, to determine the spin vector and rotation period by comparison with observed coma structures, and to map the comet's dust production activity.

This chapter tries to explain the computational method and to outline the applications of the results.

2 BASIC MODEL PARAMETERS

All dust models, so far, assume that the cometary dust is emitted radially from the region around the nucleus, with a velocity depending on the size and mass of the particles (Finson & Probstein 1968, Fertig & Schwehm 1984, Divine *et al.* 1985). As the dust is accelerated by the cometary gas, this assumption is true from about 20 nuclear radii outward (Finson & Probstein 1968, Gombosi *et al.* 1983, Gombosi 1987). At this distance the coupling of gas and dust ends, and the dynamical behaviour of the particles is controlled by a limited set of forces such as solar gravitation, solar radiation pressure, and electromagnetic forces. The various coma models differ mainly in the number and nature of the forces on the dust particles that they allow for. These forces will be discussed later in this chapter. Nearer to the nucleus than the 20 radii mentioned above, the motion of the dust particles is not necessarily radial, as the gas flow may develop a lateral component due to a day–night anisotropy of the gas production, as discussed by Wallis & Macpherson (1981) and Huebner *et al.* (1987). The velocity function of the dust particles can be deduced from astronomical observations. This approach was used by Bessel (1836), Bobrovnikoff (1931), and Finson & Probstein (1968). Alternatively, the terminal velocities of the dust grains can be obtained from hydrodynamical calculations. The most recent values, used in this chapter, are taken from Gombosi (1987). Examples of emission velocity functions from different authors are given in Fig. 1.

Another model parameter influencing the particle dynamics is the radiation pressure coefficient β, which describes the influence of the radiation pressure on the particle. The coefficient β is defined as the ratio of the acceleration g_r by radiation pressure and the acceleration g_g by gravitational forces, $\beta = g_r / g_g$. Its value depends on the complex refraction index of the particle material and the size and shape of the dust particle. Burns *et al.* (1979) give a good description of the radiation forces on small particles. The values most widely used are results from Mie calculations, such as those from Schwehm & Rohde (1979). They are strictly valid for spherical particles only. The cometary dust particles are most certainly not spherical, but only these calculations can provide the radiation pressure coefficient as a function of the particle size. Most laboratory measurements of scattering properties for irregularly shaped particles are performed with only a few particle sizes. Weiss-Wrana (1983) presents an overview of the available data. However, the Mie calculations form a

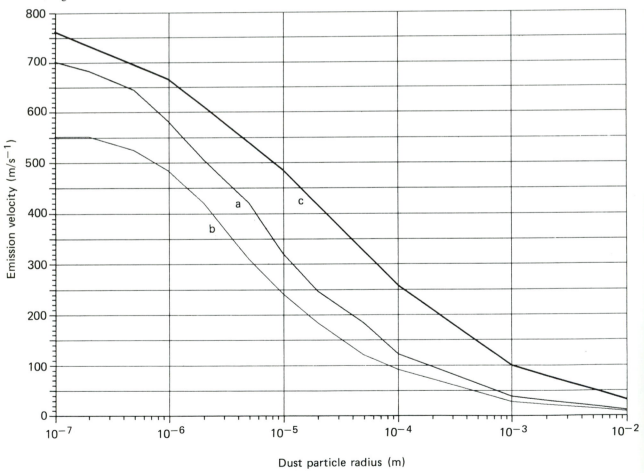

Fig. 1 — Dust particle emission velocities as function of the particle radius.
a: from Finson & Probstein (1968)
b: from Gombosi *et al.* (1983)
c: from Gombosi (1987)

good approximation to the real scattering behaviour of the cometary dust particles. In Fig. 2 the β functions of two materials are given. Olivine is a representative of a dielectric, and magnetite a representative of an absorbing material.

Some models, e.g. that of Mendis *et al.* (1985), take the interaction of the particles with the interplanetary magnetic field into account. The dust particles will be charged by impinging plasma particles which produce secondary particles and by the photoeffect of solar illumination. Böhnhardt (1987) computes the charging of small grains in the cometary environment, depending on their material and exposure time to the solar wind. To describe the Lorentz forces on the dust particles, some assumptions on the structure of the magnetic field in the vicinity of the comet are needed, as well as data on the charging of dust particles. This

charging effect is influenced by the plasma environment which itself depends on the magnetic field structure. Therefore the electromagnetic forces on dust particles are rather difficult to model in a realistic way.

3   PRINCIPLE   OF   DUST   DENSITY COMPUTATION

The spatial dust density at a given location results from the volume change of a volume element along the particle trajectory to the requested location. As the dust particle number in the volume element is constant, the density along the trajectory can be computed if the dust source function at the emission of the volume element is known. The time-dependent volume dV of the infinitesimal volume element is

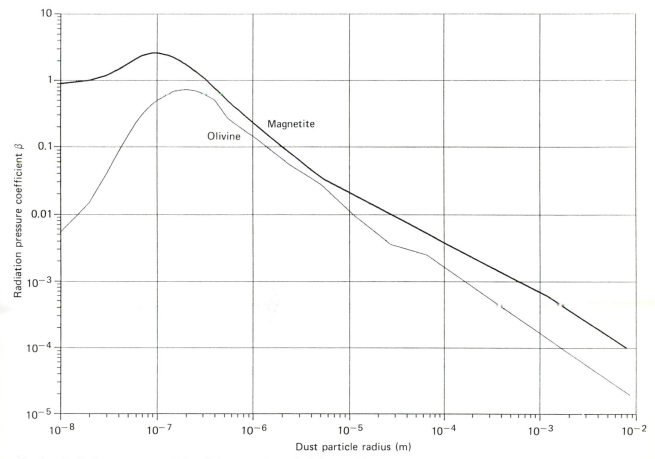

Fig. 2 — Radiation pressure coefficient β for magnetite and olivine particles as a function of particle size.

$$dV(t) = [R(\theta + d\theta), F, t - R(\theta, \phi, t)$$
$$\times \; R(\theta, \phi + d\theta, t) - R(\theta, \phi, t)]$$
$$\delta R(\theta, \phi, t)/\delta t \cdot dT. \tag{1}$$

The vector R gives the coordinate of the volume element in the CBS-system of coordinates (Massonne 1984, 1987). In this system the Sun–comet line remains fixed and forms the z-axis. The x-axis is located in the orbital plane of the comet, and is counted positive in the direction of motion. The y-axis is perpendicular on the other axes, and forms with them a right-handed system. The angles θ and ϕ define the emission direction of the dust particles, as given in Fig. 3.

The symbol dT describes the infinitesimal time interval in which the dust particles with the velocity $v = \delta R \; (\theta, \; \phi, \; t_E)$ have been emitted in the volume element at the emission time t_E.

Equation (1) is equivalent to

$$dV = \frac{\delta R}{\delta \theta} \times \frac{\delta R}{\delta \phi} \cdot \frac{\delta R}{\delta t} \cdot d\theta \cdot d\phi \cdot dT. \tag{2}$$

The source function $f(\theta, \; \phi, \; t)$ specifies the number dn of dust particles emitted in the direction $(\theta, \; \phi)$ per volume angle element and per time interval dT. It is

defined as

$$f(\theta, \phi, t) = \frac{dn(\theta, \phi, t)}{dT \cdot \sin \theta \cdot d\theta \cdot d\phi}. \tag{3}$$

The spatial dust density $\rho(\theta, \phi, t)$ results from particle number and volume to

$$\rho(\theta, \theta, \phi, t) = \frac{dn(\theta, \phi, t)}{dV(t)}. \tag{4}$$

With the definition (3) of the source function, insertion of (2) into (4) leads to

$$\rho(\theta, \phi, t) = \frac{f(\theta, \phi, t_E)}{\left| \dfrac{1}{\sin \theta} \cdot \dfrac{\delta R}{\delta \theta} \times \dfrac{\delta r}{\delta \phi} \cdot \dfrac{\delta R}{\delta t} \right|}. \tag{5}$$

The time t_E is the emission time, and $(\theta, \; \phi)$ are the emission coordinates of the dust particles which reach the requested location at the time t. As the emission is radial, the emission coordinates transform directly into the emission direction. The equations for the partial derivatives $\delta R/\delta\theta$, $\delta R/\delta\phi$ and $\delta R/\delta t$ depend on the assumed particle dynamics. Examples will be given in the next section.

From equation (5) it is clear that the computation of the spatial dust density at the location $R(t_A)$ at the time t_A requires the determination of the emission

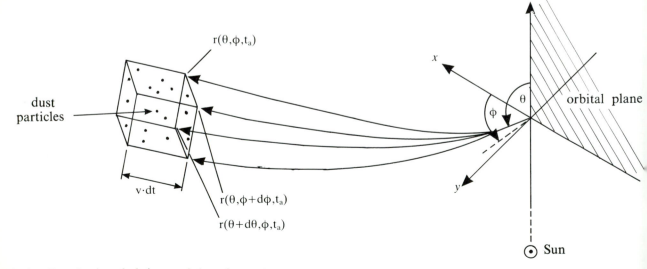

Fig. 3 — Dust density calculation, symbols, and geometry.

times and directions of all particles reaching that location. If there exist multiple trajectories to $R(t_A)$, their individual contributions to the spatial density are computed from equation (5) and are added to obtain the total dust density. This purely mathematical derivation will now be explained in detail.

4 DUST DENSITY COMPUTATION FOR DIFFERENT PARTICLE DYNAMICS

4.1 Radial expansion

The simplest model of dust dynamics assumes a radial outflow of the dust from the nucleus with constant velocity. In this case the equations from section 3 give fairly simple results. The particle location R as a function of the emission coordinates θ and ϕ is

$$R(t) = v \cdot (t - t_E) \cdot \begin{cases} \sin\theta \cdot \cos\phi \\ \sin\theta \cdot \sin\phi \\ \cos\theta. \end{cases} \quad (6)$$

The particle velocity is v, the emission time is t_E. The radial distance R of the dust location from the nucleus is then

$$R = v \cdot (t - t_E). \quad (7)$$

The three partial derivatives in equation (5) result from (6) as

$$\frac{\delta R}{\delta \theta} = v \cdot (t - t_E) \cdot \begin{cases} \cos\theta \cdot \cos\phi \\ \cos\theta \cdot \sin\phi \\ -\sin\theta \end{cases} \quad (8.1)$$

$$\frac{\delta R}{\delta f} = v \cdot (t - t_E) \cdot \begin{cases} -\sin\theta \cdot \sin\phi \\ \sin\theta \cdot \cos\phi \\ 0 \end{cases} \quad (8.2)$$

$$\frac{\delta R}{\delta t} = v \cdot \begin{cases} \sin\theta \cdot \cos\phi \\ \sin\theta \cdot \sin\phi \\ \cos\theta \end{cases} \quad (8.3)$$

Assuming a constant dust production f all over the nucleus surface, insertion of equations (8) into equation (5) leads to the spatial dust density

$$\rho = \frac{f}{|v^3 \cdot (t - t_E)^2|} = \frac{f}{|v \cdot R^2|} \quad (9)$$

Regardless of the rather crude assumption of constant particle velocity and dust production, this model gives an astonishingly good description of the dust environment near the nucleus. Many authors use it as a 'zero-order' approximation for data reduction, or to enhance structures due to higher order effects, i.e. dust production anisotropies. Grün et al. (1986) use this approach to analyse cometary images, and Vaisberg et al. (1986) evaluate some VeGa spaceprobe dust flux measurements. For the reduction of cometary images this dynamical approach is particularly useful, as the brightness distribution of the resulting dust cloud can be computed analytically. The brightness of the dust cloud is proportional to the integral over the dust density along the line of sight. The dust density ρ is a function of the distance R to the cometary nucleus only:

$$\rho(R) = c/R^2, \quad (10)$$

c being a constant that can be computed from equation (9). The density integral $I(d)$ along a line of sight passing the nucleus at a distance d (see Fig. 4) is

$$I(d) = \int_{-\infty}^{+\infty} \rho(x, d) \, dx. \quad (11)$$

With $R^2 = x^2 + d^2$ and (10), equation (11) can be transformed to

$$I(d) = c \cdot \int_{-\infty}^{+\infty} \frac{dx}{x^2 + d^2}. \quad (12)$$

This integral can be solved:

$$I(d) = \frac{\pi \cdot c}{d}. \quad (13)$$

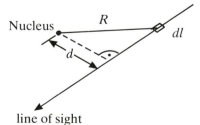

Fig. 4 — Dust density integral through a spherical dust density distribution.

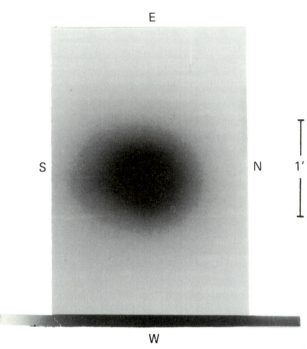

The brightness distribution resulting from a spherical dust cloud with the density distribution from equation (10) has radial symmetry and is inversely proportional to the distance from the nucleus image. Massonne (1987) developed an enhancement algorithm for the processing of coma images. An example of this is given in Figs 5 to 7. Fig. 5 shows an observation of Halley's Comet from 12 November 1985. It was taken on Calar Alto by Grün *et al.* (1986) with an IHW continuum filter, showing mainly the contribution of the cometary dust. The crosses in Fig. 6 represent the radial brightness profile of this observation, computed as the mean brightness on circles around the brightness centre. The profile follows closely the derived $1/d$ law from equation (13), represented by a line. The innermost part of the profile is affected by atmospheric effect, and at the outer boundary the effects of background light are visible. If the original brightness value of every picture element is divided by the value from a radial profile, the parts of the image which differ from this

Fig. 5 — Coma of Halley's Comet, as observed by Grün *et al.* (1986).

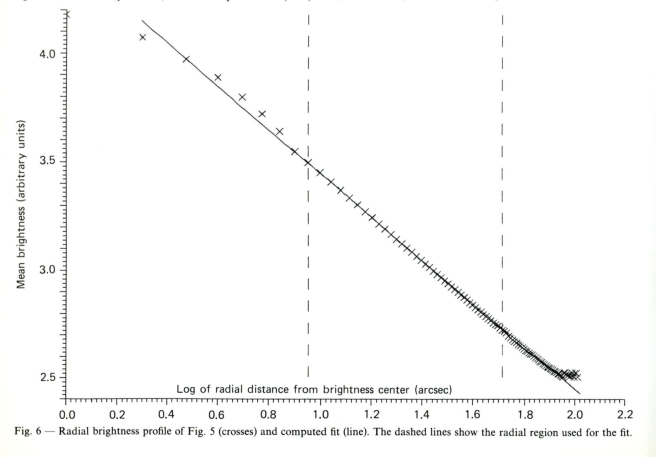

Fig. 6 — Radial brightness profile of Fig. 5 (crosses) and computed fit (line). The dashed lines show the radial region used for the fit.

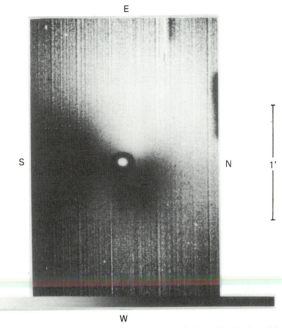

E

S   N   1'

W

Fig. 7 — Fig. 5 image, normalized to the radial profile depicted in Fig. 6.

symmetric distribution are enhanced. Fig. 7 shows the same observation as Fig. 5, but normalized to the radial profile from equation (13). Enhanced dust emission in southern and northwestern direction is clearly visible.

The northwestern emission feature coincides with the general tail direction, and can be a perspectively foreshortened view of the dust tail; the southern emission may be due to dust jets. This image normalization to the radial profile is a useful tool to enhance coma features. It can be extended to more complicated radial profiles which take other effects, i.e. gas emission light, into account.

4.2 The fountain model

The next refinement of dust dynamics leads to the fountain model. Bessel (1836) had proposed a repulsive force from the Sun on the dust particles. Bessel himself computed the particle motion under the influence of solar gravitation and a repulsive force, now known as radiation pressure. The pure fountain model, as described by Eddington (1910), considers only the radiation pressure, and not the orbital movement of the cometary nucleus. Therefore its

validity is limited to short time scales and small particles emitted from the nucleus with high velocity. The dust particles are emitted from the nucleus with the velocity v and an angle θ to the Sun–nucleus line. The coordinates of a dust particle after time t are

$$z(t) = v \cdot \cos \theta \cdot t + g_r \cdot t^2 / 2 \qquad (14.1)$$

$$R(t) = v \cdot \sin \theta \cdot t. \qquad (14.2)$$

The coordinates z and R are defined in Fig. 8. The acceleration of the particles due to radiation pressure is g_r. The maximal sunward distance from the nucleus that the dust particles can reach is the apex distance E:

$$E = v^2 / (2 \cdot g_r). \qquad (15)$$

The time t_E that the dust particles need to reach this distance is

$$t_E = v / g_r \qquad (16)$$

If the spatial coordinates are normalized to the apex distance A and the time to the apex-time t_A, equations (14) with the new variables lead to

$$\omega = z/A, \ \rho = R/A, \ \tau = t/t_A \qquad (17)$$

where

$$\omega = 2 \cdot \cos \theta \cdot \tau + \tau^2 \qquad (18.1)$$

$$\rho = 2 \cdot \sin \theta \cdot \tau. \qquad (18.2)$$

From equations (18) the emission time τ_E and direction θ_E of the dust particles reaching the coordinate (ω, ρ) can be computed.

$$\tau_{E1/2} = \sqrt{(\omega + 2)} \pm \sqrt{(\omega + 1 - (\rho/2)^2)} \quad (19.1)$$

$$\theta_{E1/12} = [-1 \pm \sqrt{(1 + \omega - (\rho/2)^2)}] / \sqrt{[2 + \omega \pm 2 \cdot \sqrt{(1 + \omega - (\rho/2)^2)}]}. \quad (19.2)$$

The emission time is counted negative from the arrival of the particle at the location (ω, ρ). For every

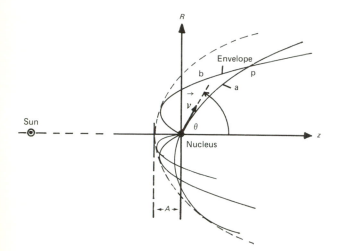

$$\delta R/\delta\theta = \begin{cases} v\cdot\cos\theta\cdot\cos\phi\cdot t \\ v\cdot\cos\theta\cdot\sin\phi\cdot t \\ -v\cdot\sin\theta\cdot t \end{cases}$$

$$\delta R/\delta\phi = \begin{cases} -v\cdot\sin\theta\cdot\sin\phi\cdot t \\ v\cdot\sin\theta\cdot\cos\phi\cdot t \\ 0 \end{cases} \qquad (22)$$

$$\delta R/\delta t = \begin{cases} v\cdot\sin\theta\cdot\cos\phi \\ v\cdot\sin\theta\cdot\sin\phi \\ v\cdot\cos\theta + a_{RP}\cdot t. \end{cases}$$

Fig. 8 — Fountain model, coordinates and symbols.

spatial coordinate reached by the cometary dust two trajectories exist, denoted (a) and (b) in Fig. 8. They are the direct trajectory (a) and the indirect trajectory (b). In equations (19) the upper sign denotes the indirect trajectory which first runs in the sunward direction and is then turned by the radiation pressure before reaching the target coordinates. The lower sign denotes the direct trajectory. Only at the envelope are the two trajectories identical. The envelope coordinates are

$$\omega_{env} + {}^1 = \rho_{env}{}^2/2 \qquad (20.1)$$

or in the coordinates from Fig. 8

$$z_{env} = R_{env}{}^2/(4 \times A) - A. \qquad (20.2)$$

As the z-axis is the symmetry axis, the envelope forms a rotation paraboloid. Outside the envelope no dust particles with emission velocity v and radiation pressure acceleration g_r are found.

From equations (14) one can compute the dust density. The dust particle's spatial location $R(t)$ in the CBS-system of coordinates is

$$R(t) = \begin{cases} x(t) & = & v\cdot\sin\theta\cdot\cos\phi\cdot t \\ y(t) & = & v\cdot\sin\theta\cdot\sin\phi\cdot t \\ & & v\cdot\cos\theta\cdot t + \frac{1}{2}\cdot g_r\cdot t^2 \end{cases} \qquad (21)$$

The three partial derivatives needed to solve equation (5) are easily found:

Inserting equations (22) in equation (5) leads to the spatial dust density ρ at a certain coordinate R:

$$\rho(R) = \frac{f}{v^3\cdot t_E(R)^2\cdot(1 + g_r\cdot t_E(R)\cdot\cos\theta_E(R)/v)}. \qquad (23)$$

The relevant emission times t_E and emission angles θ_E for the dust particles to reach the coordinate (R) can be computed from equations (19). The dust source function f may also depend on the emission time and direction.

4.3 Keplerian motion of nucleus and dust

Apart from the so-called non-gravitational forces which, as Whipple (1950) first proposed, result from the repulsion of the cometary gas, the nucleus moves in a two-body orbit around the Sun. Its motion can be described by the well-developed formalism of celestial mechanics. This formalism can also be applied to the motion of the dust particles, if only the radiation pressure force is taken into account. Its direction is opposite to that of the solar gravitation. The magnitude of the radiation pressure force is inversely proportional to the square of the distance to the Sun, as is gravitation. Therefore the effect of the radiation pressure can be modelled as a reduction of the Sun's gravitation, as Bessel had noted in 1836. The radiation pressure coefficient β describes the influence of this force. If β is greater than unity, the radiation pressure force is larger than gravitation, and the central force on the dust is repulsive. The heliocentric trajectories of the dust particles emitted from the nucleus, as well as the nucleus orbit, are conic sections. The orbital elements of the dust particles are known, as their location at emission time is identical with the nucleus position, and their velocity is the vector sum of the heliocentric velocity of the nucleus and the cometocentric emission velocity. In this way the cometocentric trajectory of the dust particles can be computed as the difference of the heliocentric trajectories of dust and nucleus. The partial derivatives needed for equation (5) cannot be solved analytically, and numerical methods have to be used. As in the case of the fountain model, at first the emission times and directions of dust particles reaching particular coordinates at particular time have to be determined. Using the full orbital motion, more than two solutions are possible. In Fig. 9 three possible dust particle trajectories are plotted which reach the point P on the 9 March 1986 at 7:13 a.m. The emission times and angles are given in the figure caption. The emission velocity was 81 m s^{-1}; the value of β was 0.006. The

curves A and B correspond to the direct and indirect trajectory of the fountain model, respectively. The contribution of the third trajectory, C, to the total dust density at P is small—about 2%.

The main difference between the fountain model and the full theory is not the existence of more possible trajectories, but the change in the particle dynamics. In Fig. 10 the cometocentric trajectories of 10 μm dust particles in the comet's orbital plane are shown as full lines. The dotted line represents the parabolic envelope according to the fountain model. The difference is clearly visible.

4.4 Approximations to Keplerian motion

The search for possible trajectory solutions to reach a given location is time-consuming if the full Keplerian theory is used. Therefore some authors have tried to find approximate descriptions of particle motions which lead to faster computation. The approach of Fertig & Schwehm (1984) was later used by Massonne (1986) in his dust model. Richter & Keller (1987) also published an approach to this problem, concentrating on the celestial mechanics and the dynamics in the tail

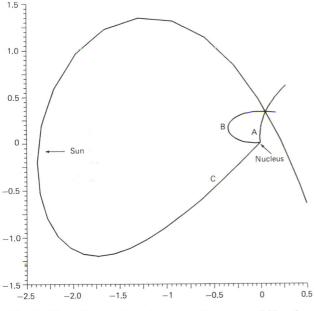

Fig. 9:   Three dust particle trajectories to the same spatial location. View on the orbital plane of Halley's Comet. Arrival at P on the 9 March 1986, 7:13 a.m.

Trajectory	Emission time	Contribution to total density
A	4 March 1986 8:38 a.m.	90%
B	15 February 1986 0:33 a.m.	8%
C	19 November 1985 4:06 p.m.	2%

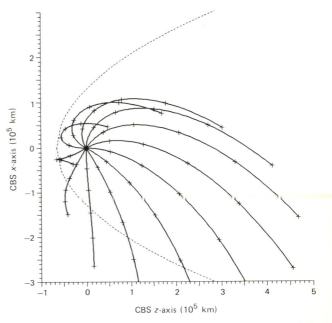

Fig. 10 — Trajectories of 10 μm particles computed with the full dynamical theory. The dotted line represents the fountain model envelope.

region. The approximation used here describes the particle's motion by transformation matrices ϕ and ψ. A full description is given in Massonne's PhD. thesis (1987). They depend on the dust particle's emission and arrival times, and give the particle's location $R(t)$ as a function of the departure velocity $v(t_E)$ at emission time and the particle's radiation pressure coefficient β:

$$R(t) = \phi\,(t_E, t)\cdot v(t_E) + \beta\cdot\psi(t_E, t) \qquad (24)$$

This equation can easily be inverted to find the departure velocity required to reach a certain location.

A simple loop can test the departure velocities needed at various departure times to find the physically relevant solution. The difference from the full dynamical theory of dust particle motion is negligible in the coma region compared with the influence of other effects, e.g. electromagnetic forces or changes of the dust particle size by electrostatic disintegration of charged fluffy particles. The coma model described here has been developed by Fertig and Massonne at the European Space Operations Centre (ESOC), and was used in the Giotto mission operations. Grün *et al.* (1986) and Massonne & Grün (1987) also apply the model to groundbased and Giotto data.

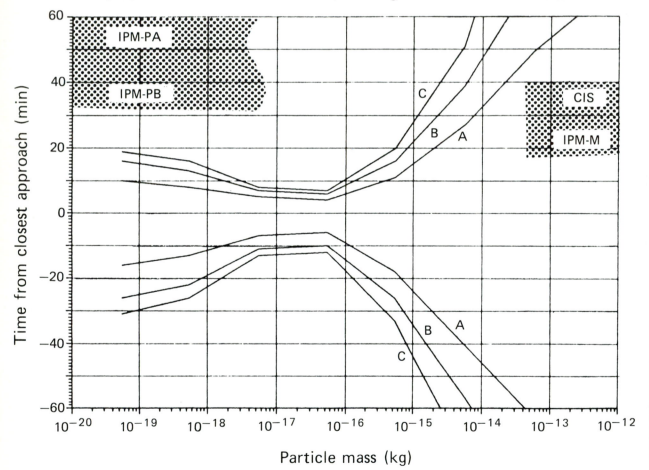

Particle mass (kg)

Fig. 11 — Envelope locations on the Giotto trajectory for magnetite particles with different emission velocities as a function of their mass. The emission velocities are taken from

A: Gombosi *et al.* (1983)
B: Finson & Probstein (1968)
C: Gombosi (1987) (see also Fig. 1)

The shaded regions represent Giotto DIDSY – results from McDonnell (1986).

5 USING THE MODEL

The model was used to compute the dust fluxes along the Giotto spacecraft trajectories. The effects, on the model results, of different particle materials and dust production distributions at the nucleus surface have been studied by these means; especially, the times at which the spaceprobes enter and leave the dust cloud of the respective particle size are a first means to derive the particle properties, as computed by Massonne (1986, 1987). Fig. 11 shows the envelope locations on the Giotto trajectory for magnetite particles at three different emission velocities as a function of the particle mass.

The computed times can be compared with the first and last observations of dust particles of the respective mass, as in the review by Grün *et al* (1987).

Fig. 12 shows a computed flux profile along the Giotto trajectory for 0.6 µm olivine particles.

The modelled dust production was set to zero on the night side of the nucleus. This results in a drop of the total dust flux to the indirect particle flux rate on that part of the trajectory where the direct particles would

come from the night-side of the nucleus. If this side is inactive, similar features should be observed in the spacecraft data.

The introduction of active areas on the model nucleus complicates the results. Fig. 13 shows the modelled dust flux for the nucleus with four active areas and a rotation axis perpendicular on the orbital plane depicted in the figure. The features in the resulting flux profile heavily depend on the spin axis orientation. Therefore an evaluation of observed features is not possible if the spin axis is uncertain. On the other hand this dependence can be used to test different spin axis orientations against observed features.

The same arguments are valid for synthetic image generation. With knowledge of the three-dimensional dust distribution the image of the dust cloud can be constructed by integration of the scattered sunlight along lines of sight. Fig. 14 shows the principle of the synthetic image generation. Each line of sight forms one picture element (pixel) of the synthetic image. As the scattering geometry of the sunlight at the dust particles is constant in the coma, numerical integra-

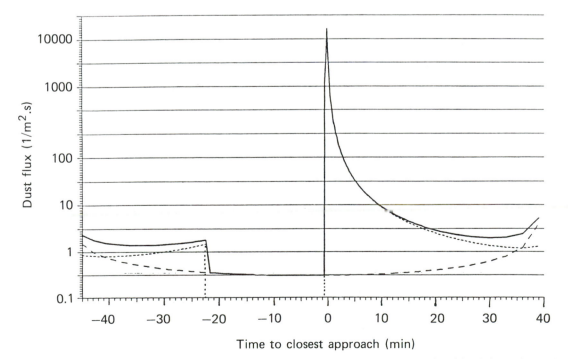

Fig. 12 — Flux of 0.6 µm olivine particles along the Giotto trajectory. Dust production on the day-side of the nucleus only.

tion of the dust densities computed with the coma model leads to the image.

Fig. 15 shows a set of six synthetic images of Halley's Comet at successive times as seen from Earth. In Münch (1986) these images are reproduced in a false colour representation, thus enhancing some details in the coma structure. Two arbitrarily chosen jets are active. The position of these jets is an assumption; all other model parameters are the most recent values known at the time of the image generation. As the dust production inside the jets has been set to a value 50 times higher than on the rest of the nucleus surface, the evaluation of the spiral jets with time is clearly visible. The Sun is at the bottom of the images. A part of the dust envelope is also visible.

Such images can now be compared with astronomical observations, and can be used to test derived spin axis orientations and active area positions, as tried by Grün *et al.* (1986). As the results may be ambiguous, model parameters derived from image analysis must

be treated with caution. Sekanina & Larson (1984, 1986) performed excessive analysis on images of Halley's Comet taken in 1910. Their results clearly depict the possibilities as well as the limitations of this method.

6 CONCLUSION AND OPEN QUESTIONS

The coma models as they stand at present provide a useful tool to test derived cometary parameters against measurements and observations. However, the ambiguity of the results limits the value of the method. The investigation of basic relations between nucleus parameters and coma structure is of great value for our understanding of the particle dynamics and the measured data.

The detailed modelling of the observed coma features leads to some still open questions:

– The spin orientation and rotation period of the

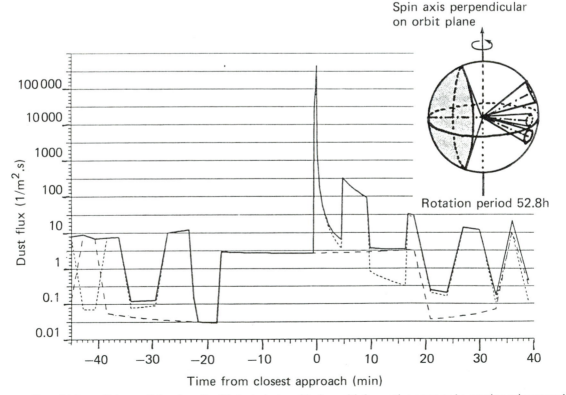

Fig. 13 — Flux of 0.6 μm olivine particles along the Giotto trajectory. Nucleus with four active areas and a rotation axis perpendicular on the orbital plane.

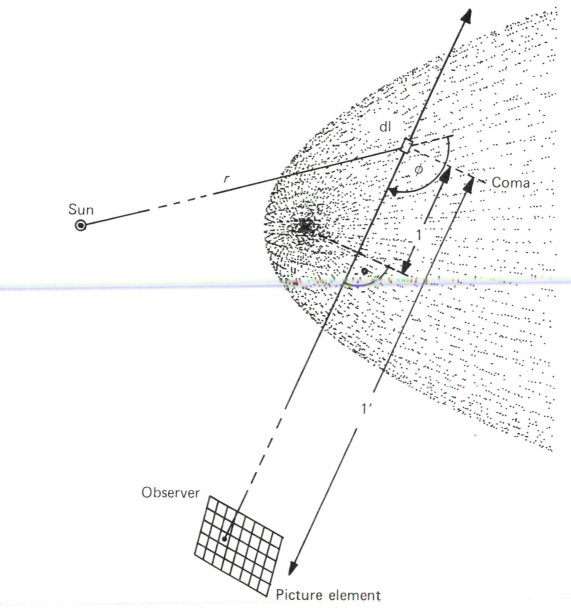

Fig. 14 — Principle of synthetic image generation.

nucleus are still under discussion, thus limiting the value of all nucleus feature modelling.

– The activity distribution on the nucleus surface is still rather uncertain. The strong near-nucleus jet emission visible on the spacecraft camera observations may be smeared out by non-radial acceleration forces on the dust particles, leading to more uniform dust distribution at larger nucleus distances.

– Dust particle parameters such as the radiation pressure coefficient β are known only for a restricted number of materials, such as olivine and magnetite. Laboratory measurements of the optical properties over a large wavelength interval are needed to enable the modelling of interesting dust particle materials such as graphite or organic compounds.

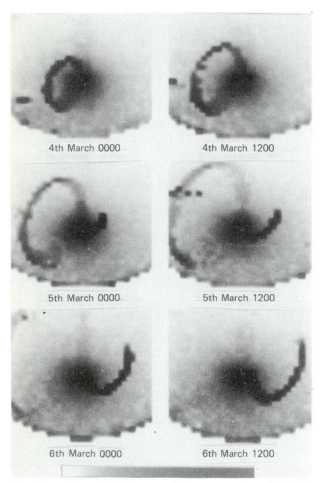

4th March 0000   4th March 1200

5th March 0000   5th March 1200

6th March 0000   6th March 1200

Fig. 15 — Six synthetic images of the coma region of Halley's Comet. Two dust jets are active. The times for which the images have been simulated are given below the images.

REFERENCES

Bessel, F. (1836) Beobachtungen über die physische Beschaffenheit des Halley'schen Kometen und dadurch veranlaßte Bemerkungen, *Astron. Nachr.* **13** 185–232

Bobrovnikoff, N. (1931) Halley's Comet in its apparition of 1909–1911, *Pub. Lick Obs.* **17** 309–482

Böhnhardt, H. (1987) The charge of fluffy dust grains of silicate and carbon near P/Halley and P/Giacobini–Zinner, *ESA SP-250* **II** 207–213

Burns J., Lamy P., & Soter S. (1979) Radiation forces on small particles in the solar system, *Icarus* **40** 1–48

Divine, N., Fechtig, H., Gombosi, T., Hanner, M., *et al.* (1985) The Comet Halley dust and gas environment, *Cometary Science Team Preprint Series* No. **72** JPL D-2823

Eddington, A.L (1910) The envelopes of Comet Morehouse (1908c), *Mon. Not. R. Astr. Soc.* **70** 442–458

Fertig, J., & Schwehm, G. (1984) Dust environment models for Comet P/Halley: support for targetting of the Giotto space-craft *Adv. Space Res.* **4** 9, 213–216

Finson, M., & Probstein, R. (1968) A theory of dust comets. I. model and equations, *Astroph. J.* **154** 327–352

Gombosi, T., *et al.* (1983) Gas dynamic calculations of dust terminal velocities with realistic dust size distributions, In: *Cometary Exploration II*, Gombosi, T.I., ed. Budapest, 99–111

Gombosi, T. (1987) A heuristic model of the Comet Halley dust distribution, *ESA SP-250* **II** 167–171

Grün, E., Massonne, L., Schwehm, G., (1987) New properties of cometary dust, *ESA Sp-278*, 305–314

Grün, E., Graser, U., Kohoutek, L., Thiele, U., Massonne, L., & Schwehm, G., Structures in the coma of Comet Halley, *Nature* **321** 144–147 (1986)

Huebner, W., Keller, H., Wilhelm, K., Whipple, F., Delamere, W., Reitsema, H., Schmidt, H., (1987) Dust-gas interaction deduced from Halley multicolour camera observations, *ESA SP-250* **II** 363–364

Larson, S., & Sekanina, Z. (1984), Coma morphology and dust emission pattern of Periodic Comet Halley. II. Nucleus spin vector and modelling of major dust features in 1910, *Astron. J.* **89** 1408–1425

Massonne, L. (1984) P/Halley's jet topology: an interface proposal, *Giotto Study Note* **49** ESA/ESOC/OAD, Darmstadt

Massonne, L. (1986) Coma morphology and dust emission pattern of Comet Halley, *Adv. Space Res.* **5** 12, 187–196

Massonne, L. (1987), Modellierung einer kometaren Staubkoma. PhD Thesis, Heidelberg 1987

Massonne, L., & Grün, E. (1987) Structures in Halley's dust coma, *ESA SP-250* **III** 319–321

McDonnell, J.A.M., Alexander, W.M., Burton, W.M., Bussoletti, E., Clark, D.H., Grard, R.J.L., Grün, E., Hanner, M.S., Hughes, D.W., Igenbergs, E., Kuczera, H., Lindblad, B.A., Mandeville, J.-C., Minafra, A., Schwehm, G.H., Sekanina, Z., Wallis, M.K, Zarnecki, J.C., Chakaveh, S.C., Evans, G.C., Evans, S.T., Firth, J.G., Littler, A.N., Massonne, L., Olearczyk, R.E., Pankiewicz, G.S., Stevenson, T.J., & Turner, R.F., (1986), Dust density and mass distribution near Comet Halley from Giotto observations, *Nature* **321** 338–341

Mendis, D., Houpis, H., & Marconi, L. (1985) The physics of comets, *Fundamentals of Cosmic Physics* **10** 1–380

Münch, R. (1987) Heading for the Giotto encounter with Comet Halley, *ESA Bulletin* **45** 14–20

Richter, K., & Keller, H. (1987) Density and brightness distribution of cometary dust tails, *Astron. Astrophys* **171** 317–326

Schwehm, G., & Rohde, M. (1977) Dynamical effects on circumsolar dust grains, *J. Geophys.* **42** 727–735

Sekanina, Z. (1986) Nucleus studies of Comet Halley, *Adv. Space Res.* **5** 12, 307–316

Sekanina, Z., & Larson, S. (1986) Coma morphology and dust-emission pattern of Periodic Comet Halley. IV. Spin vector refinement and map of discrete dust sources for 1910, *Astron. J.* **92** 462–482

Vaisberg, O.L., Smirnov, V.N., Gorn, L.S., Iovlev, M.V., Balikchin, M.A., Klimov, S.I., Savin, S.P., Shapiro, V.D., & Shevchenko, V.I. (1986) Dust coma structure of Comet Halley from SP-1 detector measurements, *Nature* **321** 274–276

Wallis M., & Macpherson, A. (1981) On the outgassing and jet thrust of snowball comets, *Astronom. Astrophys.* **98** 45–49

Weiss-Wrana, K. (1983) Optical properties of interplanetary dust: comparison with light scattering by larger meteoric and terrestrial grains, *Astron. Astrophys.* **126** 240–250

Whipple, F. (1950), A comet model 1. The acceleration of Comet Encke, *Astrophys, J.* **111** 375

7

Moment analysis of the near-nucleus morphology of comets

J. Watanabe

1 INTRODUCTION

It is important to monitor the near-nucleus coma structure when studying the physics and kinematics of comets. Many bright comets show a coma structure that changes from night to night. The coma features, which include jets, fans, arcs, and halos, are called near-nucleus phenomena. These phenomena are products of directed emission of material, and provide evidence of a non-uniform surface structure of the cometary nucleus. Recently, it was demonstrated quantitatively that the structure of the dust coma near the nucleus contains information on the rotational motion of the cometary nucleus and on the distribution, number, size, activity, and other characteristics of the active regions (Sekanina 1981).

Interest in near-nucleus imaging, therefore, has considerably increased. High-speed computers and digital image processing techniques allow us to analyze the coma morphology quantitatively. Moreover, the situation has been further improved by the introduction of highly sensitive solid state detector arrays, such as charge-coupled devices.

Morphological analysis of astronomical images are broadly divided into two methods.

One is the method of contrast enhancement, which is often applied to comets in order to make the low-contrast features appear more prominent. Sekanina & Farrell (1978) applied this method to the photographs of the dust tail of Comet West 1976 VI, with a digital low-pass filter for the reduction of the noise. Many plates of Comet Halley taken during the 1910 apparition were analyzed by using a linear shift-difference algorithm of the contrast enhancement by Wood & Albrecht (1981) and by Klinglesmith (1981). The visibility of images processed by linear shift-difference, however, depends strongly on the direction of the shift. Larson & Sekanina (1984) devised a new algorithm of the contrast enhancement, which contains the radial and the rotational shift differences about the centre of the light (nucleus), and applied it to the near-nucleus images of Comet Halley taken at Mount Wilson Observatory during the 1910 apparition.

When using contrast enhancement, great care must be exercised to identify processing artefacts that may be misleading in the analysis.

Another method of morphological analysis of astronomical images is based on moment analysis, which is often applied to galaxies in order to determine basic morphological parameters (for example, the position of the centre, size, ellipticity, position angle of the major axis etc.). Watanabe (1987) applied this method to the gaseous coma of Comet IRAS–Araki–Alcock 1983 VII, and discussed the rotational motion of its nucleus. Near-nucleus images of Comet Halley were also analyzed by this method (Watanabe et al. 1987a, b).

It is true that moment analysis is less effective than contrast enhancement in the study of the fine structure of the coma of bright comets. However, moment analysis can be applied to comets of which the coma does not develop sufficiently for analysis by contrast enhancement.

This paper introduces moment analysis for comets and its applications.

2 MOMENT ANALYSIS OF THE MORPHOLOGY OF COMETS

2.1 The method of moment analysis

Quantifying the morphology of astronomical images requires some preliminary data reduction. Photographic plates must be digitized into matrix data by a scanning microdensitometer. The sensitometric calibration for transforming the data of density to those of intensity is not always necessary if the interest of the study is concentrated on the morphology. If some kind of solid state detector array, such as CCD, is used in the observations, we can obtain the data in digital form. In this case, however, additional data reduction processes consisting of dark current subtraction and non-uniform sensitivity calibration (flat-field calibration) must be carried out.

To reduce the noise, especially the grain noise of the photographic data, some appropriate methods of image-processing, for example smoothing methods or digital filtering methods, can be applied.

The last step of preliminary data reduction is to remove stars, plate faults, and pixel faults from the image data. Although an automated object removal algorithm has been devised by M. Watanabe (1983) and Ichikawa et al. (1987), we must be careful to remove star images superimposed on the coma.

After these reduction processes we can advance to the moment analysis of the morphology. I shall describe the method briefly, on the basis of development by Stobie (1980).

The position of a pixel is represented by the matrix coordinates (I, J). The position of the nucleus is expressed by (I_N, J_N) in the matrix coordinates. The position of the point of maximum density (or intensity) in the coma is sometimes assumed to be that of the nucleus.

Once a threshold level is determined, its contour can be drawn on the image data. Its effective radius R of the contour is determined by

$$R = \sqrt{an/\pi}, \qquad (1)$$

where n is the number of pixels whose values are greater than the threshold, and a the area of one pixel.

The centre of the contour (I_C, J_C) is determined by

$$I_C = \Sigma\, I/n, \qquad (2)$$

$$J_C = \Sigma\, J/n, \qquad (3)$$

The deviation of this position (I_C, J_C) from the nucleus (I_N, J_N) contains information on the asymmetric mass ejection from the nucleus. It is possible to determine a rough structure of the extended coma by calculating (I_C, J_C) at each threshold. Figs 1a, b show an example of Comet IRAS–Araki–Alcock 1983 VII, which had a developed fan-like structure in the coma. We can easily see the direction of the asymmetric mass ejection from the nucleus (Watanabe et al. 1987a).

In the next step, the contour is fitted by an ellipse centred at (I_C, J_C) by the following formulas:

$$\tan 2\theta = 2U_{IJ}/(U_{II} - U_{JJ}), \qquad (4)$$

$$A^2 = 2(U_{II} + U_{JJ}) + 2([U_{II} + U_{JJ}]^2 + 4U_{IJ}{}^2) \qquad (5)$$

$$B^2 = 2(U_{II} + U_{JJ}) - 2([U_{II} - U_{JJ}]^2 + 4U_{IJ}{}^2) \qquad (6)$$

where, A and B are the lengths of the semi-major and semi-minor axes, θ the position angle of the major axis of the ellipse measured counterclockwise from the I-axis, and U the second moment defined as

$$U_{II} = \Sigma\,(I - I_c)^2/n, \qquad (7)$$

$$U_{JJ} = \Sigma\,(J - J_c)^2/n, \qquad (8)$$

$$U_{IJ} = \Sigma\,(I - I_c) \quad (J - J_c)/n, \qquad (9)$$

The quadrant of θ is determined by
$$4U_{IJ} = (A^2 - B^2)\cos\theta\sin\theta. \qquad (10)$$

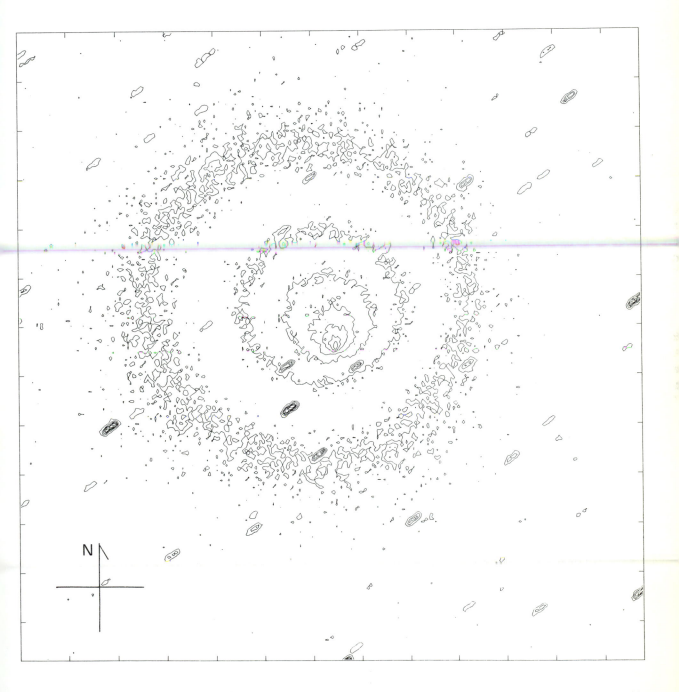

Fig. 1a — The image of the coma of Comet 1983 VII IRAS–Araki–Alcock taken at 1404 UT on 8 May 1983, using 50 cm Schmidt telescope at the Dodaira Observatory. The emulsion is Kodak 103a-0 (no filter), exposed 4 minutes. The field of view is 35′ × 35′; north is at the top.

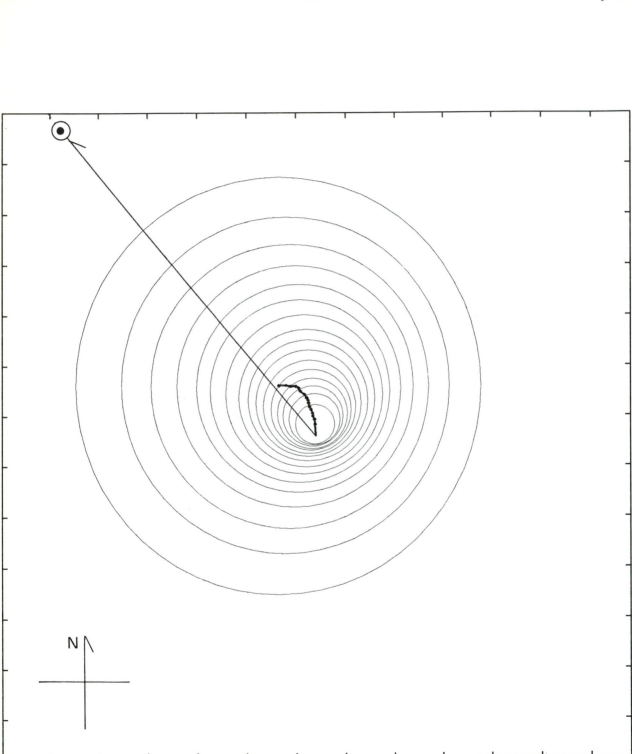

Fig. 1b — The reconstructed image of Fig. 1a after star removal. The points of the centre (I_C, J_C) at each contour are connected by the curve.
⊙ expresses the Sunward direction. It is clear that the coma expanded toward the Sun.

These values of A, B, and θ also contain information on the asymmetric mass ejection from the nucleus. When there is an elongated jet feature in the coma, we can estimate the position angle of its jet direction by that of the major axis of the fitted ellipse θ. The ratio R_{AB} of major to minor axes is defined by

$$R_{AB} = A/B. \qquad (11)$$

This can be a rough indicator of the jet activity. Figs 2a, 2b*, 3a, 3b*, 4a, 4b* show contour maps of the near-nucleus images of Comet Halley taken on 10, 11, and 12 December 1985 respectively. The inner contours of Fig. 2*and 3*have fairly round shapes, while those of Fig. 4*show elongated jet features. These situations are quantitatively represented in Table 1.

To estimate the error of the ellipsoidal fit to a contour, we can make use of the standard deviation SD of the density (or intensity) values of pixels which exist onto the curved line of the reconstructed ellipse on the image. The ratio S of the SD to the average M may be the relative indicator of the error of fitting. These values are calculated as

$$M = \Sigma f(I, J)/N_e, \qquad (12)$$

$$SD = (N_n \cdot \Sigma f^2(I, J) - (\Sigma f(I, J,))^2/N_e \cdot (N_e - 1), (13)$$

$$S = SD/M, \qquad (14)$$

* to be found in the colour illustration section in the centre of the book.

where N_e is the number of pixels on a curved line of the reconstructed ellipse, $f(I, J)$ the density (or intensity) value of the pixel (I, J), M the average value of those pixels. This value S will indicate quantitatively how well the ellipse represents a contour.

2.2 Application to Comet Halley

In the 1985/86 apparition of Comet Halley, a large number of high-resolution groundbased images were obtained for near-nucleus studies coordinated by the IHW. In Japan, the group of the Tokyo Astronomical Observatory played a central role in performing a sequential near-nucleus imaging observation. The data of the near-nucleus imaging were taken from 8 December 1985 through 20 January 1986, at the Dodaira observatory, at the Okayama Astrophysical Observatory, and at the Kagoshima Space Center. On 28 nights about 300 CCD images were obtained without filters. Although moment analysis applied to those data has been reported elsewhere in detail, it will be appropriate to show some results briefly as examples of moment analysis for comets.

Figs 5 and 6 show the day-to-day variation of the axial ratio R_{AB} of the contours of four different effective radii (10″, 20″, 30″, 40″) during the period of observation; Fig. 5 preperihelion, Fig. 6 postperihelion. We can see that the elongated feature appeared on 8, 12, 20, 24 December and 8 January. For the event of 12 December see Fig. 7, the outer contours, of which effective radii are larger than 20″, did not

Table 1. The axial ratio and the position angle of the major axis of each contour

Date	Dec. 10 10:43 UT			Dec. 11 10:45 UT			Dec. 12 12:26 UT		
Effective Radius	(1) R_{AB}	(2) θ	(3) S	(1) R_{AB}	(2) θ	(3) S	(1) R_{AB}	(2) θ	(3) S
6″	1.08	216°	5.8	1.07	202°	4.4	1.64	179°	7.0
10″	1.10	194°	3.4	1.08	193°	3.1	1.41	180°	4.8
15″	1.12	200°	2.6	1.09	189°	2.7	1.25	176°	3.6
20″	1.10	206°	2.3	1.11	188°	2.1	1.14	180°	3.3
25″	1.09	216°	2.3	1.11	187°	2.5	1.08	198°	3.4
30″	1.06	219°	2.5	1.10	187°	2.8	1.08	235°	3.5

(1): The ratio of the major to minor axes.
(2): The position angle of the major axis.
(3): The ratio of the standard deviation to the average of the intensity of pixels onto the fitted ellipse $S = SD/M$.

Table 2. The position angle of the major axis θ on the day of the jet event

Date	Dec. 8	Dec. 12	Dec. 20	Dec. 24	Jan. 8
10″	205°	181°	188°	200°	151°
15″	208°	180°	191°	198°	145°
20″	214°	180°	201°	209°	145°
25″	215°	240°	225°	204°	138°
30″	216°	250°	240°	210°	133°
◎	249°	247°	246°	245°	241°
Mot.	249°	248°	248°	248°	248°

◎: The position angle of the direction to the Sun.

Table 3. The position angle of the major axis θ on the day of the jet event.

Date	Mar. 7	Mar. 12	Mar. 23	Mar. 24	Apr. 6	Apr. 24
10″	72°	140°	60°	142°	114°	200°
15″	87°	142°	61°	156°	111°	211°
20″	102°	149°	60°	161°	108°	217°
25″	107°	240°	58°	166°	106°	223°
30″	113°	232°	64°	175°	114°	219°
◎	78°	78°	79°	80°	107°	257°
Mot.	227°	226°	229°	230°	254°	313°

◎: The position angle of the direction to the Sun.

elongate. This indicates that we caught the beginning of the jet ejection at this time (Watanabe *et al.* 1986, 1987b). On the other hand, there is an extended elongation over 40″ from the nucleus on 8, 20, 24 December.

Tables 2 and 3 show the position angle of the major axis θ at five effective radii (10″, 15″, 20″, 25″, 30″) on those days when the elongated feature appeared. The direction of the elongation seems to be near 180°–210° except for the event of 8 January. These

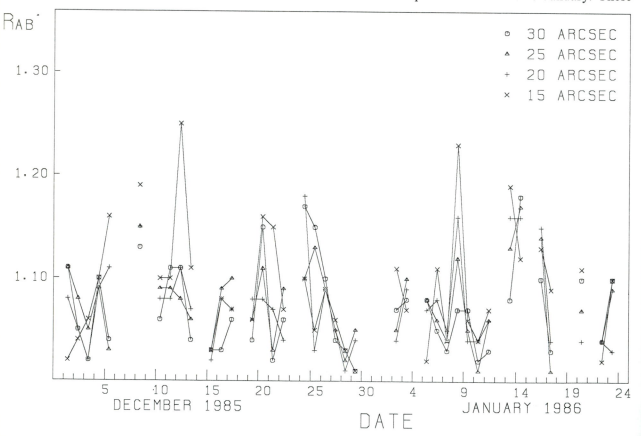

Fig. 5 — Day-to-day variation of the axial ratio R_{AB} of the contours of 4 different radii (15″, 20″, 25″, 30″) before the perihelion passage of Comet Halley.

southward jet phenomena were frequently reported in the IAUC etc. Thus moment analysis allowed us to observe those phenomena quantitatively.

3 THE FUTURE PERSPECTIVE OF THE MOMENT METHOD

In the above section we have described the concept of moment analysis for morphology, and we have given a few examples of its application. It can be seen that moment analysis is effective in finding out the morphological change of a comet quantitatively.

Although the method of contrast enhancement is not described in this paper, that method is very effective for finding out the fine structure in the coma and for studying the rotational motion of the nucleus. In this apparition of Comet Halley, many investi-

gations were carried out by various types of contrast enhancement methods (Parson et al. 1986, Sekanina & Larson 1986, Cosmovici et al. 1986, Chakaveh et al. 1986). A'Hearn et al. (1986) discovered strong CN jets by making use of contrast enhancement. This method is adequate for the morphological analysis of comets which are active and show complex coma structure such as Comet Halley after the perihelion passage.

On the other hand, moment analysis is not appropriate when the comet becomes active enough to eject materials by several jets in different directions simultaneously. For example, the scattering of the value of the axial ratio R_{AB} at each effective radius becomes large in January of 1986 in Fig. 5, compared with that of December 1985. This reflects the increasing activity of the comet. The occurrence of Lyman-

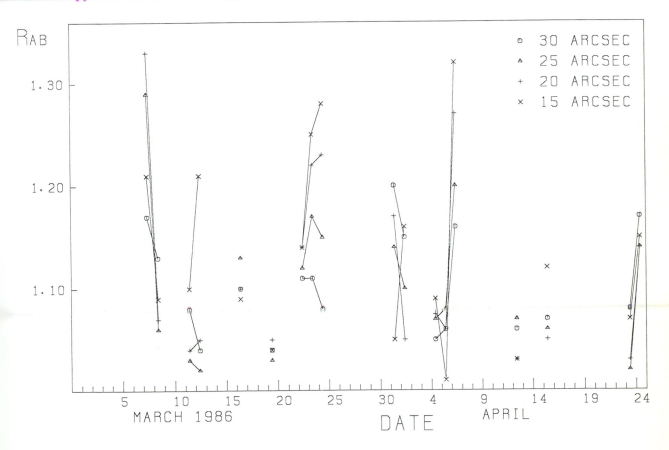

Fig. 6 — Same as Fig. 5 but the post-perihelion phase.

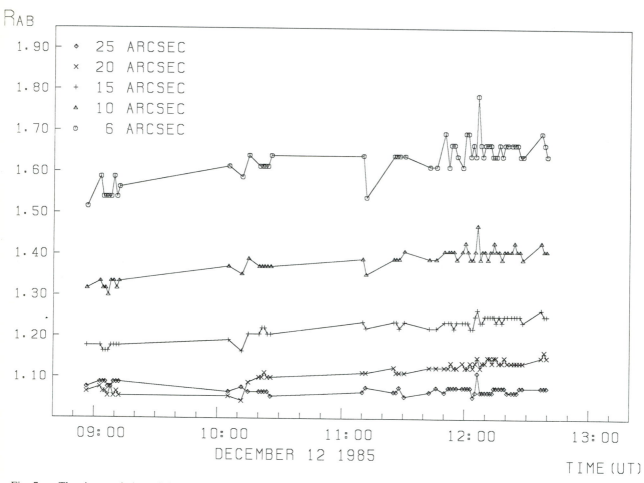

Fig. 7 — The time variation of the axial ratio R_{AB} on 12 December in 1985. The R_{AB} increases with time, which indicates the jet expansion.

α outbursts and dust jet phenomena in December 1985 is shown in Fig. 8.

In spite of this disadvantage, moment analysis seems to be effective for finding out quantitatively the morphological change. Its easy processing algorithm is a great advantage, with little processing artefacts that may be misleading in the analysis. It must be emphasized that the method of moment analysis can be applied to a comet even if its coma does not develop enough to be analyzed by contrast enhancement. Almost all comets, especially the periodic comets, are considerably less active than Comet Halley. Even if the comet does not show the remarkable near-nucleus phenomena, it is important to observe its morphological features for the study of the rotational motion of the nucleus (Sekanina 1979). Therefore, it will be worth trying to use moment analysis for morphological study.

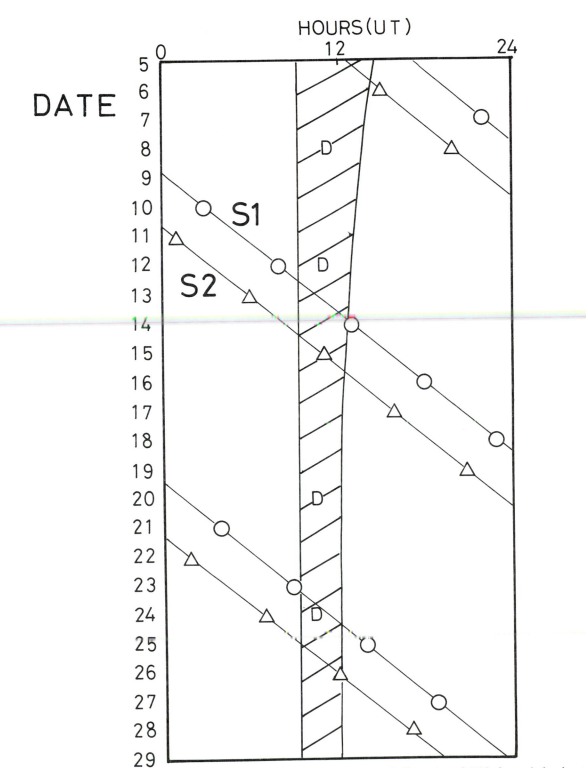

Fig. 8 — Occurrence diagram of Ly-α outbursts and observed dust jets phenomena in December of 1985. Open circles denote Ly-α outbursts of the S1 source, and open triangles those of the S2 (see Kaneda *et al.* 1986b). D represents the observed dust jets phenomena recognized by the moment analysis. The hatched region is the time when Comet Halley could be observed from groundbased telescopes in Japan.

REFERENCES

A'Hearn, M.F., Hoban, S., Birch, P.V., Bowers, C., Martin, R., & Klinglesmith III, A. (1986) *Nature* **324** 649

Chakaveh, S.C., Green, S.F., Ridley, J.K., McDonnell, J.A.M., & Hughes, D.W. (1986) *Proc. 20th ESLAB Symp.* (ESA SP-250) **2** 163

Cosmovici, C.B., Mack, P., Craubner, H., & Schwarz, Z. (1986) *Proc. 20th ESLAB Symp.* (ESA SP-250) **2** 151

Ichikawa, S., Okamura, S., Watanabe, M., Hamabe, M., Aoki, T., & Kodaira, K. (1987) *Ann. Tokyo Astron. Obs., 2nd Series* **21** 285

Klinglesmith, D.A. (1981) In: *Modern observational techniques for comets*, JPL Publ. 81–68, ed. Brandt, J.C., Donn, B., Greenberg, J.M., & Rahe, J. (U.S. Gpo, Washington, D.C.), 223

Larson, S.M., & Sekanina, Z. (1984) *A.J.* **89** 571

Larson, S.M., Sekanina, Z., Levy, D., Tapia, S., & Senay, M. (1986) *Proc. 20th ESLAB Symp.* (ESA SP-250) **2** 145

Sekanina, Z. (1979) *Icarus* **37** 420

Sekanina, Z. (1981) *Ann. Rev. Earth Planet. Sci.* **9** 113

Sekanina, Z., & Farrell, J.A. (1978) *A.J.* **83** 1675

Sekanina, Z., & Larson, S.M. (1986) *A.J.* **92** 462

Stobie, A.S. (1980) *J. British Interplanet. Soc.* **33** 323

Watanabe, M. (1983) *Ann. Tokyo Astron. Obs., 2nd Series* **19** 121

Watanabe, J. (1987) *Publ. Astron. Soc. Japan* **39** (in press)

Watanabe, J., Kawakami, H., Tomita, K., Kinoshita, H., Nakamura, T., & Kozai, Y. (1986) *Proc. 20th ESLAB Symp.* (ESA SP-250) **3** 267

Watanabe, J., Kawakami, H., Tomita, K., Takagishi, K., Kinoshita, H., Nakamura, T., & Kozai, Y. (1987a) *Proc. of Symp. on Diversity and Similarity of Comets,* ESA SP-278, in press

Watanabe, J., Kawakami, H., Tomita, K., Kinoshita, H., Nakamura, T., & Kozai, Y. (1987b), *A. & Ap.* in press

Wood, H.J., & Albrecht, R. (1981) In: *Modern observational techniques for comets*, JPL Publ. 81–68, Brandt, J.C., Donn, B., Greenberg, J.M., & Rahe, J. eds (U.S. Gpo, Washington, D.C.), 216

8

The composition of dust particles in the environment of Comet Halley

D.E. Brownlee
J. Kissel

1 INTRODUCTION

Three mass spectrometers carried on the Giotto and Vega spacecraft produced the historic first direct analyses of the composition of cometary dust. This information is of key scientific interest because of the likelihood that particles released from Halley and other active comets are preserved samples that previously existed at the outer fringes of the planetary region of the solar nebula. The particles may be composed of nebular materials that formed in the solar system as well as materials that are preserved interstellar solids and predate the planetary system. The properties of cometary solids provide constraints on models of nebular and presolar processes and environments. The comparison of comet dust with asteroidal meteorites provides insight into the processes that produced dust and planetesimals over an extreme range of radial distance from the centre of the solar system.

Before the Halley flybys, compositional information on bona-fide comet dust was obtainable only by analysis of meteors and by telescopic measurements of comets. The detection of the 10 μm and 18 μm 'silicate' emission features in cometary comae indicate that silicates are a major component of comet dust (Ney 1982). Coma measurements in the optical and infrared indicate that comet dust contains an absorbing component and is not pure silicate. The characteristically low albedos of coma dust (Hanner 1984) are presumably due to a carbonaceous component. The elemental composition of comet dust was largely unknown before the Halley encounters but spectral analysis of cometary meteors (Millman 1972) and rocket measurement of mesospheric ion enhancements during meteor showers (Goldberg & Aiken 1973) were consistent with the suspicion that cometary solids have chondritic elemental abundances. Extensive laboratory analyses have been conducted on collected samples of interplanetary dust, but at present it is not possible to prove definitively that any one of these meteoritic particles has a cometary origin (Bradley *et al.* 1988). It is nearly certain that cometary particles have been studied, but among the particles studied it has not been possible to tell which ones are asteroidal and which ones are cometary.

2 MEASUREMENTS AT HALLEY

The direct dust composition measurements at Halley were made with a unique mass spectrometer design developed by Dr Jochen Kissel, the principal investigator for the instruments flown both on Giotto and on Vega (Kissel 1986). The mass analyser part of the instruments is similar to a design used for laser microprobe laboratory instruments, but the ion source and operation of the Halley instruments are unlike those used for any previous instrument. The Giotto instrument is named PIA (particle impact analyser) and the Soviet constructed Vega instruments are called PUMA 1 and 2. The instruments have extraordinary sensitivity and they provided ion mass spectra for individual comet particles down to femtogram mass. In all, the three mass spectrometers provided mass spectra for over 5000 individual particles in the 0.1 to 1 μm range.

Ions measured by the instruments were produced by the hypervelocity impact of dust particles onto a target of AgPt alloy for PIA or pure Ag for PUMA. A fraction of the positive ions produced by a dust particle impact were extracted by an electric field and accelerated to an energy of 1 keV. The ions then travelled down a flight tube where their time of flight was used to separate ions with different charge/mass ratio. The flight path contained an electrostatic reflector that increased the flight time and focused ions that were generated with different initial energies. The flight time for singly charged Ag was approximately 40 μs, and the resolution (at Ag) for the best spectra exceeds 150 ($M/\Delta M$). At the end of the flight tube, ions were detected by an ion multiplier whose dynodes were coupled to operational amplifiers to give a logarithmic output. The nominal dynamic range of five orders of magnitude of signal amplitude was transmitted as a 7-bit word giving a minimum uncertainty of peak amplitude of approximately 10% or 15% for the ratio of two amplitudes. The measured spectra were transmitted in either of four different modes. Four percent of the spectra were transmitted in 'mode zero', the highest fidelity mode, in which the mass spectra were sampled at 67 ns intervals, and all data were recorded. In the other modes the data were compressed by transmission only of peaks and valleys in the signal. The mode zero spectra have the best resolution and are the most reliable data for minor element and isotopic analysis.

Although all three instruments were of nearly identical design, there were differences in hardware and operation that made each unique. The apparent failure of two stages of the five stage logarithmic on PIA resulted in reduced amplitudes and truncation of low-amplitude signals from the Giotto data. The PIA spectra are excellent but there is some uncertainty in interpretation of relative peak amplitudes. On Vega-2, power problems on the spacecraft resulted in lower

amplifier gain, and the dynamic range was 3.5 decades on PUMA-2 instead of the planned 5 decades. The operation of PUMA-1 was normal. On PUMA-1 and -2 the reflector voltage was intentionally switched between two values every 30 s. In one mode (long) only ions with <50 eV thermal energy from the impact were reflected by the electrostatic mirror, while in the 'short' mode, ions with initial energies up to 150 eV were reflected to pass on to the multiplier. The short mode spectra from PUMA have higher mass 12 and 16 peaks than the long mode, while the heavier mass peaks in both spectra are similar. This indicates that most of the low mass ions picked up appreciable energy from the impact process (Krueger & Kissel 1987).

3 CALIBRATION

The instrumental data consist of time of flight spectra containing peaks of ions with the same charge/mass ratio. The relationship between the spectrum and the original particle composition depends on many factors that are not well understood and cannot be calibrated in a conventional manner. Factors involved include ion production efficiency, recombination, ionization state, molecule formation, extraction efficiency, linearity, and transmission through the entire system to the multiplier. Ideally, these effects could be studied in the laboratory, but the high velocity of the particles and their small size makes simulation experiments extremely difficult. During development, calibration tests were made with dust particles accelerated with the Heidelberg 2MV Van de Graff accelerator (Krueger & Kissel 1984). Most of this work was done with submicrometre particles at 10 to 25 km s^{-1}, but one iron grain was successfully launched at 64 km s^{-1}. One of the interesting results from these tests was that there was a general trend that spectra became simpler at increasing velocities. Low-velocity shots had many lines, while the high-velocity particles produced spectra that contained prominent lines of singly ionized atoms. These experiments provided data for estimating particle mass from the signal amplitudes. The laboratory calibrations also showed that the ratio of silver target ions to the projectile ions was a function of projectile density. Extrapolation of these findings provides a basis for estimating particle density (Kissel & Krueger 1987b).

Unfortunately, it was not possible to run the calibration experiments with particles that had velocities, compositions, and morphologies that closely match the particles encountered at Halley. In addition to the problems of accelerating submicrometre dust particles to 70 km s^{-1} a factor complicating laboratory calibration is the difficulty of keeping the surfaces of the targets and the projectile clean. For example a 3nm contaminant coating on a 100nm particle constitutes 10% of the particle mass and possibly a much higher fraction of the projectile mass that actually produces ions that are drawn into the time of flight tube. Spectra from the laboratory tests often contained major lines that could not be attributed to either the projectile or target. Although future efforts could yield adequate calibration data for the instruments, the present situation is that there are no laboratory studies with the instruments that can be confidently used to convert spectral peak ratios to accurate elemental abundance ratios. Future calibration work could provide critical factors for quantifying abundance ratios. They could also provide essential information on the possibility that ion yields vary with particle composition and structure. For example, if most of the liberated ions originate from the particle surface, then the measured composition may not be representative of the bulk composition.

The theory of ion production from impacts on the PIA and PUMA targets has been discussed by Krueger & Kissel (1987). The major source of ions is not a quasi-equilibrium thermal plasma generated by the impacts. The specific kinetic energy of impacting Halley, dust, approximately 30 eV/amu, is sufficient to generate a high degree of ionization. If hot thermal plasma in Saha equilibrium was the major source of ions, then one would expect many multiply charged ions in the spectra. Multiply charged ions are not apparent in Halley spectra, and thermal ionization is apparently not an important ion source under these conditions. Owing to high density it is possible that ions in a thermal plasma would recombine before they could be extracted into the instrument's flight tube. Kissel & Krueger (1987a) suggest that the character of dominant ion emission in these high velocity impacts is similar to that created by ion sputtering in secondary ion mass spectrometry (SIMS). The simi-

larity in the processes involves the rapid dissipation of energy near surfaces. They argue that ion emission arises from the interaction of the rear surface of the particle and the shock wave generated at the projectile–target interface. Ionization occurs at energies corresponding to the shock velocity in the particle. The relative ion yields deduced from SIMS measurements have been used by Jessberger et al. (1986, 1988) for conversion of ion peak amplitudes to abundances.

4 RESULTS

The many spectra returned from the flyby missions are of variable quality, and most investigators have selected sets of suitable spectra. This selection probably does not produce a serious bias in the results, because most of the unsuitable spectra have obvious shortcomings such as high background, spurious peaks, or low signal strength. Roughly a third of the spectra are of good quality from which elemental peaks can be unambiguously identified. Many of the spectra are excellent, and isotopic effects can easily be resolved. The flight spectra are remarkable in that molecular and multiply charged ions are rare second-order effects. This, and the near lack of obvious contaminant lines from the substrate, greatly simplify interpretation of the data. The flight spectra are much better than the laboratory calibration spectra, partly because the target metal on the missions was sputter-cleaned by coma gas. Most spectra consist of a line or doublet at mass 107–109 (Ag) and lines at mass 12 (C), 16 (O), 24 (Mg), 28 (Si), and 56 (Fe). On good spectra, lines corresponding to singly charged N, Na, Al, S, Ca and other elements can be seen. Other than Ag, the only line that is likely to be associated with the target is a line at 35 (Cl) that was seen on PUMA-1 but not on the PIA spectra.

4.1 Bulk Composition

Without any correction for the efficiency of ion production, the spectra show a remarkable similarity to solar abundances. An average of 39 spectra is shown in Fig 1. As can be seen from the average of 5000 spectra shown in Table 1 (from Langevin et al. 1987), there is good agreement between PIA and

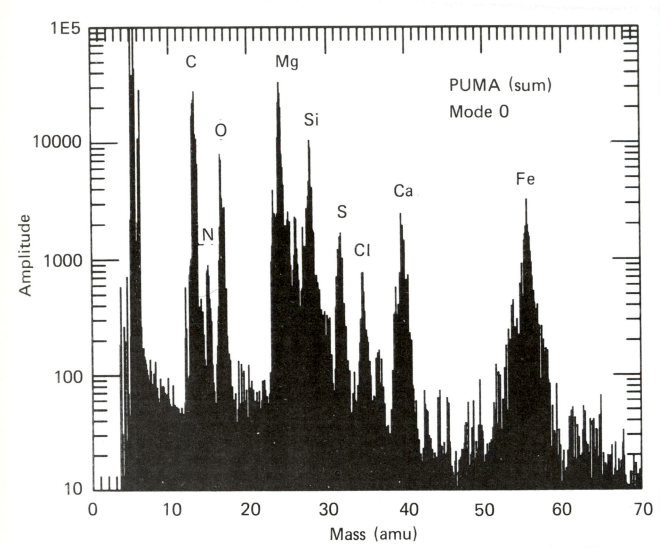

Fig. 1 — The sum of 39 high quality mode 0 spectra. The individual spectra were corrected for gain (mass) and zero shifts.

PUMA for the heavy elements, and there is general qualitative agreement with chondritic abundances. Assuming the bulk elemental composition of Halley is chondritic for the heavy elements, then the agreement of the raw spacecraft data with solar composition suggests that factors for conversion of ion abundances to elemental abundances are typically less than 4 for the major elements heavier than oxygen. Jessberger *et al.* (1986, 1988) used the best Giotto and Vega spectra to quantitatively compare the mean Halley composition with chondrites. Using ion production efficiencies deduced from laboratory SIMS work, they estimated that the 40 picograms of Halley dust that

was analysed matched type 1 (CI) chondrites within a factor of 2 for O, Na, Mg, Al, Si, S, Ca, Cr, Mn, Fe, and Ni. The only dramatic deviation from CI abundances is carbon and nitrogen. As shown in Fig. 2, the C and N abundances are a closer match to solar abundances than to chondritic composition. With the correction factors from SIMS work, they estimate that the mean carbon/silicon ratio in their set of particles is six to eight times higher than CI values. This would indicate that carbon is the second most abundant element in Halley dust after oxygen. Assuming that all other elements have CI abundances, this would imply that, on an atom basis, mean Halley dust contains

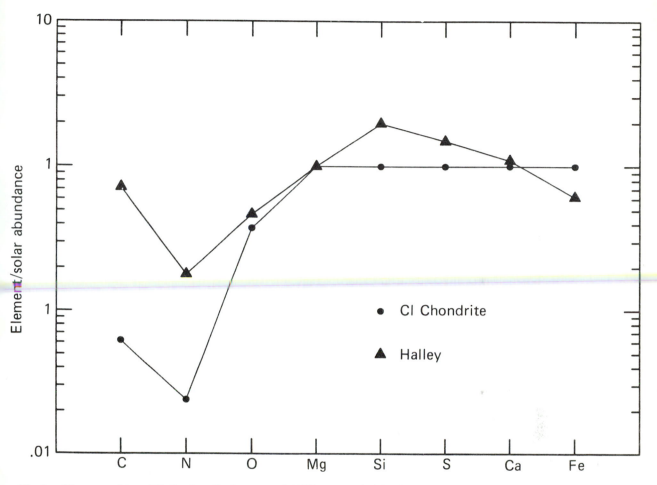

Fig. 2 — The composition of Halley dust (Jessberger *et al.* 1988) compared with solar and CI chondrite abundances (Anders & Ebihara 1982). In concordance with Jessberger *et al.* the meteorite and Halley data have been normalized at Mg instead of Si, the traditional normalization element.

approximately 30% carbon and 50% oxygen. On a weight percent basis the Halley dust would be 20% carbon. The C/O implied by these results is in general agreement with the solar value of 0.6. Sagdeev *et al.* (1986) have shown that the high carbon abundance is not concentrated in any particular size range. Most of the other studies are in general agreement with peak height ratios used by Jessberger. There is little doubt that the Halley dust contains high carbon abundances, but the precise magnitude of the enrichment is uncertain. Quantitation of the spacecraft data relies on the use of ion yield factors that have not been

Table 1 — Relative ion abundances for 5000 PUMA spectra. These are raw numbers uncorrected for ion yield. Table adapted from Langevin *et al.* 1987.

	H	C	N	O	Na	Mg	Si	S	Ca	Fe
PIA	5.0	3.0	0.15	4.1		2.1	1	0.16	0.05	0.20
PUMA-1 (short)	2.0	8.0		4.6	0.12	1.8	1	0.30	0.17	0.33
PUMA-1 (long)	4.0	1.0		0.7	0.10	2.5	1	0.25	0.18	0.28
CI (atom)	5.9	0.76	0.06	7.3	0.12	1.0	1	0.48	0.07	0.86

directly determined. The factors deduced from SIMS may be appropriate, but until real impact ionization calibrations can be done with dust particles at 70 km s^{-1} there will be uncertainty in the quantitative abundance of carbon measured at Halley. There may be effects in ion production that are not at all well understood. For example, Sagdeev *et al.* (1986) point out that an anomalous number of spectra do not contain oxygen. In many cases this must be some instrumental effect because the particle contains elements such as Mg and Si that are very unlikely to occur in the reduced state.

4.2 Individual grain composition

Although the average composition of Halley solids appears to be chondritic with enhanced carbon, there is a high degree of compositional variability between individual particles. The most conspicuous heterogeneity is the abundance of the low atomic number (Z) elements. One of the most significant early findings from the PIA and PUMA spectra was the discovery that approximately 30% of the particle spectra are dominated by low Z elements. These particles have been called 'CHON' (Kissel *et al.* 1986 a, b). In the PUMA-1 spectra, where a large dynamic range can be utilized, it is usually seen that the CHON particles are not pure low Z, but they also contain minor amounts of heavy elements such as Mg, Si, and Fe. In a study of 5000 spectra, Langevin *et al.* (1987) determined that 30% were CHON, 35% were a mixture of CHON and silicates (Mg, Si, and Fe) and that 35% contained no appreciable content of low atomic number elements except oxygen. Such high abundance of submicrometre organic particles is not seen in meteorites, and the frequency of the CHON component indicates an apparent distinction between Halley dust and the meteorite parent bodies. This conclusion is fortunately independent of uncertainty in ion yields. For comparison with meteorites, however, there is concern about the stability of CHON and the true composition of the CHON particles. CHON particles could be stable in the nucleus and coma but might sublime or disintegrate in other environments. It is also not known if the CHON particles are really low Z throughout their interiors or whether they could be a mantle coating over a core of different composition.

The low Z particles detected by PIA were subdivided by Clark & Mason (1986) into classes dominated by the following element groups; C-H-O-N, C-H, C-H-O, C-H-N, H-O, and H-C. The abundance frequency of these types relative to the silicate particles varied during the Giotto flyby, indicating compositional heterogeneity in the coma. This heterogeneity could be due to differences in particles emanating from different jets on the nucleus, or they could be due to sublimation processing of grains in the coma. The CN jets discovered by A'Hearn *et al.* (1986) appear to be produced by CN compounds lost from grains during their transit through the coma. The biggest effect noted by Clark *et al.* was a decrease in the relative abundance of the C-H-O composition particles with increasing distance from the nucleus. They suggested that this could be the result of sublimation of an intermediate volatility compound such as formaldehyde or formaldehyde polymer. Support for this assertion is given by evidence for the existence of polymerized formaldehyde (polyoxymethylene) in gas composition analyses made at Halley (Mitchell *et al.* 1987, Huebner 1987).

The elemental composition of the 'silicate' elements in the PIA and PUMA data has been used to compare Halley with meteorites and interplanetary dust (Brownlee *et al.* 1987, Lawler *et al.* 1989). Even though all of these materials have bulk chondritic elemental compositions at the level of accuracy of the Halley data, there are considerable spatial variations at the submicrometre level. The variations are due to the mineralogical and structural differences related to the origin and evolution of mineral grains contained in the different bodies. The Halley particles are generally dominated by the major chondritic elements, but there is a high degree of variability from one grain to the next. The compositional dispersion seen at the submicrometre level suggests that the (mineral) grain size in Halley is similar to the typical size (submicrometre) particles analysed by PIA and PUMA. The range of Mg, Fe, and Si compositions clearly distinguishes Halley from the carbon-rich types of carbonaceous chondrites. At the submicron scale, Halley contains abundant pure Mg silicates and high Fe phases as well. In comparison, submicrometre Mg silicates are exceedingly rare in the CI and CM meteorites and high Fe grains are also comparatively rare relative to Halley. The fine–grained material in these meteorites is composed of hydrated silicates that have constrained Mg/Fe ratios that rarely approach nearly pure Mg or Fe end members. Halley is more compositionally diverse than the carbonaceous chondrites at the submicrometre level. The dispersion of abundances in Halley is also distinct from that found in the fine–grained matrix of the unequilibrated ordinary chondrites and some classes of interplanetary dust. Some classes of interplanetary dust do show submicrometre compositional variations similar to the Halley data. The tentative best match is with the porous particles that are composed of anhydrous minerals. These particles are mineralogically distinct from the carbonaceous chondrites that are dominated by hydrated minerals. In qualitative agreement with Halley, the anhydrous/porous interplanetary particles are also composed of three types of submicrometre grains: pure silicates, pure low Z and mixed, low Z plus silicates.

4.3 Isotopic effects

The quality of many of the mode zero spectra is sufficient for isotopic analysis of some of the major elements. The elements most suitable for meaningful analysis are Mg and C. The observed isotopic ratios for these elements are often consistent with terrestrial values, but there are cases where large deviations are seen. Variations of a thousand in the 12/13 ratio were reported by Solc $et\ al.$ (1987), and equally large effects are sometimes seen for the 24/25 ratio. It is unlikely that effects as large as this are actual isotope ratios. While it cannot be ruled out that the large effects are real, it is more likely that they are instrumental in nature. Owing to the number of uncertainties with the spectra and the lack of detailed laboratory study of the process involved, it is not possible to rigorously show that any real isotopic effects were observed in Halley dust. The conservative conclusion at the present time is that with large uncertainty the Mg and C isotopic composition of Halley dust is terrestrial. It is likely that continued work on the data and laboratory simulations will provide more confidence in the isotopic data contained in the Halley spectra. Carbon is perhaps the most likely element where isotopic effects might be resolved. Recent observation of

C12/C13 ratios near 10 for selected micrometre–sized grains from carbonaceous chondrites are nearly a factor of 10 lower than the terrestrial ratio, and it is possible that similar effects could be resolvable with PUMA and PIA if they exist in cometary material. The meteoritic effect is associated with SiC grains and a correlation between this composition and isotopically heavy carbon in Halley would be strong evidence for a true isotopic effect.

4.4 Molecular composition

The PIA and PUMA spectra are dominated by singly charged atomic ions. No single spectrum has been published that shows prominent lines that can confidently be attributed to molecular ions. Molecules are apparently normally dissociated during hypervelocity impact, and only trace amounts are likely to survive. Surviving molecules and fragments may have been emitted from the rear projectile surfaces by a process analogous to desorption (Kissel & Krueger 1987a). In a statistical study of 43 high-quality spectra, Kissel and Krueger present evidence that molecules and fragments do exist in the spectra as minor components. In their study they summed spectra, they subtracted the atomic lines and looked for coincidences in the mass of molecular fragments in the residual data. From this study they infer the existence of a variety of organic molecules in the Halley particles. Among other compounds that they suggest are present are butadeine, toluene, adenine, formaldehyde, and xanthine. Most of the inferred molecules do not contain oxygen, and the authors estimate that the C/O atom ratio in the organic material is 5:1.

5 CONCLUSIONS AND COMMENTS

The particle composition data from three similar instruments flown on three spacecraft have produced an intimate view of the composition of Comet Halley. They show that the bulk elemental composition of Halley dust is consistent with that of chondrites, except that Halley appears to have much higher carbon and nitrogen contents than any meteorite. Roughly a third of the micrometre and smaller particles from Halley are dominated by carbon and other low atomic weight elements. Dependent on resolution of the present uncertainty in the gas/dust ratio for Halley, it is apparent that dust is an important and probably the dominant form of carbon escaping from Halley. From the observed correlation with other elements it is evident that most carbon is in the form of organic compounds and not graphite.

A major goal in cometary research is to investigate links between comets, interstellar grains, and primitive solar system materials. Accordingly, it is important to relate the properties of Halley dust with those of meteorites and interplanetary dust. It is apparent from the data that the mean carbon abundance in Halley dust is higher than found in chondrites. The most carbon–rich chondrite is Ivuna, a type I carbonaceous chondrite that contains 4.8% wt carbon with a C/Si atomic ratio of unity. This is 50% higher that in other CI meteorites and a factor of 40 above that in ordinary chondrites. The carbon abundance in laboratory samples of interplanetary dust has not yet been accurately measured for a significant number of particles. While some particles may have carbon abundance as high as implied by the Halley results, it appears that typical particles have carbon compositions either similar to the CI chondrites or intermediate between Halley and CI (Schramm *et al*. 1989). The apparent difference in the carbon abundance if taken at face value implies that Halley dust is different from meteorites and most interplanetary dust. The complication with this comparison is the uncertainty in the stability of carbon in the Halley particles. It is possible that a substantial fraction of the carbon in Halley dust may be volatile and would not survive post-coma environments. For this to be an important effect the carbon would have to be in a form that can be retained in particles for a day, but lost on time scales of 10 000 years or more. Evidence for carbon loss from Halley dust comes from the observation of cyanogen jets that evidently are produced by release of carbonaceous material from solid grains in dust jets (A'Hearn *et al*. 1986). Another line of evidence that is consistent with volatile loss from dust grains is the dust impact rate and size distribution data that suggests that particles are disintegrating in Halley's coma (Simpson *et al*. 1986, Vaisberg *et al*. 1987). Other lines of evidence for loss of organic material from coma dust are described by Wallis *et al*. (1986).

The submicrometre compositional variability in Halley dust clearly indicates that Halley material is different from and more heterogeneous than the carbon-rich chondrites. These meteorites are dominated by hydrous minerals, they are not porous, and they have experienced aqueous alteration early in the history of the solar system. While they are mineralogically different from the Halley material it is possible that aqueous alteration of Halley-like material coupled with loss of carbonaceous material and compaction could produce rocks similar to the CI and CM meteorites. The material that appears to most closely match the fine–scale heterogeneity of Halley is the highly porous class of interplanetary dust that is composed largely of anhydrous minerals. These particles contain pure carbonaceous, mixed carbon–silicate, and pure silicate submicron components (Bradley & Brownlee 1986) that qualitatively are similar to the three major components in the Halley particles. If the analogy with this particle type is correct, this implies that Halley is also dominated by anhydrous minerals. It is interesting that many of these silicates are high-temperature minerals such as enstatite and forsterite that obviously never equilibrated with a low-temperature solar composition gas. If Comet Halley were constructed of this interplanetary dust type, it would be a black mixture of anhydrous silicate, carbonaceous matter, and ice, all finely mixed on a scale of micrometres.

There is a strong possibility that comets contain presolar interstellar grains, and a considerable amount of excitement has surrounded the Halley particulate data as a means of testing for a link between comets and interstellar dust. Several authors have interpreted their results as confirmation of the Greenberg model (Greenberg 1982) of comets being mixtures of elongated core–mantle interstellar grains assembled into a 'bird's nest' structure. The low nuclear albedo of Halley and the fine grain size of its dust are consistent predictions of the model. The most important agreement with the Greenberg model is the high carbon content of the grains. In the model, most of the carbon is contained in radiation-processed mantles that surround silicate cores. The observed high carbon abundance, however, does not prove a link between Halley and interstellar grains, nor does it show that these grains have core–mantle structure predicted by the model. The requirement that interstellar grains be carbon rich is common to all models of interstellar matter. A third of the Halley particles are composed almost entirely of silicate materials that clearly do not have significant organic mantles. A similar fraction of the particles shows no evidence for elements heavier than oxygen, and they probably do not have cores. The rest of the particles do have compositions that are consistent with particles with silicate and substantial organic components. Survival of unmantled grains and the formation of coreless organic particles are not predicted by the Greenberg model, but these difficulties could perhaps be reconciled as products of processing during or after the formation of Halley. A major question for future work is whether the considerable range of organic compositions seen in Halley dust is compatible with accreted interstellar mantles that have been extensively processed by irradiation.

Some of the observations of Halley dust are consistent with the Greenberg model, but to state that Halley is actually composed of interstellar grains all having core–mantle structure would be a gross overinterpretation of the data. A more convincing connection between the data and the model would be evidence for radiation processing of the organic component or evidence that the organic component actually occurs as thick mantles over silicate cores. If a bird's-nest structure of core–mantle grains can be proven for Halley, then it could be shown that Halley is structurally unlike any of the hundreds of interplanetary dust particles that have been studied in the laboratory by electron microscopy.

The chondritic composition, high carbon content, heterogeneity, and submicrometre intermixing of organic and silicate components are all consistent with the notion that comets are composed of extremely primitive materials. Numerous parent body processes, such as the aqueous alteration that occurred for the CI chondrites, could have produced more compositional homogeneity than is apparent in the Halley data. Although there are many uncertainties in the Halley results, the mission data suggest that cometary solids are different, more complex, and in a sense more 'primitive' than meteorites. The term 'primitive' is used here to emphasize that the fine-grained, heterogeneous Halley materials more closely resemble the likely properties of the submicrometre

interstellar grains that were in the protosolar nebula than the larger more homogeneous 'secondary' grains that dominate the meteorites. The Halley results support the widely held belief that the ice–dust grain conglomerates in comets are likely to carry chemical and isotopic records of the processes and environments that formed or influenced them. These environments include the outer regions of the solar nebula, and in the interstellar and circumstellar environments that preceded it. The 1986 Halley missions provided the initial characterization of cometary matter, but extraction of the full range of information recorded in cometary solids will require more detailed studies. The link between comets, interstellar grains, and core–mantle structures can be investigated more fully on a future rendezvous mission. The ultimate utilization of cometary solids as a solar system/interstellar medium 'Rosetta stone' will require a sample return mission such as the one currently planned by ESA that has aptly been named Rosetta.

REFERENCES

A'Hearn, M.F., Hoban, S., Birch, P.V., Bowers, C., Martin, R., & Klinglesmith, D.A. (1986) Cyanogen jets in Comet Halley, *Nature* **324** pp 649–650

Anders, E., & Ebihara, M. (1982) Solar system abundances of the elements, *Geochim. Cosmochim. Acta.* **42** 2363–2380

Bradley, J.P., & Brownlee, D.E. (1986) Cometary particles: thin sectioning and electron beam analysis, *Science*, **231** 1542–1544

Bradley, J.P., & Sandford, S.A., Walker, R.W. (1988) Interplanetary dust particles, In: *Meteorites and the early solar System*, Kerridge, J., ed. Univ. of Arizona press, 861–898

Brownlee, D.E., Wheelock, M.M., Temple, S., Bradley, J.P., and Kissel, J. (1987) *Lunar and Planet. Sci.*, **18** 133–134

Clark, B., & Mason, L.W. (1986) Systematics of the 'CHON' and other light-element particle populations in Comet Halley. In *ESA SP-250*, 353–358

Goldberg, R.A., & Aikin, A.C. (1973) Comet Encke meteor metallic ion identification by mass spectrometry, *Science*, **180** 292–296

Greenberg, M. (1982) *What are comets made of?* A model based on interstellar dust, in *Comets*, Wilkening, L.L., ed. U. Arizona press 131–163

Hanner, M. (1985) Dark grains in Comet Crommelin, *Astron. Astrophys*, **152** 177–181

Huebner, W.F. (1987) First polymer identified in Comet Halley, *Science*, **237** 628–630

Jessberger, E.K. Kissel, J., Fechtig, H., & Krueger, F.R. (1986) *Eur. Space Agency Spec. Publ.*, **249** 27–30

Jessberger, E.K., Christoforidis, A., & Kissel, J. (1988) Aspects of the major element composition of Halley's dust, *Nature*, **332** 691–695

Kissel, J. (1986) The Giotto particulate impact analyzer, *Eur. Space Agency Spec. Pub.*, **1077** 67–68

Kissel, J., Sagdeev, R.Z., Bertaux, J.-L., Angarov, V.N., Audouze, J., Blamont, J.E., Büchler, K., Evlanov, E.N., Fechtig, H., Fomenkova, M.N., von Hoerner, H., Inogamov, N.A., Khromov, V.N., Knabe, W., Krueger, F.R., Langevin, .Y., Leonas, V.B., Levasseur-Regourd, A.C., Managadze, G.G., Podkolzin, M.N., Shapirov, V.D., Tabaldyev, S.R., & Zubkov, B.V. (1986a) Composition of Comet Halley dust particles from Vega observations, *Nature*, **321** 280–282

Kissel, J., Brownlee, D.E., Büchler, K., Clarke, B.C., Fechtig, H., Grün, E., Hornung, K., Igenbergs, E.B., Jessberger, E.K., Krueger, F.R., Kuczera, H., McDonnell, J.A.M., Morfill, G.M., Rahe, J., Schwehm, G.H., Sekanina, Z., Utterback, N.G., Völk, H.J., & Zark, H.A. (1986b) Composition of Comet Halley dust particles from Giotto observations, *Nature*, **321** 336–337

Kissel, J., & Krueger, F.R. (1987a) Ion formation by impact of fast dust particles and comparison with related techniques, *Appl. Phys. A*, **42** 69–85

Kissel, J., & Krueger, F.R. (1987b) The organic component in dust from Comet Halley as measured by the PUMA mass spectrometer on board Vega 1, *Nature*, **326** 755–760

Krueger, F.R., & Kissel, J. (1984) *Eur. Space Agency Spec. Pub.*, **244** 43–48

Krueger, F.R., & Kissel, J. (1987) The chemical composition of the dust of Comet P/Halley as measured by 'PUMA' on board VeGa-1, *Naturwissenschaften*, **74** 312–316

Langevin, J., Kissel, J., Bertaux, J-L., & Chassefiere, E. (1987) First statistical analysis of 5000 mass spectra of cometary grains obtained by PUMA 1 (Vega-1) and PIA (Giotto) impact ionization mass spectrometers in the compressed modes. *Astron. Astrophys.*, **187** 761–766

Lawler, M.E., Brownlee, D.E., Temple, S., & Wheelock, M.M. (1989) Iron, magnesium and silicon composition of dust from Comet Halley, *ICARUS* **80** 225–242

Mitchell, D.L., Lin, R.P., Anderson, K.L., Carlson, C.W., Curtis, D.W., Korth, A., Reme, H., Sauvauvd, J.A., D'Uston, C., & Mendis, D.A. (1987) *Science,* **237** 626–628

Millman, P.M. (1972) Cometary meteoroids, In: *From plasma to planet*, Nobel Symp. No. 21, Elvius, A., ed. 157–168

Ney, E. (1982) Optical and infrared observations of bright comets in the range 0.5μ to 20μ, In *Comets*, Wilkening, L.L., ed. 323–340

Sagdeev, R.Z., Kissel, J., Evlanov, E.N., Mukhin, L.M., Zubkov, O.F., Prilutskii, O.F., & Formenkova, M.N. (1986) Elemental composition of the dust component of Halley Comet: preliminary analysis, *ESA SP-250*, 349–352

Schramm, L.S., Brownlee, D.E., & Wheelock, M.M. (1989) The major element composition of stratospheric micrometeorites, *Meteoritics* **24** 99–112

Simpson, J.A., Rabinowitz, D., Ksanformality, L.V., & Sagdeev, R.Z. (1986) Halley's Comet coma dust particle mass spectra, flux distributions and jet structures derived from measurements on the Vega-1 and Vega-2 spacecraft, *ESA SP-250* 11–16

Solc, M., Vanysek, V., & Kissel, J. (1987) Carbon isotope ratio in PUMA-1 spectra of P/Halley dust, *Astron. Astrophys.*, **187** 385–387

Vaisberg, O.L., Smirnov, V., Omelchenko, A., Gorn, L., & Iovlev, M. (1987) Spatial and mass distribution of low mass dust particles in Comet P/Halley's coma, *Astron. Astrophys.*, **187** 753–760

Wallis, M.K., Rabilizirov, R., & Wickramasinghe, N.C. (1986) Evaporating grains in Halley's coma, *ESA SP-250*, 251–254

9

The evidence that comets are made of interstellar dust

1 INTRODUCTION

The chemical composition and size distribution of interstellar dust has been studied by theoretical interpretation combining remote observations and laboratory experiments (Greenberg 1985, d'Hendecourt *et al*. 1985, 1986). Following the various stages of evolution (Greenberg & Hong 1974a, b) leads to a picture of grains as consisting of basically two distinct size populations: tenth micrometre and hundredth (or less) micrometre. The larger particles are core–mantle structures with cores of silicates, inner mantles of a complex organic refractory material, and outer mantles of various ices dominated by frozen H_2O. The very small particles are again of two types; one type is mostly carbon, the other had earlier been presumed to be made of silicate material but may in fact be very large carbon molecules called PAHs (polycyclic aromatic hydrocarbons). The key to the evolution of grain mantles is solid-state photoprocessing while fully taking into account the gas phase and surface reactions of interstellar atoms and molecules. After a mean lifetime of several thousands of million years most grains are consumed by star formation. Some, however, remain as the basic building material of Solar System bodies, and are brought to us at the present time in various stages of preservation since our solar system formed about $4\frac{1}{2}$ thousand million years ago.

Section 2 reviews the basic size, shape, and composition parameters of interstellar dust. Section 3 considers the direct evidence for the cold comet aggregation process and the implication for the comet nucleus density. The question of chemical and physical evolution of the comet nucleus while in the Oort cloud is taken up in section 4. Section 5 discusses a number of Comet Halley results and their interpretation in terms of the aggregated interstellar dust model. Section 6 shows some connections and relationships between comets and other small solar system bodies.

2 SIZE, SHAPE, AND COMPOSITION OF INTERSTELLAR DUST

Recent studies of the observations of so-called diffuse cloud dust (dust not in molecular clouds) in the ultraviolet have revealed the fact that there are three populations of dust (Greenberg & Chlewicki 1983, Chlewicki 1985). There are elongated 'large' grains of $\sim 0.12\ \mu m$ in mean radius which provide the major blocking of starlight in the visual. The polarization of starlight in the visual which is associated with extinction implies that these particles are elongated by at least a factor three (Greenberg 1968). There are also very small carbonaceous particles of $\lesssim 0.01\ \mu m$ in radius which produce a strong absorption feature at about 220 nm (Stecher 1965). In addition there is an independent population of $\sim 0.01\ \mu m$ silicate–type particles.

How the large grains form and evolve is a complex physical and chemical story. Basically, they start with silicate particles forming in the atmospheres of cool stars and, after being ejected into space and cooling down to 10–15 K, providing nucleation cores for the growth of mantles of ices. These mantles result from accretion of gas phase atoms and molecules of the abundant atomic species, oxygen, carbon, nitrogen and sulphur along with the hydrogen (Greenberg 1982a) which may also undergo subsequent surface reactions. Since the silicate cores provide an extra polarization at 10 μm—associated with their Si-O stretch infrared absorption—they must be non-spherical and aligned. The non-sphericity is apparently not a result of crystallinity, because the shape of the 10 μm absorption is characteristic of amorphous silicates. It is probably due to small (more or less spherical) particles stuck together: the elongation of two connected equal-sized particles is obviously 2:1. Although the accretion of the gases is random it does not, as expected, make the particles more spherical. Rather, because the particles must be spinning suprathermally in order to produce the required alignment (Greenberg & Chlewicki 1987), the extra centrifugal force causes the atoms and molecules which hit and stick, to tend to *slide* outward from the centre of rotation; i.e. the surface diffusion is slightly enhanced by the centrifugal energy difference (Greenberg & Aanestad, unpublished). This differential sticking leads to *enhanced* non-sphericity. The observation of an extra degree of linear polarization at 3.07 μm shows that the solid H_2O in interstellar mantles is also elongated, thus confirming this effect. In addition to the H_2O, many other simple molecules—and even ionic-species—like CO, CO_2, H_2S, CH_3OH, OCS, OCN^-, NH_4^+, H_2CO, etc. have been observed in interstellar dust. These ices are always being photoprocessed by ultraviolet radiation from either distant stars or by UV created by cosmic rays or arising from local hot stars and/or stellar winds. The result of such photoprocessing is not only a change in the basic composition of the ices but also the production of complex organic refractory residues which have been studied in the Leiden laboratory and compared with astronomical observations (Greenberg 1982b, Agarwal *et al.* 1985, Schutte 1988).

The direct evidence for organic residues in space,

although predicted as early as 1971 (Greenberg *et al.* 1971), was finally exhibited in the observation of a 3.4 μm feature seen towards objects in the galactic centre. The first such observation was of SgrAW (Willner *et al.* 1979), and, since then, better observations of this feature taken with higher resolution have confirmed its presence in a number of different galactic centre sources (Willner & Pipher 1982, Allen & Wickrama-singhe 1981, Butchart *et al.* 1986). Laboratory produced residues have been successful in recreating the *essential* shape of the 3.4 μm feature which consists of contributions from the C-H stretch in CH_2 and CH_3 groups in complex organic molecules (Greenberg 1984a, Schutte & Greenberg 1986). Although the *precise* shape of the interstellar 3.4 μm feature (or features) has not been matched, there are several good reasons for this: first, no laboratory residue has yet been subjected to the complete radiation history of an interstellar grain, because of present laboratory limitations; second, line of sight effects by grains with varying histories create averages of spectra of various complex organic residues (Greenberg 1984a).

The spectral correspondence between the laboratory residue and the galactic centre is supported quantitatively by the measured strength of the laboratory 3.4 μm feature. The major component of the dust in low-density (diffuse) clouds is the organic refractory material.

The evolutionary picture of dust which is emerging is a cyclic one in which the particles find themselves alternately in diffuse clouds and in molecular clouds. A small silicate core captured within a molecular cloud gradually builds up an inner mantle of organic refractory material which has been produced by photoprocessing of the volatile ices. Within the dense clouds critical densities lead to star formation and subsequent ejection of some of the cloud material back into the surrounding space. Much of this material, finding itself in a very tenuous low-density environment, expands to the diffuse cloud phase. During the intermediate stage of passage from high to low density, the volatile grain mantles are heavily photoprocessed, producing the major part of the organic refractory mantles which are seen on diffuse cloud grains. Dust particles in the diffuse medium are subjected to numerous destructive processes which rapidly erode refractory material after evaporating the volatiles. It is important to note that without their organic refractory mantles the silicate cores could not survive. The rate of destruction of pure silicate grains leads to a maximum lifetime of $\tau_{Sil} \simeq 4 \times 10^8$ yr which converts to a galactic average mass loss rate of $d\rho_{Sil}/dt = -5 \times 10^{-43}$ g cm^{-3}s^{-1}. Assuming a mass loss rate from M stars of 1 $M_\odot$ yr^{-1} and a full cosmic abundance silicate production leads to a production rate for silicates of $d\rho_{Sil}/dt < 10^{-45}$ g cm^{-3}s^{-1}, which is 100 times lower (Greenberg 1985) than the destruction rate. On the other hand, the mean production rate of the O.R. of $d\rho_{O.R.}/dt \simeq 10^{-41}$ g cm^{-3}s^{-1} is adequate to replenish the mantle material lost in the diffuse cloud phase even if the O.R. is somewhat less tough than the silicates (Greenberg 1982a). Therefore silicate core–organic refractory mantle grains survive the diffuse cloud phase to re-enter the molecular cloud phase, having lost a fraction of the protective organic refractory 'shield' which permits the silicate cores to survive.

The mean star production rate of 1–2 $M_\odot$ yr^{-1} implies an interstellar medium turnover time of $\sim 5 \times 10^9$ yr, so that this is the absolute maximum lifetime of a dust particle no matter how resistant to destruction. If we use a mean molecular cloud–diffuse cloud period of 2×10^8 yr (10^8 years in each based on a mean time spent in and out of spiral arms) then a typical grain anywhere in space will have undergone at least 20 cycles, so that, for example, the typical diffuse cloud dust particle age is $\gtrsim 10^9$ yr and consists of a mix of particles which have undergone a *wide variety* of photoprocessing. Using a mean cloud lifetime of 5×10^7 years for the cycle time may be more representative.

It will be useful to calculate the total ultraviolet energy absorbed by a grain in standard energy dose units of rads (1 rad = 100 erg g^{-1}). The dose rate absorbed by a tenth micrometre typical core–mantle particle exposed to the diffuse cloud flux of ultraviolet photons of energy $E = h\nu \geq 6$ eV (a conservative lower bound for photoprocessing) is 0.8 rad s^{-1}, while the dose rate on a very small, $a = 0.01$ μm, particle is about ten times higher. Note that the organic refractory mantles are subjected to the highest photoprocessing rates in the diffuse cloud phase— higher by factors of 10 000 or so than in the molecular cloud. This would imply that the organic refractory

mantle on a grain is not a homogeneous substance but rather layered like an onion. Sequential organic mantle formation (in the molecular cloud phase) and intense photoprocessing (in the diffuse cloud phase) would lead to a structure in which the innermost layers have been the most irradiated, and the outermost layer in the most recent molecular cloud phase is first generation organic refractory surrounded by lightly photoprocessed ices. Because of this kind of layering, and the fact that the grains are of various ages, leads one to expect average homogeneity of diffuse cloud grains both in size and structure which is observed as a uniformity in the visual extinction curve and a rather structureless 3.4 μm feature. In other words, diffuse cloud grains represent a steady-state average of grains of a multiplicity of chemical and physical histories. Since further photoprocessing of organics leads to a greater and greater depletion of O, N, and H, the innermost layers are the most 'carbonized' and the most non-volatile. This property will be reflected in the emission characteristics of heated comet dust, owing to the loss of the more volatile organic components after the dust leaves the comet surface.

We arrive at a picture of the 'large' grains as being multiple-layered particles with a mean silicate core radius of 0.05 μm (elongated by a factor of about 3). In diffuse clouds there is a mean mantle thickness of $\lesssim 0.05$ μm (total radius about 0.1 μm) and an elongation of 3 or 4 to 1. In molecular clouds there is an additional mantle (outer mantle) of more volatile species dominated by H_2O but generally containing CO, NH_3, H_2CO etc. (see Table 1). This outer mantle may grow to a thickness of 0.01 in only 10^5 years in a cloud with hydrogen density $n_H = 10^3$ cm^{-3} and, neglecting desorption, in $\gtrsim 10^6$ years it would deplete all the available gas (excluding H and He) in a cloud of density $n_H = 10^4$ cm^{-3} (Greenberg 1985). There are normally desorption mechanisms provided by ultraviolet processes which lead to grain explosions (d'Hendecourt et al. 1985, Greenberg 1978) thus inhibiting such complete grain growth. In *very* dense clouds such processes are turned off by lack of ultraviolet radiation and the grains then may deplete all the remaining gas. Such a situation would prevail in the latest stage of the molecular cloud correspond-ing to earliest stages of the protosolar system *and* comet formation.

In addition to the large particles there are numerous much smaller ($a \lesssim 0.01$ μm) particles which are responsible for key ultraviolet features of the interstellar extinction: the 220 nm 'hump' particles and the particles needed for the extra far ultraviolet extinction beyond $\lambda \simeq 170$ nm (FUV particles).

Graphite was the first candidate suggested for the 220 nm hump (Stecher & Donn 1965). One of the basic difficulties with this material has been the fact that the shape and position of the hump depend strongly

Table 1.   Molecules directly observed in interstellar grains and/or strongly inferred from laboratory spectra and theories of grain mantle evolution

Molecule	Comment*	
H_2O	O	M_2
CO	O	M_2
H_2S	O	M_2
NH_3	O	M_2
H_2CO	O	M_2
$(H_2CO)_n$	I	M_2
OCN^-	O	$M_2(M_1)$
NH_4^+	O	M_2
CH_3OH	O	M_2
OCS	O	M_2
CO_2	I,O	M_2
CH_4	I	M_2
S_2	I	M_2
complex organic	O	M_1
'silicate'	O	C
'carbonaceous'	(O, I)	B

*O = observed, M_1 = inner mantle, M_2 = outer mantle, B = small bare, I = inferred, C = core

on the size and shape of the particles whereas the observed hump is highly uniform in structure (Savage & Mathis 1979). Another problem has been the fact that, although the hump and the visual extinction are remarkably well correlated, the sources of these two dust components are independent in the sense that the visual extinction particles are basically created in the interstellar medium, while the graphite is (presumed to be) formed in stellar atmospheres (Czyzak & Santiago 1973).

We shall not discuss here the question of how graphite carbon can form in the first place. However, if interstellar dust provides the basic components of the protosolar nebula, as is indicated by the evidence

of their presence in comets, then there should be some remnants in primitive meteorites. The lack of well-crystallized carbon in meteorites (Nuth 1985) would seem to preclude graphite as a substantial component of the interstellar medium; certainly not as much as the $\sim 32\%$ of all carbon required in the form of graphite to give the 220 nm hump (Chlewicki 1985, Hong & Greenberg 1980, Mathis et al. 1977). There is also a problem of replenishing the graphite at a rate sufficient to counter its erosion and destruction rate (Greenberg 1986a, Draine & Salpeter 1979).

A *purely interstellar process* leading to the presence of a carbon component follows from ultraviolet photoprocessing of grain mantle ices (Van der Zwet et al. 1986, Greenberg et al. 1987). A factor in favour of interstellar produced organic particles contributing to the 220 nm hump is the degree of correlation between the visual extinction and the 220 nm hump (Savage & Mathis 1979, Savage 1975). In other words we might expect a visual-hump correlation if the 'visual particles' can be the *source* of the '220 nm particles'.

It has been proposed that certain linear unsaturated carbon chain molecules have extremely strong absorption in the 200 nm region. We have, in matrices, produced these absorptions by ultraviolet photolysis of a number of starting organic molecules—either hydrocarbons or those containing oxygen or nitrogen. The stable end products appear from infrared and ultraviolet spectra to be linear chain molecules of the polyacetylene and cyano-polyacetylene type. The assumption here is that some of the organic refractory mantles are either broken off by some grain destruction process, such as shocks, or ejected by grain explosions, and appear in space *initially* as very small carbonaceous particles which are small enough ($2\pi a/\lambda \ll 1$) to produce the hump (Greenberg 1979, Van de Hulst 1957).

As a cyclic phenomenon the 220 nm particles are accreted along with the icy mantles in the molecular cloud phase (Greenberg 1982a), and as a result of oxidation and other photochemically induced chemical reactions in the solid which is oxygen rich they are *incorporated* into the mantle material. The observed 220 nm decrease in strength in molecular clouds (Snow & Seat 1980) is consistent with reincorporation in the mantles (Greenberg 1978). Regeneration of the hump particles then occurs when the particles reappear in the diffuse cloud phase.

Other forms of small carbon particles (or perhaps bare molecules) similar to polycyclic aromatic hydrocarbons are suggested by their contribution to certain interstellar infrared emission features (Leger *et al.* 1987). These particles appear to use up of the order of 5% of the cosmically available carbon, and may finally wind up either as a part of the organic refractory component or as imbedded in the volatile mantles. It is not clear what happens to them in the molecular cloud phase.

Grain modelling of the far ultraviolet part of the extinction curve has been successfully performed, using very small silicate particles and PAHs.

The optical properties of interstellar dust are a very important ingredient not only in theories of extinction and polarization in space but also in theories of how they act as aggregates. For application to comet and comet dust modelling the important parameters are the absorptivity of the core–organic refractory grains and their scattering properties. Observationally, it is seen that the diffuse cloud grains absorb about 60% of the incident (visible) radiation (albedo $\equiv \alpha = 0.6$) and that they scatter about 80% of the radiation in the forward direction (asymmetry factor $g = < \cos \theta_{sc} > = 0.8$) (Savage & Mathis 1979). The laboratory-created organic refractory mantles are strongly absorbing in the ultraviolet, and experiments, supplemented by theory, lead one to expect that strongly photoprocessed organics have a complex index of refraction in the visual of $m \simeq 1.7 - 0.15\,i$ (Chlewicki & Greenberg 1988), which implies a dark material when small particles are aggregated. The laboratory first-generation organic residues start out yellow (UV absorption) and become darker (brown) with radiation. Theoretical calculations of core–mantle particles with such mantle optical properties have been shown to match the observed extinction and polarization as well as the albedo of interstellar dust (Chlewicki & Greenberg 1988, Greenberg & Chlewicki 1987).

A schematic representation of grains in the various regions of space is shown in Fig. 1. In the final stage of cloud condensation we may expect that all remaining (condensable) molecules will have accreted onto the dust. In addition, the very small ($\leq 0.01\ \mu m$) particles will be collected and trapped within the outer volatile icy mantle. An alternative route for the very small particles is that they may *themselves* accrete icy mantles in the dense cloud after the ultraviolet has been 'turned off', so that some may accrete as small protuberances on the outer parts of the large grains enclosed in their own small mantles (Greenberg & Hong 1974b).

3 DUST AGGREGATION AND MORPHOLOGY

In our solar system all the planets and satellites have incorporated into their bodies at least the most refractory components of the interstellar dust which existed in the presolar nebula. In general, many of the direct connections to the dust have been lost. Comets are the most likely bodies to have preserved their original composition, but the first important question is did their original composition include the volatile as well as the non-volatile components of dust? The evidence is that the volatile molecule S_2 may be traced back to the photochemical evolution of the interstellar dust ice (predominantly H_2O) mantles which are *known* to contain sulphur-bearing molecules (Grim & Greenberg 1987). Secondly, an upper limit on the temperature of formation of the comet is provided by the fact that if the predominantly H_2O grain mantles evaporate they release the S_2 which cannot reform in the protosolar nebula and is therefore lost to the comet. Thirdly, even if where the comets form the temperature is low enough to keep the ice from evaporating, the formation must occur far enough out so that volatiles brought by turbulence from inner regions of the solar nebula (where grains evaporate) are not incorporated into the nucleus, thus diluting the S_2 concentration below what is observed (Greenberg *et al.* 1985). This additional constraint on the region of comet formation is placed by the fact that the mean concentration of S_2 relative to H_2O in interstellar dust, as deduced from laboratory experiments, does not allow for much dilution by the addition of non-S_2 containing H_2O. Another piece of evidence is given by the fact that the fraction of CH_4 observed in comets Halley and Wilson is consistent with the chemistry of photoprocessed interstellar dust 'but not with either equilibrium or disequilibrium condensation sequences in the solar nebula' (Larson *et al.* 1989).

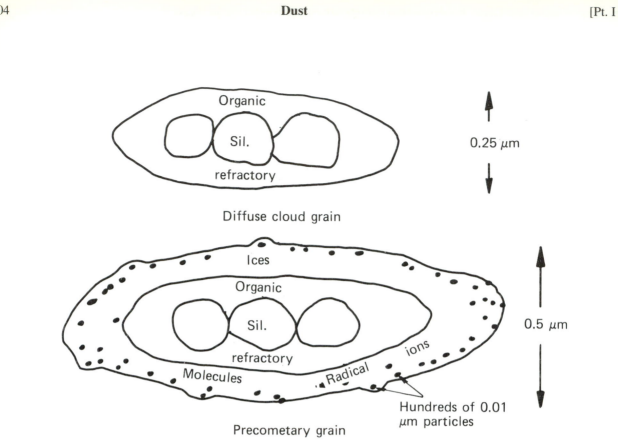

Fig. 1 — Schematic illustration of an average diffuse cloud grain and of an average fully accreted precometary grain.

The temperature in the presolar nebula, as calculated by Ruznaikina & Maeva (Ruznaikina & Maeva 1988), supports the low temperature cometary accretion concept. According to their theory of the process of formation of a protoplanetary disk, the maximum temperature at a distance of 1 AU lies in the range $400\,K \leq T \leq 500\,K$, so that at distances 2–4 AU from the sun, where the asteroid belt is now located, the temperature did not exceed 250–300 K. At distances between 20 and 30 AU (Uranus–Neptune) the temperature was $T < 67\,K$, so the only significant source of heating during grain coagulation would have been by collisions. Totally inelastic collisions at turbulent-induced speeds of $\sim 0.1\,km\,s^{-1}$ would raise the grain temperatures by only ~ 15 degrees (Greenberg 1979).

Yamamoto & Kozasa (1987) even suggest that most of the aggregates and planetesimals formed outside the planetary region at around a few hundred AU.

As a first approximation, therefore, we consider a

Table 2. Suggested mass distribution of the principal chemical constituents of a cometesimal based on the dust model. Parentheses refer to very small particle components (a $\leq 0.01\,\mu m$)

Component	Mass fraction
Silicates	0.20
Carbon (carbonaceous)	(0.06)[a]
Non-volatile Complex	
Organic Refractory	0.19
H_2O	0.37[b]
CO	0.03[c]
CO_2	0.02[d]
Other Molecules + Radicals (H_2CO, NH_3, OCN^-, HCO, S_2, CH_3OH ...)	0.13

[a] This amount of carbon in small particles is derived by using graphite. Carbonaceous particles which may *more* strongly absorb than graphite (per C atom) at 220 nm could bring this percentage down.
[b] Based on *70%* of volatile interstellar grain mantles as H_2O (Greenberg 1982a, Van de Bult et al. 1984).
[c] Based on an average $[CO/H_2O]$ in interstellar grain mantles $\leq 15\%$ (Grim & Greenberg 1987).
[d] Based on photoprocessing in grain mantles.

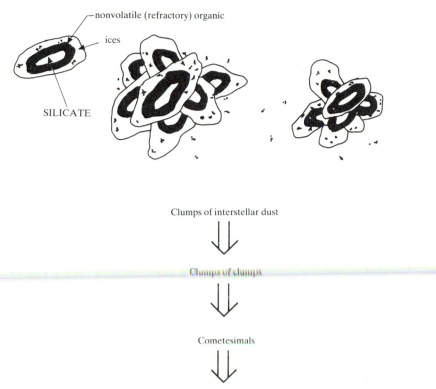

Fig. 2 — Aggregation process from individual interstellar dust grains to comets.

comet nucleus as if its chemical composition and morphological structure are directly related to interstellar dust. Table 2 shows the relative fractions of the various chemical constituents which have been obtained by an extrapolation from the molecular cloud dust phase (Greenberg 1983a). In all cases the full cosmic abundances of the elements are assumed.

In forming the nucleus we assume that (see Fig. 2) first clumps of grains form and then clumps of clumps and so on until finally we reach the size of the comet nucleus. The question is, does this lead to a fully compact structure, or does it lead to an open structure. Donn & Hughes (1986) have suggested fractal aggregation and also some compaction by collisions. With the highly elongated particles of interstellar dust, the rotation after collision has not been simulated in fractal models. However, the elongation must provide a basis for fluffy aggregation with a substantial portion of the tensile strength due to the tangle of long dust grains, just as in a bird's nest of

twigs (Greenberg & Gustafson 1981, Greenberg 1980). The evidence we have for an open structure is the low density of meteors, which are known to be comet debris (Verniani 1969, 1973, Millman 1976).

We define the porosity, P, of an aggregate of interstellar dust as

$$P = 1 - \frac{V_{\text{solid}}}{V} = 1 - \rho \qquad (1)$$

where V_{solid} is the volume of the solid material inside the aggregate, and V is the total volume of the aggregate, its volume being defined by some appropriate boundary. We have $0 \le P < 1$, where $P = 0$ corresponds to a solid aggregate and $P \approx 1$ corresponds to a cloud of independent interstellar dust particles.

If the aggregation into a comet were to lead to a fully compacted structure (i.e. $P = 0$), the nucleus density would be

$$\rho_{nucleus} = \frac{M_{sil} + M_{or} + M_c + M_{ice}}{V_{sil} + V_{or} + V_c + V_{ice}}, \qquad (2)$$

where M and V refer to the mass and volume fractions of the various materials in the protosolar nebula dust. Using the mass fractions of the protosolar nebula dust given in section 2 and densities of $\rho_{sil} = 3.5$, $\rho_{or} = 1.8$, $\rho_{ice} = 1.2$ and $\rho_c = 2$, in units of grams per cubic centimeter, one finds $\rho_{nucleus} = 1.54$ g cm^{-3}. This is about 15% higher than previously derived (Greenberg, 1986b), because of the present assumption of higher material densities. The density and porosity of the coma dust may now be derived by assuming that the only difference between comet material and coma dust is that the volume filled originally by the ices and carbonaceous particles in the nucleus material is taken up by an equal volume of vacuum inside the coma dust. This corresponds to taking $M_c = M_{ice} = 0$ in equation (2), resulting in $\rho_{dust} = 0.6$ g cm^{-3}. The porosity of the coma dust derived in this way would be

$$P_{dust} = 1 - \frac{V_{sil} + V_{or}}{V_{sil} + V_{or} + V_c + V_{ice}} = 0.75. \qquad (3)$$

At this point we note that the mean density of meteors, some of which are clearly of cometary origin, is generally less than 0.6 g cm^{-3}. According to Verniani (1969, 1973), those meteors whose mean aphelion distance is highest and which are therefore least affected by the sun have the lowest densities of about 0.2 g cm^{-3}. If we assume that the density of coma dust equals that of these meteors, this implies that the coma dust must have a higher porosity than derived above. The implied density change by a factor of three $(= 0.2/0.6)$ now gives

$$P_{dust} = 1 - \frac{V_{sil} + V_{or}}{3 \times (V_{sil} + V_{or} + V_c + V_{ice})} = 0.92. \qquad (4)$$

Consequently, by reconstitution of the ices and carbonaceous particles into the coma dust, the porosity and density of the comet itself may be derived:

$$P_{nucleus} = 1 - \frac{V_{sil} + V_{or} + V_{ice} + V_c}{3 \times (V_{sil} + V_{or} + V_c + V_{ice})} = 0.67 \qquad (5)$$

$$\rho_{nucleus} = \frac{M_{sil} + M_{or} + M_c + M_{ice}}{3 \times (V_{sil} + V_{or} + V_c + V_{ice})} = 0.51. \qquad (6)$$

An even lower dust packing factor of about 0.2 is derived from the mean comet density of $\rho = 0.25$ g cm^{-3} derived, from the tidal splitting of P/Comet Brooks 2, by Sekanina & Yeomans (1985) and by the latest interpretation of meteor densities as being even less than 0.1 g cm^{-3} (Olsson-Steel 1987, 1989). A model of such an open aggregate of 100 typical precometary grains is shown in Fig. 3a*. Each particle as represented corresponds to an interstellar grain $\frac{1}{2}$ μm thick and about $1\frac{1}{2}$ μm long. The mean mantle thickness corresponds in reality to a size distribution of thicknesses starting from zero. The packing factor of the particles is about 0.2 (80% empty space) and leads to a mean mass density of 0.3 gm cm^{-3} and an aggregate diameter of 5 μm.

Another possibility is to take the comet density of 0.6 g cm^{-3} (as derived by Sagdeev et al. 1988) as a starting point and use the same volume and mass fractions as above, to derive the comet and coma dust porosities. These then turn out to be $P_{dust} = 0.90$ and $P_{nucleus} = 0.61$. A summary of these results is contained in Table 3.

Table 3. Comet nucleus densities and porosities, with corresponding coma dust densities and porosities. The superscripts denote $M_{sil} : M_{or}$ in the dust. The case 1 : 1 is the original cometary ratio according to the model and the case 2 : 1 corresponds to the situation where half of the organic refractory has evaporated

$\rho_{nucleus}$ (g cm^{-3})	$P_{nucleus}$	$P_{dust}^{1:1}$	$P_{dust}^{2:1}$	$\rho_{dust}^{1:1}$ (g cm^{-3})	$\rho_{dust}^{2:1}$ (g cm^{-3})
1.54	0.	0.75	0.83	0.60	0.45
0.60	0.61	0.906	0.933	0.23	0.18
0.51	0.67	0.92	0.94	0.20	0.16
0.26	0.83	0.96	0.975	0.10	0.08

4 EVOLUTION OF THE NUCLEUS IN THE OORT CLOUD AND LATER

Comets, once formed, become objects of the Oort cloud circulating about the Sun in near interstellar space. Insofar as further ultraviolet processing is concerned, even $4\frac{1}{2}$ thousand million years can lead only to changes in the outer few micrometres (!) of the nucleus. Furthermore, the temperature of the nucleus, barring possible internal heating by radionuclides, is even lower than the 10–15 K of interstellar grains.

* to be found in the colour illustration section in the centre of the book.

There is no evidence of a ^{26}Mg excess in Comet Halley dust mass-spectra (Jessberger & Kissel 1987), as has been found in some meteorites, so that internal heating by the radionuclide ^{26}Al did not occur. However, cosmic ray protons are another matter, and their effects have been considered by several authors (Johnson *et al.* 1983, Whipple 1977, Moore & Donn 1982, Strazzulla 1986). Their principal effect is to change the chemical composition of the material they penetrate in a way similar to the changes induced by ultraviolet processing of dust in interstellar space. The question is how significant is this additional processing?

It had already been shown by Donn & Moore that only the outer fraction of a metre of the nucleus could be chemically modified by the MeV cosmic ray protons, so that the higher energy ones must first be slowed during penetration before they are effective. A recent calculation of the radiation dosimetry of a cometary nucleus by cosmic ray protons with energies ranging from 1 MeV to 10^{10} GeV and including all secondary cascade effects (Ryan & Draganic 1986) will be used as a comparison with the ultraviolet radiation dosimetry of interstellar grains. According to the new calculation the absorbed C.R. dose after $4\frac{1}{2}$ thousand million years at about 1 m depth in a comet of density 1 g cm^{-3} is 10^{11} rad, which is much smaller than the typical ultraviolet interstellar photoprocessing dose of 6×10^{16} rad on a core–mantle grain in its lifetime. Thus the ultraviolet dose on grains during only one of the several thousand million years they spend *before* aggregation is about 2.5×10^{16} rad, which is 100 000 times more than the cosmic ray dose at the surface *after* aggregation and even more in the interior. A range of comparisons is shown in Table 4. Since the outer few metres of a comet are lost already in its first apparition, there is no reason to expect to find evidence of cosmic ray effects inside a comet like Halley which has been around many times. However, the production of a significant 'crust', some by cosmic ionizing particle effects, some by evaporation of the outer volatiles without loss of the dust, and some by recapture of previously emitted dust, appears plausible and some interpretations of its effects are suggestive of what is observed (Johnson *et al.* 1987a, b). With a structure like that of Fig. 3 in mind, one might picture releasing the volatile ices and some of the more volatile organics, thus leaving an even more open collection of dust grains behind. Noting that the inner organic refractory mantle is the most absorbing component of interstellar dust, the albedo of the non-volatile crust may even be lower than the average one for the nucleus (see section 5a).

The next question is the modification of the nucleus by internal heating after entering the inner solar system. Heat conduction of porous structures is substantially lower than that of solid material. First of all we start with the fact that the grain ices are amorphous, which means that their thermal conductivity is a few orders of magnitude lower than crystalline ice at $T \simeq 20$ K and 100 times less at $T = 40$ K (Klinger 1980). Next, I have estimated the thermal conductivity of the structure shown in Fig. 3 by letting each grain have about 4–5 grains touching it and by letting each contact surface be about 4×10^{-15} cm^2 (limited by local heat dissipation after collisions during aggregation). Using a mean grain area of 2×10^{-11} cm^2 (fully 'grown' presolar grain) means that only about 0.1% of the surface of each grain can conduct heat so that the net heat conductivity of a comet is expected to be $\lesssim 1 \times 10^{-5}$ that of crystalline ice; i.e. the heat conductivity is reduced by a factor of 100 000! This leads to a very high temperature gradient at the nucleus surface, and at the same time, because of the very open structure, a very high evaporation rate which leads to efficient cooling. A detailed calculation incorporating both of these effects has not yet been done, but it leads to the expectation that while the surface of such a comet would be hotter than had been expected (as is in fact observed), the amount of internal heating may still be low. The diffusion of vapour into the interior is limited by the very small pore size (recondensation occurs quickly), so that the morphological structure of the aggregated dust tends to be self-preserving. A factor which had not earlier been considered as contributing to thermal evolution of comets while still in the Oort cloud was the effect of passing hot stars and supernovae. Stern & Shull (1988) have shown that the frequency of occurrence of such accidental passages is significant, and have calculated the depth to which significant heating below the comet surface may occur. However, they have apparently overestimated the heat conductivity by a factor of 100 or

Table 4.

The cosmic-ray ionizing particles energy dose on a cometary nucleus after formation compared with the ultraviolet radiation dose on interstellar dust in diffuse clouds and molecular clouds before aggregation into a cometary nucleus. The figures for molecular clouds assume a reduction in the ultraviolet flux by a factor of 100, which corresponds to a cloud of density $n_H = 5 \times 10^3$ cm^{-1} and 1 pc thick.

Depth (metres)		CR dose 4.5×10^9 yr (Oort cloud) (Mrad)	Oort cloud CR dose/interstellar UV dose			
$\rho = 1.0$ (g cm^{-3})	$\rho = 0.2$ (g cm^{-3})		2.5×10^9 yr (Diff. cloud)	2.5×10^9 yr (Mol. cloud)	1.0×10^8 yr (Diff. cloud)	1.0×10^8 yr (Mol. cloud)
0.1	0.5	3×10^5	5.0×10^{-6}	5.0×10^{-4}	1.25×10^{-4}	1.25×10^{-2}
1.0	5.0	1×10^5	1.7×10^{-6}	1.7×10^{-4}	4.00×10^{-5}	4.00×10^{-3}
5.0	25.0	1×10^4	1.7×10^{-7}	1.7×10^{-5}	4.00×10^{-6}	4.00×10^{-4}
10.0	50.0	3×10^3	5.0×10^{-8}	5.0×10^{-6}	1.25×10^{-7}	1.25×10^{-5}

more. Their estimate of the heat conductivity of amorphous ice at, say $T = 40$ K, is at least ten times higher than that of Klinger (1980), and their estimate of the contact surface is also a hundred times too high. The net result is that their estimate of internal heating is considerably greater (by *at least* a factor of ten) than it actually is. The conclusion appears to be that while in the Oort cloud comets are really quite unaffected by their environment and the first time a comet enters the solar system, the only modification to be expected relative to its initial composition is limited to a rather thin surface layer. (This conclusion is confirmed by Stern in a private communication.) Thermal processing of periodic comets is another matter, but even at an internal temperature of 80 K the overall thermal conductivity of the fluffy comet is still less than $\simeq$ 0.0004 that of crystalline ice, a fact which implies that, for comets with a mean solar distance as large as that of Comet Halley, their interior should be very close to its original composition—again with the exception of an outer 'crust' which I will discuss later in the context of the smallness of the active surface of Comet Halley and probably other comets, as well.

5 COMPARISON OF THE INTERSTELLAR DUST MODEL WITH COMET HALLEY RESULTS

(a) Albedo and surface properties
Both Vega and Giotto measures of the brightness of the surface of Comet Halley outside the jets indicated very low albedos. According to Sagdeev *et al.* (1986) the television system (TVS) aboard Vega-1 and -2 measured the bidirectional reflectance factor ρ_{max} to

be 0.02 at a phase angle of 32°. Assuming the phase function of the nucleus surface to be similar to that of the Moon, this gave a value for the geometric albedo A $\sim 0.04^{+0.02}_{-0.01}$. The images obtained by the Halley multi-colour camera aboard Giotto gave $A < 0.04$. These values are consistent with the albedo derived for the open structure shown in Fig. 3* (Greenberg 1986b). Because the individual grains have strongly forward-throwing scattering phase functions and are somewhat absorbing, light penetrates the comet surface and is absorbed by multiple scattering within. A very good approximation to the surface albedo of a loosely agglomerated structure of particles with scattering asymmetry factor g (g = mean value of the cosine of the scattering angle $= \langle \cos \theta \rangle$), when $g \to 1$, and individual particle albedo α (α = scattering/absorption + scattering) is $A \simeq \alpha(1 - g)/2$ (42). For the particles shown in Fig. 3a one expects $g \simeq 0.8$–0.9 and $\alpha \simeq 0.5$–0.6, so that one gets $A \simeq 0.06$–0.025, which is about what is observed. It is likely that not all of the surface microstructure of a comet is simply as pictured in Fig. 3a, but, whether or not a crust has formed, the lowering of the albedo is certainly enhanced by microparticle structures.

Only about 10% of the surface of Halley is 'active', i.e. releasing gas and dust. If the interstellar dust structure is evacuated of its volatiles and does *not* come off as dust, the remaining material consists of a crust of fluffy aggregated silicate core–organic refractory mantle dust grains. It is not obvious that this is intrinsically more (or less) dark than the original aggregated dust *with* ices, but the former seems the more likely. It does provide a 'cover' through which the volatiles underneath are inhibited from evaporat-

* to be found in the colour illustration section in the centre of the book.

ing because of the very low heat conductivity of the crust and the subsequent *lack* of heating to the depths where the material is volatile. In any case with the open structure suggested by the dust model, the *entire* surface is predicted to be dark, and whether or not there are available areas through which the volatile dust may evaporate bringing with it the refractory components would be random and variable in time perhaps because of buildup and release of pockets of excess pressure.

As noted by Sekanina (1986), the nucleus of Comet Halley appears to be an extremely dynamic body, with the number of activity centres opening and closing on a time scale of one day or a fraction thereof. In view of the ~52 hour rotation period confirmed by Suizei (Kaneda *et al.* 1986, Sekanina & Larson 1986a,b) and the erratic behaviour of even the more enduring vents (Simpson *et al.* 1986), a yielding surface is more likely than a rigid one. This is in line with the result of the calculation in which we have shown that whatever cosmic ray effects there may have been in the Oort cloud, they were limited to an outer layer which has long since been eroded away. See Johnson *et al.* (1987a,b) for some discussions of this.

The evidence that Comet Halley activity is confined to the sunlit side (Sekanina & Larson 1986b) also suggests that the heat is limited to the surface, and that little residual rise in temperature persists in the shadow region. The low thermal conductivity and low albedo are also supported by the high surface temperature of 300–400 K (Sagdeev *et al.* 1986).

(b) Particle sizes

The SP-2 experiment carried by the Vega spacecraft (Sagdeev *et al.* 1986, Mazets *et al.* 1987) as well as the dust impact detection system (DIDSY) on the Giotto spacecraft (McDonnell *et al.* 1987, 1989) gave evidence for large numbers of previously 'unexpected' submicrometre-sized particles down to the limit of detection at 10^{-17} g, as well as particles in the 10^{-14} g range, although the data on the smaller ones is difficult to be sure of because of problems with deployment of the detector. Both of these sizes are expected and were indeed 'predicted' from the size distribution of interstellar grains (Greenberg 1982b, 1984b). As a result of heating, the ices evaporate and release the previously trapped hundredth-micrometre

sized particles, some of which themselves may be heated and evaporated or eroded if they are of the carbonaceous type proposed here for the 220 nm hump of the interstellar extinction. A typical mean mass for the larger (0.12 μm) individual interstellar grains (*after* ice mantle evaporation) is about 0.65×10^{-13} g (see next section), and for the small (~0.01 μm) carbonaceous or silicate grains is about 10^{-17} g.

The cumulative particle fluxes obtained by McDonnell *et al.* (1989) and by Mazets *et al.* (1987) may be used to deduce the number density n_m by first differentiating and then dividing the result by the spacecraft velocity which is much larger than the particle ejection speed from the nucleus. From the data of McDonnell *et al.* on 14 March one deduces n_m as shown in Fig. 4. We note that owing to a change in calibration of the instrument these are a factor of up to 25 times larger in the range 10^{-15} g $< M < 10^{-10}$ g than would be obtained from McDonnell *et al.* (1987). The size distributions from the Vega results shown in Fig. 4 are similar to that of McDonnell *et al.* up to $M \lesssim 10^{-10}$ g but are as much as two orders of magnitude higher for higher masses. The continued rise in all measured size distributions for $M < 10^{-13}$ g is indicative of a large number of individual interstellar size grains.

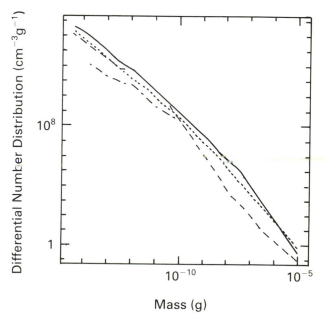

Fig. 4 — The number density of dust particles as a function of particle mass within the range 10^{-15} g $\leq M \leq 10^{-5}$ g.

The fact that there are subsubmicrometre sized particles (10^{-17} g) which can be released from the icy grain mantles seems to preclude a strong chemical modification and polymerization and grain sticking produced by cosmic rays after grain aggregation. If such extra radiation were to occur, the carbonaceous hundredth micrometre particles, rather than simply being trapped in ices, would, instead, be chemically bonded in a polymerized material and, having lost their individuality, would not reappear as observed in the size spectrum of Halley dust.

The spatial and temporal distribution of the masses and flux of dust particles measured by the dust counter and mass analyser (DUCMA) on VeGa-1, -2 showed, among other things, that the lowest masses were the first particles encountered at the fringes of the coma. One of the explanations for this phenomenon by Simpson et al. (1986) is that some of the dust particles are 'comprised of much smaller particles' from which pieces are shed which appear at great distances as the material which binds them sublimes; i.e., the initial dust is clumpy.

The density of the dust particles deduced by Krasnopolsky et al. (1986) is about 0.35 g cm^{-3}. Although this also supports the clumpy structure it is not quite as low as the density of the clumps shown in Fig. 3a which *without* the volatiles is $\lesssim 0.1$ g cm^{-3}. The dust densities deduced by Kissel and Krueger (1987) are $\rho \simeq 0.3$ g cm^{-3}; however, Krueger (1987) allows that significantly lower densities are possible. In fact (see next section) lowering the estimates of their dust masses by only a factor of 2 (well within their stated uncertainty) would bring their density into line with the fluffy comet prediction.

(c) Mass spectra of dust

The dust impact mass analysers on Vega-1, -2 (PUMA) and on Giotto (PIA) showed a predominance of the light elements H, C, O, N (organics) relative to the heavier elements Si, Mg, Fe (rockies) in the dust (Kissel et al. 1986a, b). There is considerable variety among the mass spectra, although 80% of the particles were indeed dominated individually by the organics. Many of these are like those which would consist of the relative proportion of organics to silicate in the core–mantle interstellar grains. One should bear

in mind that the organic refractory mantles have a range of possible thicknesses from zero to perhaps 0.2 μm, so that the light to heavy element ratio for *individual* (10^{-14} g) grains ranges from ~ 1 to significantly *greater* than 10:1. However, clumps of dust ($\gtrsim 10^{-13}$ g) which better represent an average composition would give a light to heavy ratio of about 3–5:1 by *number* and about 1:1 by mass in this model. Note, however, that the organics which are refractory in interstellar space conditions ($T \lesssim 100$ K) contain species (certainly in the outer first-generation organic refractory mantles) which will evaporate at the dust temperatures achieved at 1 AU. The optical absorptivity of the organic mantles raises the dust temperature to as high or higher than 600 K at 1 AU. This would bring the mean light to heavy atom ratio down, from its pre-cometary value, and, in fact, this is borne out by the infrared emission spectra of the dust (see next section).

There are some spectra which could be characterized as pure H, C, N, O, and others as pure silicates (Si, Mg, Fe, O).

Kissel & Krueger (1987) have attempted a molecular analysis of the comet dust, and in particular its organic component. Masses between 2×10^{-15} and 10^{-11} were measured with the masses of most of the particles estimated to be in the range 10^{-12}–10^{-13} g with 'systematic error within an order of magnitude'. We note that *one* typical silicate core–organic refractory interstellar dust grain has a *predicted* mass of $\sim 0.65 \times 10^{-13}$ g where I have used a silicate core consisting of three 0.05 μm radius silicate particles with an organic refractory mantle of 0.12 μm outer radius and length 0.75 μm (elongation = 3:1). Thus within the experimental uncertainties it appears that Kissel & Krueger are looking at particles as small as individual interstellar grains as well as small clumps. Their typical total relative atomic abundances in their molecules (of the organic refractory) show a significant lack of oxygen *just* as is predicted by the interstellar dust model (see Tables 5, 6). This is typical of moderately to fully photoprocessed interstellar grain organics as deduced from ultraviolet processed residues. As shown by Greenberg (1982 a, b) the C:O ratio in the organics is expected to be $\sim 1:2$; i.e. a four-fold enhancement of carbon is expected relative to oxygen. Most of the oxygen is in the volatile components with a substantial amount also in the silicates (see Table 5). It should be pointed out that the silicate core-organic refractory mantle morphology is substantiated by the time of flight mass spectroscopy of the CHON ions with respect to the rockies (Jessberger 1989).

Insofar as the smallest masses measured by Kissel & Krueger are concerned, they appear to be more massive than the very small bare particles which form a part of the interstellar dust size distribution (but note the quoted uncertainty). These particles would be expected to be either pure silicates or pure carbonaceous (mostly C with significantly less O and N) materials, and were predicted as providing a 'fine mist' of comet dust (Greenberg 1984b) with masses $< 10^{-16}$ g. More analysis is perhaps required to resolve whether these 'bare' interstellar particles are in the comet dust populations.

(d) Dust contributions to the comet atmosphere

An apparent anomaly exists when one compares the observed dust to gas production ratio of $Q_{dust}/Q_{gas} \simeq 0.25-0.1$ (Sagdeev et al. 1986, McDonnel et al. 1986) with that predicted (Greenberg 1982b, 1983a) for dust which is predominantly organic. The latter gives a value of $Q_{dust}/Q_{gas} \simeq 0.82$, which is consistent with comets Arend Roland 1957III (0.8 ± 0.2) and with Comet Bennet 1970II (0.6 ± 0.4) (Delsemme 1982). Even *without* an organic refractory component the predicted value is about 0.25, but then why is the organic component of dust predominant in the mass spectra (see next section)? Are we missing a part of the dust, or is it that *at* the nucleus the dust to gas ratio is indeed higher than where it is measured? One of the key questions is how far should one extend to mass limit in the dust spectrum. Extending it gives higher dust to gas ratios (McDonnell 1987 at Leiden, Fluffy Structures Workshop).

Photographic observations of Comet Halley's faint Sunward spike have been analysed by Sekanina (1986) to show that it indicates the presence of extremely small ($\ll 0.1$ μm) dielectric or nearly dielectric grains. These he associates with the prominent spectral signature of CN seen in the jet even before the scattered light by the spike was detected. Although the formation mechanism for the cyano radical is not certain, it is virtually certain that the CN

jets were formed *in situ*, which suggests that they arise from molecules which are sublimating from the very small grains. The carbonaceous grains which we have proposed for the 220 nm hump have both the size and the chemical composition (cyanopolyenes, in part) to be the source of such molecules. While they absorb strongly at 220 nm, the carbonaceous particles need not do so in the visible where they could appear dielectric. Another possible source of the CN jets is the presence of OCN⁻ and associated salts in interstellar dust mantles. It is observed in some instances that about 1% of the volatile outer mantles is the molecular ion OCN⁻ (Grim & Greenberg 1987). Some of this could also be tied up in organic salts as part of the organic refractory inner mantles. Both the refractory sources are consistent (small bare particles or organic refractory mantles) with the interpretation of A'Hearn et al. (1986) as '. . . jets of submicrometre particles, perhaps the CHON particles detected from spacecraft . . .'

The ion mass spectrometer on board Giotto detected, among others, a surprisingly large abundance of the C⁺ ion throughout the coma (Balsiger et al. 1986). The suggestion made by the authors was that small carbonaceous particles heated in the coma could be releasing carbon atoms or highly unsaturated molecules providing a distributed source for photolysis and ionization leading to the C⁺. The source of the carbon molecules could be, in part, the hundredth micrometre 220 nm hump interstellar (carbonaceous) particles *or* the relatively volatile part of the organic refractory mantle material on the core–mantle grains.

(e) Infrared evidence for silicate core–organic mantle dust

The infrared sounding of Comet Halley from Vega-1 (Mazets et al. 1986) revealed spectral emission at 3.3 μm and 7.5 μm attributed to C-H and C-C bonds which could be related to the presence of carbonaceous material. Groundbased infrared emission spectra of the 3.4 μm band by a number of observers was obtained (Baas et al. 1986, Wickramsinghe & Allen 1986, Knacke et al. 1986, Danks et al. 1986), and have been interpreted as being caused either by very small solid organic refractory particles or by solid particles plus large molecules. A point to be remembered here is that some of the organic matter which is refractory at

Table 5. Atomic constituents of various comet components model as fractions of cosmic abundance based on dust.

	Silicates	O.R.	Small carbonaceous	H_2O	CO	Other
H*		1.7 (−4)		4.7 (−4)		4.4 (−4)
C		0.45	≤0.27a		0.10	0.17
N		0.25[b]				0.75
O	0.09	0.13[b]		0.65	0.05	0.8
Mg	1.00					
Si	1.00					
Fe	1.00					

* The hydrogens are estimated as follows: 1 for each C in the O.R., 2 for each O in H_2O, 1 for each C in 'other', 1 for each N in 'other', 1 for each 0 in other.
[a] Based on graphite which we know is not valid.
[b] Nitrogen is relatively more abundant in the organic refractory than oxygen, so that the fractional (by cosmic abundance) nitrogen value could be significantly higher and that of oxygen significantly lower. These changes affect 'other' accordingly.

normal interstellar temperatures is volatile at comet dust and surface temperatures. The detailed structures shown in Baas et al. (1986) probably provide the best available means of identifying the organics. It is important to note the need for the sources of emission 'structures' to be very small in order to make the 3.4 μm emission possible. The 3.4 μm emission must be due to much hotter ($T \gtrsim 480$ K at 1 AU) particles (Greenberg & Zhao 1987) than the temperatures ($T \sim 320$ K at 1 AU) which produce the continuum. Particles with radii < 1 μm will *generally* be much hotter than large ones (or black bodies) (Greenberg 1971). The 10 μm emission (Gehrz & Ney 1986, Bregman et al. 1987) also requires such hot particles. However, *isolated* silicate particles, *even very small ones*, do not absorb solar radiation enough to be much hotter than black bodies, and therefore are too cool to emit sufficiently at 10 μm to show up as they do. On the other hand, the organic refractory mantles *do* absorb solar radiation strongly because the imaginary part of their index of refraction (Greenberg &

Chlewicki 1987, Chlewicki & Greenberg 1989) in the visual is about 0.1 to 0.15 (at least 10 times higher than that of the silicates). The *mantles* are thus the source of *heating* the silicate cores, and, in fact, the expected temperatures of individual core–mantle grains are as high as *700 K at 1 AU* (Greenberg and Hage 1989). Fluffy aggregates of such particles up to a certain overall size, which is limited by the fact that their overall optical depth must be substantially less than unity, will be as hot as their *individual* components. Larger aggregates approach black body temperatures (or temperatures like that of the comet surface), so that there is expected to be a distribution of dust temperatures from much hotter than black body down to about black body. The larger particles give rise to the continuum emission, while the smaller ones produce the *excess* emissions at 3.4 μm and at 10 μm. The conclusion is that to produce the 3.4 μm and 10 μm excess emission, the particles must: (1) be fluffy, (2) have individual components ∼0.1 μm in size, and (3) have silicates heated by contact with (mantles of)

Table 6. Mean abundances of the light elements in the cometary constituents

	Volatiles relative to 0		Relative to cosmic abundance		
Element	CI chondrite	Int. dust model	Volatiles	O.R.	Silicates + carbon
H	1.5	1.75	9.1 (−4)	1.7 (−4)	
C	0.2	0.18	0.28	0.45	0.27
N	0.1	0.16	0.75	0.25	
O	1	1	0.77	0.13	0.09

organic refractory absorbing material which emits at 3.4 μm.

Detailed sophisticated calculations of the emission/absorption properties of porous aggregates provided quantitative evidence to show that, in order to satisfy simultaneously such independent properties of coma dust as its (1) 3.4 μm emission (2) 9.7 μm emission (amount and shape), (3) coma dust mass distribution and (4) coma dust mass spectroscopic composition, the comet nucleus must be a low density aggregate of interstellar dust (Greenberg and Hage 1989). The basic units of the coma dust are the submicron silicate core-organic refractory mantle particles characteristic of diffuse cloud interstellar dust which are aggregated into very porous structures. These core–mantle unit particles have lost their original ices and the more volatile part of their organic refractory material after leaving the comet nucleus. The comet dust porosity derived, assuming about half the organics have evaporated, is $P \gtrsim 0.95$. The comet nucleus porosity deduced from the coma dust porosity is then most likely in the range $0.6 < P_{nucleus} < 0.8$ which leads to a comet density of 0.6 g cm^{-3} $> \rho_{nucleus} > 0.3$ g cm^{-3} and probably no more than ~ 0.4 g cm^{-3}.

(f) Molecule and ion components in the gaseous coma
Of course, H_2O was the most abundant molecule deduced in the coma of Comet Halley. The next most abundant species is CO. For example, Krankowsky et al. (1986) found a ratio $Q_{CO}/Q_{H_2O} \simeq 0.03$, and infrared data (Mazets et al. 1986) gave $Q_{CO_2}/Q_{H_2O} \simeq 10^{-2}$, while IUE observations (Festou et al. 1986) gave $Q_{CO}/Q_{H_2O} \simeq 0.1$–$0.2$. These values are more or less within the ranges suggested by the volatile composition of interstellar dust (Greenberg et al. 1985, White et al. 1985, Greenberg 1983b). There are two possible sources of CO. One of these is, of course, as part of the ice which evaporates from the grains. Another is the photodissociation of the more volatile molecules of the organic refractory component. The existence of carboxylic acid groups in laboratory first-generation organic residues and, by inference, in the outer parts of the organic dust mantles makes such a source highly plausible. The existence of an extended CO source in Halley associated with the dust provides support for the fact that a large fraction of CO comes off as the dust fragments and releases small grains

from which the not-so-refractory organics evaporate and are photodissociated (see section (d) for other gas components from dust). There is no definite evidence for the presence of NH_3 in the ion mass spectra (Balsiger et al. 1986), and there may even be a lack of nitrogen in the coma gas. This is yet to be definitely confirmed, but one possible reason could be that nitrogen is strongly bound in the organic refractories and is rather part of the dust than directly in volatile forms like NH_3 and N_2. Although NH_3 had earlier been suggested to be a substantial component of interstellar dust, the observational evidence (van de Bult et al. 1984), as well as theoretical arguments, lead to generally rather small amounts of NH_3 in grain mantles and possible more N_2 (d'Hendecourt et al. 1985).

It was noted by Balsiger et al. (1986) that the C/O ratio is about half of the cosmic abundance ratio. This had earlier been called the missing carbon mystery by Delsemme (1982) and had been attributed to the 'hiding' of a large fraction of the carbon in the organic refractory component (Greenberg 1983a). The dust mass spectra where the carbon to oxygen ratio is much higher than cosmic abundance confirm this prediction (Kissel & Krueger 1987).

(g) Density of Comet Halley
Now that the nucleus of Comet Halley has finally been directly seen, it has become possible to make proper estimates of its mass density. A detailed investigation had been reported by Rickman, at the comet nucleus sample return meeting, of the mass of Comet Halley which, given its size, leads to a mean bulk density as low as 0.1 g cm^{-3} and no higher than 0.5 g cm^{-3}, which is in reasonable agreement with the prediction by the fluffy structure in Fig. 2. See Chapter 13 by Rickman in this volume for densities of comets more generally. Rickman et al. (1986) lean toward lower densities than Sagdeev et al. (1988) who claim that densities ~ 0.6 g cm^{-3} may be more reasonable. Meteor densities, if as low as claimed (Verniani 1969, 1973, Millman 1976, Olsson-Steel 1986) generally appear to be more consistent with cometary densities $\lesssim 0.3$ g cm^{-3}, i.e. packing factors no higher than 0.2, or porosities $\simeq 0.8$. The comet dust porosity of $P \lesssim 0.95$ suggests that the comet nucleus density is < 0.5 g cm^{-3} (see Table 3).

6 THE COSMIC DUST CONNECTION

Since all solar system bodies formed out of the original dense interstellar cloud, answering the question of whether and to what extent any memory of the interstellar dust remains may provide clues to the original aggregation processes. Of course, subsequent metamorphosis must also be considered. The inner planets represent the extreme case where the original dust signatures are totally lost both as a consequence of high initial condensation temperature and gravitational differentiation. How do asteroids, meteorites, meteors, and interplanetary dust fit into the hierarchy?

Interplanetary dust has classically been observed via its scattering of visible and near ultraviolet radiation—the zodiacal light. The addition of infrared observations has revealed some significant physical distinctions between particles as a function of distance from the Sun. Those which are within 1 AU scatter visible light much more effectively than those which are beyond 1 AU. At the same time, those which are farther out are more effective emitters of infrared radiation. This implies a difference in *kind* as well as number with increasing solar distance. This problem has been addressed by a number of authors (Hong 1988, Deul 1988, Cook 1978, Dumont & Levasseur-Regourd 1987). The most obvious explanation of this phenomenon is that the radial decrease of the albedo of the zodiacal light particles is produced by a decrease in material density, just as the albedo of cometary dust is decreased because of its fluffiness. The interplanetary particle probe results of Pioneer 10/11 were also interpreted in terms of a radial decrease of particle density (Fechtig 1984).

It has been suggested that the zodiacal light is predominantly produced by particles which started out as comet dust. The alternative point of view is that interplanetary particles result from asteroidal collisions. Probably something in between may be true, although, if some asteroids are just inert comets, the distinction may be academic. According to Dohnanyi (1976), asteroids play only a minor role as a dust source, and this appeared to be confirmed by the Pioneer 10/11 data which did not show any dust increase in the asteroidal belt (Humes *et al.* 1974, 1975). With the assumption that most interplanetary particles start out as comet dust whose morphological structure consists of bird's-nest-like ensembles of interstellar dust particles (like that in Fig. 3), Mukai & Fechtig (1983) proposed and calculated the effects of an ingenious mechanism by which solar heating would lead to a gradual compaction of the initially fluffy dust by evaporation of the volatiles in what they called 'Greenberg particles'. They suggested that one could account in this way for the appearance of the small interplanetary dust particles (IDPs)—commonly known as 'Brownlee particles'—as having evolved from Greenberg particles. Although the mean density of the relevant sample of IDPs is low, it is much higher than the initial cometary dust. But, as has been pointed out by Brownlee himself (1988), there is no evidence of a bird's-nest structure in the IDPs (Figs. 3*, 5). What we do see in Fig. 5 (which is a representative example of a low-density IDP, also called chondritic porous or CP) is an aggregate of more or less *spherical* particles of about 0.1 μm diameter whose infrared signature is that of silicates. In my initial bird's-nest model I had—but for illustrative purposes only—used simple elongated silicates as the interstellar dust cores. Because the interstellar silicates are amorphous and not crystalline, it is more likely that they form as spheres rather than as elongated structures, so that their effective non-sphericity should actually be pictured, on astrophysical grounds, rather as linear clumps of ~ 0.1 μm diameter silicates (see Fig. 1). Note that hydrated silicates are severely depleted in the CP IDPs as well as in Comet Halley dust, but are present in CI meteorites. But where are the organic refractory mantles in the IDPs? In the original (interstellar dust) comet nucleus material the ratio of O.R. mass to silicate mass is given as about 1:1. However, already in the comet dust, the loss of the more volatile O.R. molecules has led to a reduction of this ratio by about a factor of 2 to about 1:2. While the organic mantles are not 'seen' in the IDP electron micrographs, they become immediately apparent with Raman spectroscopy, and, in fact, the latter does not at all exhibit the silicate 10 μm signature (Wopenka 1988, Brownlee 1988). It appears that *every* silicate particle is covered by *some organic mantle*. If the O.R. constitutes about 5% of an IDP, a typical 0.05 μm radius silicate particle would have about a 0.001 μm (10 Å) thick layer of

* to be found in the colour illustration section in the centre of the book.

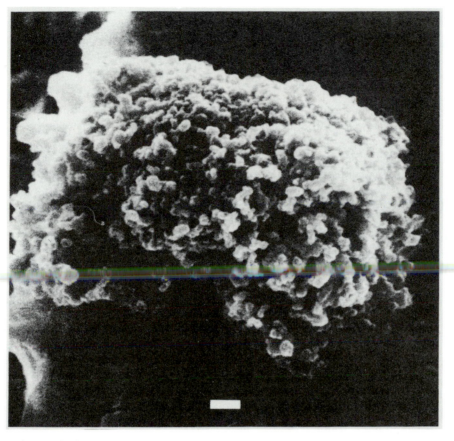

Fig. 5 — An electron microscopic photograph of a low-density interplanetary dust particle (IDP). The bar represents 1 micrometre. Such particles, called Brownlee particles, are characterized by loosely packed aggregates of small submicrometre-sized stony (silicate) particles giving a mean density of about 1 g cm⁻³. The silicate particles are thinly coated by organic refractory mantles. The packing factor is about twice as great as that used in the interstellar dust model of a comet nucleus, so that Brownlee particles morphologically resemble compressed comet dust from which a large fraction of organics has evaporated.

O.R. A uniform 50 Å organic coating on *each* particle would imply a 33% concentration in the IDPs. This seems perhaps high, but the fact that the mean silicate particle size is like that of the interstellar core pieces and *each* silicate or clump of silicates has an O.R coating is certainly suggestive of the interstellar origin, while the bird's-nest morphological structure is lost because of the removal by, for example, the Mukai–Fechtig mechanism and by other heating of a further factor of 5 of the original comet dust O.R. This begins to answer some of the objections in relating the interstellar dust model of comets and bird's-nest comet dust to the Brownlee particles. Nevertheless, it would certainly be useful to pursue further both the theory of the physical processing of interplanetary

particles and the microanalysis of the Brownlee particles to provide a more detailed comparison.

Additional evidence for the cometary–interplanetary dust evolution may be seen in the lower density of meteors whose aphelion distances are beyond 5.4 AU as compared with those which spend more time closer to the Sun (Verniani 1973). In fact, in deriving the comet density, I have used, for internal consistency, those meteors in the lower density range because of the assumption of a cometary origin for the meteors. The fragile nature of meteors is well documented, and there appears to be an increase in density for very small ones which fit into the evolutionary picture of Mukai & Fechtig (1983).

How do meteorites and their parent asteroidal

bodies fit into the cosmic dust connection? Since the formation region for the asteroids was certainly at a higher temperature than that for comets we do not expect the interstellar dust to be nearly as well preserved. Within the framework of the theory of Ruznaikina & Maeva (1988) the temperature of the pre solar nebula relevant to the asteroidal belt was 250–300 K, which was sufficient to evaporate all the dust volatiles while preserving a fraction of the organics. In fact, although in the most carbon rich meteorites the mass fraction is as much as 5%, a more characteristic value for carbonaceous chondrites is $\sim 1\%$. This would indicate that in asteroids and their meteoritic debris only a few percent of the original interstellar refractory carbon is preserved. Is it actually a remnant of the interstellar O.R. mantle material? Direct proof is difficult. However, one factor which appears to provide a basis for believing the connection lies in the preservation (Brownlee 1988, Huss & Alexander 1987) of the presolar isotopic abundances of the heavy noble gases Ar, Kr, and Xe in the carbonaceous component. These elements are presumed to have been trapped in the interstellar organic refractory mantles and retained during asteroid formation. If these mantles would have been totally evaporated before the asteroid formed, the escaping noble gases would have been mixed with all the presolar gases, distorting their isotopic ratios. Thus although meteorites may be identified with the same ancestors—interstellar dust—as comets, they are like cousins rather then siblings.

Based on the observations of the largely amorphous, carbonaceous coatings in the Allende (C3V) meteorite (Green et al. 1971, Bunch & Chang 1980, Bauman et al. 1973), Huss (1987) has suggested that the matrices in the parent bodies of the C3V, C30, and type 3 ordinary chondrites probably accreted from presolar dust that had lost the icy mantles. On the other hand he proposed that CI (C1) chondrites and the matrices of C2 chondrites probably accreted as bulk samples of presolar dust with some icy mantles intact—almost cometary. Parent body heating (not characteristic of comets) then caused the icy mantles to react with the fine-grained dust to produce the hydrothermal mineral assemblages now observed. The icy mantles in comet dust evaporate rather than melt, so that, although we should not be surprised by

seeing some resemblance between CP IDPs and CI chondrites, the differences should not be a surprise—there are no hydrated silicates in low density IDPs. If IDPs are remnants of comet dust they should more resemble the chemical and physical composition of the latter in which the H_2O evaporated rather than melted. There is a monotonic sequence of carbonaceous content from interstellar dust to comets to IDPs to meteorites.

In Fig. 6 we summarize the relationship between interstellar dust, interplanetary dust, and meteors and meteorites as conceived of here.

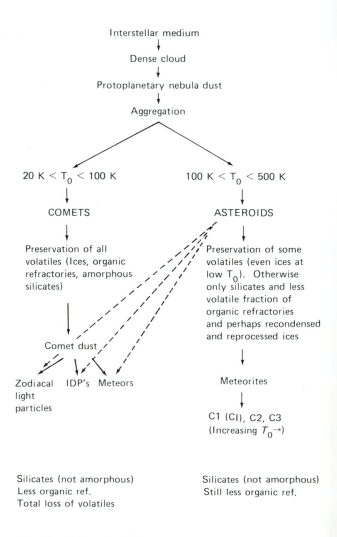

Fig. 6 — Schematic of the cosmic dust connection.

7 CONCLUSIONS

The key questions raised and discussed have been: are comets born as aggregated interstellar dust grains, and what evidence do we have that comet nuclei maintain or do not maintain their initial composition and structure? The cold formation of comets out of interstellar dust has been reinforced by the demonstration that not only is pre-aggregation ultraviolet radiation of sulphur-containing interstellar grain mantles adequate to produce the S_2 as a parent molecule, but the calculated values of cosmic ray ionizing particle penetration of the nucleus in the Oort cloud phase are not sufficient to do so except in the outer metre or so. The presence of abundant submicrometre sized particles and fluffy aggregates thereof in the dust size spectrum of Comet Halley is a confirmation of their interstellar origins.

The low density and morphological structure, on a microscale, predicted by the aggregated dust model are consistent with the low albedo and low thermal conductivity of the Comet Halley surface. The high surface temperature and the preponderance of dust/gas jets on the Sunward side provide evidence for a high surface temperature gradient and a low heat penetration. This indicates that the morphological structure of aggregated dust may be self-preserving even in the apparition phase.

The mass spectra and infrared emission spectra of the dust show that the major component of the solids is an organic refractory material similar to that observed in interstellar dust. Since the penetration of cosmic ray protons during the Oort cloud phase was only to the depth of layers which have long ago been shed, this cannot have been the cause of the present organic refractories. The evidence provided by Comet Halley results is strongly in favour of the view that comets are very much like original aggregates of interstellar dust except possibly the outermost surface layers where evaporation of the volatile components, which has occurred without dust ejection, may leave a low-density dust consisting of the *refractory components alone*. There could also be a layer of previously emitted dust which has fallen back on the surface.

There remain a number of unsolved problems in relating the aggregated interstellar dust comet model with the interplanetary particles studied in the laboratory (Bradley & Brownlee 1986), and which are presumed to have originated as comet dust. Although there are similarities, are the differences reconcilable with the ageing and chemical and morphological changes of cometary dust which have occurred during the 100 000 years spent orbiting the Sun and also those which result from impacting the Earth's atmosphere? An encouraging line of investigation shows how a combination of solar radiation and solar wind processing of fluffy particles (Strazzula 1986) consisting of small core–mantle units leads to slowly increasing density as well as chemical and physical changes of the post-cometary dust in the direction of the interplanetary dust particles collected in the atmosphere. Since the interstellar silicate cores are more likely to be beaded structures rather than needles, even the morphological as well as chemical structure of the IDPs may be derived.

We have to look to future space missions to recover comet material much more pristine than we can infer from flyby or even rendezvous missions. Ultimately, if the comet nucleus material can be retrieved from its depths and maintained intact cryogenically for laboratory studies, we may hope to study not only its atomic and molecular compositions but also its morphology. The physical and chemical nature of the presolar nebula will be seen in the most pristine form available in our solar system. Microprobes are being developed (Bradley & Brownlee 1986) which will make investigations possible of submicrometre structures. If it should turn out that the interstellar dust model is correct, individual grains whose mean lifetime before becoming part of a comet is about 5×10^9 yr will reveal cosmochemical evolution not only of the solar system but dating back a further 5 thousand million years before the Earth's beginning—back to the earliest stages of the chemical evolution of the Milky Way. Dramatic differences in isotopic abundances could be expected on scales of micrometres. The next twenty to thirty years should be exciting ones indeed for studies of our origins.

REFERENCES

Agarwal, V.K., Schutte, W., Greenberg, J.M., Ferris, J.P., Briggs, R., Connor, S., Van de Bult, C.P.E.M., & Baas, F. (1985) Photochemical reactions in interstellar grains: photolysis of CO, NH$_3$ and H$_2$O, *Origins of Life* **16** 21–40

A'Hearn, M.F., Hoban, S., Brich, P.V., Bowers, C., Martin, R., & Klinglesmith III, D.A. (1986) Cyanogen jets in Comet Halley, *Nature* **324** 649–651

Allen D.A., & Wickramasinghe, D.T. (1981) Diffuse interstellar absorption band between 2.9 and 4.0 μm, *Nature* **294** 239–240

Baas, F., Geballe, T.R., & Walther, D.M. (1986) Spectroscopy of the 3.4 micron emission feature in Comet Halley, *Astrophys. J. Lett.* **311** L97–L101

Balsiger, H., Altwegg, K., Bühler, F., Geiss, J., Ghielmetti, A.G., Goldstein, B.E., Goldstein, R., Huntress, W.T., Ip, W.-H., Lazarus, A.J. Meier, A., Neugebauer, M., Rettenmund, U., Rosenbauer, H., Schwenn, R., Sharp, R.D., Shelley, E.G., Ungstrup, E., & Young, D.T. (1986) Ion composition and dynamics at Comet Halley, *Nature* **321** 330–334

Bauman, A.J., Devany, J.R., & Bollin, E.M. (1973) Allende meteorite carbonaceous phase: intractable nature and scanning electron morphology, *Nature* **241** 264–267

Bradley, J.R., & Brownlee, D.E. (1986) Cometary particles: thin sectioning and electron beam analysis, *Science* **231** 1542–1544

Bregman, J.D. *et al.* (1987) Airborne and groundbased spectrophotometry of Comet P/Halley from 5–13 micrometers, *Astr. Astrophys.* **187** 616–620

Brownlee, D.E. (1988) Composition of comets and of interplanetary dust In: Joint Discussion IV: *The Cosmic Dust Connection*, IAU, to be published

Bunch, T.E., & Chang, S. (1980) Carbonaceous chondrite phylosilicates and light element geochemistry as indicators of parent body processes and surface conditions, *Geochim. Cosmochim. Acta* **44** 1543–1577

Butchart, I., McFadzean, A.D., Whittet, D.C.B., Geballe, T.R., & Greenberg, J.M. (1986) The 3.4 μm absorption towards the galactic center, *Astron. Astrophys.* **154** L5–L7

Chlewicki, G.C. (1985) Observational constraints on multimodal interstellar grain populations. PhD thesis, University of Leiden

Chlewicki, G.C., & Greenberg, J.M. (1990) Interstellar circular polarization and the dielectric nature of dust grains, *Astrophys. J.* in the press.

Combes, M. *et al.* (1986) Infrared sounding of Comet Halley from VeGa 1, *Nature* **321** 266–268

Cook, A.F. (1978) Albedos and size distribution of meteoroids from 0.3 to 4.8 AU, *Icarus* **33** 349–360

Czyzak, S.J., & Santiago, J.J. (1973) On the presence of graphite in the interstellar medium, *Astrophys. Sp. Sci.* **23** 443–458

Danks, A.C., Encrenaz, T., Bouchet, P., Le Bertre, T., Chalabaev, A., & Epchtein, N. (1986) Observation of an emission feature at 3.4 μm in the spectrum of Comet Halley, *20th ESLAB Symposium on the Exploration of Halley's Comet* eds Battrick, B., Rolfe, E.J., & Reinhard, R. ESA SP-250, vol. III, ESA Publications, ESTEC, Noordwijk, The Netherlands, 103–106

Delsemme, A.H. (1982) Chemical composition of cometary nuclei, In: *Comets* Wilkening, L.L., ed. University of Arizona Press, 85–130

Deul, E. (1988) Interstellar dust and gas in the Milky Way and W33, PhD thesis, University of Leiden

d'Hendecourt, L.B., Allamandola, L.J., & Greenberg, J.M. (1985) Time dependent chemistry in dense molecular clouds I. Grain surface reactions, gas/grain interactions and infrared spectroscopy, *Astron. Astrophys.* **152** 130–150

d'Hendecourt, L.B., Allamandola, L.J., Grim, R.J.A., & Greenberg, J.M. (1986) Time-dependent chemistry in dense molecular clouds II. Ultraviolet processing and infrared spectroscopy of grain mantles, *Astron. Astrophys.* **158** 119–134

Dohnanyi, J.S. (1976) Sources of interplanetary dust: asteroids In: *Lecture notes in physics*, Elsässer, H., & Fechtig, H. eds Berlin: Springer-Verlag **48** 29

Donn, B., & Hughes, D. (1986) A fractal model of a cometary nucleus formed by random accretion, *20th ESLAB Symposium on the Exploration of Halley's Comet*, Battrick, B., Rolfe, E.J., & Reinhard, R. eds ESA SP-250, Vol. III, ESA Publications, ESTEC, Noordwijk, The Netherlands, 523–524

Draine, B.T., & Salpeter, E.E. (1979) Destruction mechanisms for interstellar dust, *Astrophys. J.* **231** 438–455

Dumont, R., & Levasseur-Regourd, A.-C. (1987) Properties of interplanetary medium from infrared and optical observations, *Astron. Astrophys.* in press

Eberhardt, P. *et al.* (1987) The CO and N$_2$ abundance in Comet P/Halley *Astron. Astrophys.* **187** 481–484

Fechtig, H. (1984) The interplanetary dust environment beyond IAU and in the vicinity of the ringed planets, *Adv. Sp. Res.* **4** no. 9, 5–11

Festou, M.C., Feldman, P.D., A'Hearn, M.F., Arpigny, C., Cosmovici, C.B., Danks, A.C., McFadden, L.A., Gilmozzi, R., Patriarchi, P., Tozzi, G.P., Wallis, M.K., & Weaver, H.A. (1986) IUE observations of Comet Halley during the VeGa and Giotto encounters, *Nature* **321** 351–363

Gehrz, R.D., & Ney, E.P. (1986) Infrared temporal development of P/Halley, *20th ESLAB Symposium on the Exploration of Halley's Comets*, eds Battrick, B., Rolfe, E.J., & Reinhard, R. ESA SP-250, Vol. II, ESA Publications, ESTEC, Noordwijk, The Netherlands, 101–105

Green, H.W. III, Radcliffe, S.V., & Hever, A.H. (1971) Allende meteorite: a high voltage electron petrographic study, *Science* **172** 936–939

Greenberg, J.M. (1968) Interstellar grains, nebulae and interstellar matter, Middlehurst, B.M., & Aller, L.H. eds *Stars and stellar systems* **VII** University of Chicago Press, 221–364

Greenberg, J.M. (1971) Interstellar grain temperatures, effects of grain materials and radiation fields, *Astron. Astrophys.* **12** 240–249

Greenberg, J.M. (1978) Interstellar dust In: *Cosmic dust* eds McDonnell, J.A.M., Wiley, N.Y. 187–294

Greenberg, J.M. (1979) Grain mantle photolysis: a connection between the grain size distribution function and the abundance of complex interstellar molecules, In: *Stars and star systems*, ed. Westerlund, B.E., D. Reidel Pub. Co. 173–193

Greenberg, J.M. (1980) From interstellar dust to comets to the zodiacal light, In: *Solid particles in the solar system*, ed. Halida, I., Reidel, Dordrecht, 343–350

Greenberg, J.M. (1982a) Dust in dense clouds. One stage in a cycle, *Submillimetre wave astronomy* eds Phillips, D., & Beckman, J.E. Cambridge University Press, 261–306

Greenberg, J.M. (1982b) What are comets made of? A model based on interstellar dust, In: *Comets* ed. Wilkening, L.L., University of Arizona Press, 131–163

Greenberg, J.M. (1983a) Laboratory dust experiments—tracing the composition of cometary dust, In: *Cometary exploration*, ed. Gombosi, T.I., (Hungarian Academy of Sciences) 23–54

Greenberg, J.M. (1983b) Interstellar dust, comets, comet dust and carbonaceous meteorites, In: *Asteroids, comets and meteors*, eds Lagerkvist, C.I., & Rickman, H. Uppsala University Press, 259–268

Greenberg, J.M. (1984a) Experiments on chemical and physical evolution of interstellar grain mantles, In: *Laboratory and observational infrared spectra of interstellar dust*, eds Wolstencroft R.D., & Greenberg, J.M. Occasional Reports of the Royal Obs. Edinburgh ISSN 0309–049X, 82–92

Greenberg, J.M. (1984b) A fine mist of very small comet dust particles, *Adv. Space Res.* **4** no. 9, 211–212

Greenberg, J.M. (1985) Evolution of interstellar grains: observation, theory, laboratory experiments, *Birth and infancy of stars*, eds Lucas, R., Omont, A., & Stora, Elsevier press, 139–203

Greenberg, J.M. (1986a) Dust in diffuse clouds: one stage in a cycle, In: *Light on dark matter* ed. Israel, F.P., (Reidel), 177–188

Greenberg, J.M. (1986b) Fluffy comets, In: *Asteroids, comets and meteors* II eds Lagerkvist, C.-I., Lindblad, B.A., Lundstedt H., & Rickman, H. Uppsala University Press 221–223

Greenberg, J.M., & Chlewicki, G.C. (1983) A far-ultraviolet extinction law: what does it mean? *Astrophys. J.* **272** 563–578

Greenberg, J.M., & Chlewicki, G. (1987) Variations in ultraviolet extinction: effect of polarization revisited, *Q.J.R. Astr. Soc.* **28** 312–322

Greenberg, J.M., & Gustafson, B. (1981) A comet fragment model for zodiacal light particles, *Astron. Astrophys.* **93** 35–42

Greenberg J.M., & Hong, S.S. (1974a) Evolutionary characteristics of a bimodal grain model In: *HII regions and the galactic center*, ed. Moorwood, A.F.M., ESRO SP-105, 153–161

Greenberg J.M., & Hong, S.S. (1974b) The chemical composition and distribution of interstellar grains In: *Galactic radio astronomy*, eds Kerr, F.J., & Simonson, S.C. III Reidel, 153–177

Greenberg, J.M., & Zhao, Nansheng (1987) The 3.4 and 10 μm emissions from comet dust—Indications of many small fluffy particles, *Leiden Workshop on Fluffy Structures*.

Greenberg, J.M., Yencha, A.J., Corbett, J.W., & Frisch, H.L. (1971) Growth, distribution and chemical composition of interstellar dust, *BAAS* **3** No. 2, 250

Greenberg, J.M., Grim, R.J.A., & Van Ijzendoorn, L. (1985) Interstellar S_2 in comets, In: *Comets, Asteroids, Meteorites* II eds Lagerkvist, C.I., Lindblad, B., Lundstedt, H, & Rickman, H. Uppsala press, 218–220

Greenberg, J.M., de Groot, M.S., & Van der Zwet, G.P. (1987) Carbon components of interstellar dust, In: *Polycyclic Aromatic hydrocarbons and astrophysics*, eds Leger, A., d'Hendecourt, L.B., & Boccara, N. (Reidel) 177–181.

Greenberg, J.M., & Hage, J. (1989) From interstellar dust to comets: A unification of observational constraints, submitted to *Astrophys. J.*

Grim, R.J.A., & Greenberg, J.M. (1987a) Photoprocessing of H_2S in interstellar grain mantles as an explanation for S_2 in comets, *Astr. Astrophys.* **181** 155–168

Grim, R.J.A., & Greenberg, J.M. (1987b) Ions in grain mantles: the 4.62 micron absorption by OCN^- in W33A, *Astrophys. J.* **321** L91–L96

Hong, S.S. (1988) Distribution of interplanetary particles from infrared and visual studies?, In: Joint Discussion IV: *The cosmic dust connection*, IAU, to be published

Hong, S.S., & Greenberg, J.M. (1980) A unified model of interstellar grains: a connection between alignment efficiency, grain model size, and cosmic abundance, *Astron. Astrophys.* **88** 194–202

Hong, S.S., & Um, I.K. (1987) *Astrophys. J.* **320** 928

Humes D.H., Alvarez, J.M., O'Neal, R.L., & Kinard, W.H. (1974) The interplanetary and near Jupiter meteoroid environments, *J. Geophys. Res.* **79** 3677

Humes, D.H., Alvarez, J.M., Kinard, W.H., & O'Neal, R.L. (1975) Pioneer II meteoroid detection experiment: preliminary results, *Science* **188** 473

Huss, G.R. (1987) The role of presolar dust in the formation of the solar system, *Icarus*.

Huss, G.R., & Alexander, C. Jr. (1987) On the presolar origin of the 'normal planetary' noble gas component in meteorites, *J. Geophys. Res.* **92** no. 134, E710–E716

Jessberger, E.K., 1989 Chemical properties of cometary dust and a note on carbon isotopes, in *Comets in the Post-Halley Era*, eds. R. Newburn, M. Neugebauer, J. Rahe June 1989

Jessberger, E.K., & Kissel, J. (1987) Bits and pieces from Halley's Comet, *Lunar and Planetary Science* XVIII

Johnson, R.E., Lanzerotti, L.J., Brown, W.L., Augustiniak, W.M., & Mussil, C. (1983) Charged particle erosion of frozen volatiles in ice grains and comets, *Astron. Astrophys.* **123** 343–346

Johnson, R.E., Cooper, J.F., & Lanzerotti, L.J. (1987a) Radiation formation of a non-volatile crust, *20th ESLAB Symposium on the Exploration of Halley's Comets*, eds Battrick, B., Rolfe, E.J., & Reinhard, R. ESA SP-250, ESA Publications, ESTEC, Noordwijk, The Netherlands, 269–271

Johnson, R.E., Cooper, J.F., Lanzerotti, L.J., & Strazzulla, G. (1987b) Radiation formation of a non-volatile comet crust, *Astron., & Astrophys.* **187** 889–892

Kaneda, E. *et al.* (1986), Strong breathing of the hydrogen coma of Comet Halley, *Nature* **320** 140–141

Kissel, J., & Krueger, F.R. (1987) The organic components in dust from Halley as measured by the PUMA mass spectrometer on board VeGa-1, *Nature* **326** 755–760

Kissel, J., Sagdeev, R.Z., Bertaux, J.L., Angarov, V.N., Audouze, J., Blamont, J.E., Buchler, K., Evlanov, E.N., Fechtig, H., Fomenkova, M.N., von Hoerner, H., Inogamov, N.A., Khromov, V.N., Knabe, W., Krueger, F.R., Langevin, Y., Leonas, V.B., Levasseur-Regourd, A.C., Managadze, G.G., Podkolzin, S.N., Shapiro, V.D., Tabladyev, S.R., & Zubkov, B.V. (1986a) Composition of Comet Halley dust particles from VeGa observations, *Nature* **321** 280–282

Kissel, J., Brownlee, D.E., Büchler, K., Clark, B.C., Fechtig, H., Grün, E., Hornung, K., Igenbergs, E.B., Jessberger, E.K., Krueger, F.R., Kuczera, H., McDonnell, J.A.M., Morfill, G.M., Rahe, J., Schwehm, G.H., Sekanina, Z., Utterback, N.G., Völk, H.J. and Zook, H.A. (1986b), Composition of Comet Halley dust particles from Giotto observations (1986) *Nature* **321** 336–337

Klinger, J. (1980) Influence of a phase transition of ice on the heat and mass balance of comets, *Science* **209** 271–272

Knacke, R., Brooks, T.I., & Joyce, R.R. (1986) Observations of 3.2–3.6 micron emission features in Comet Halley, *Astrophys. J. Lett.* **310** L49–L53

Krasnopolsky, V.A., Gogoshev, M., Moreels, G., Moroz, V.I., Krysko, A.A., Gogosheva, Ts., Palazov, K., Sargoichev, S., Clairemidi, J., Vincent, M., Bertaux, J.L., Blamont, J., Troshin, V.S., & Valnicek, B. (1986) Spectroscopic study of Comet Halley by the VeGa-2 three-channel spectrometer, *Nature* **321** 269–271

Krankowsky, D., Lämmerzahl, P., Herrwerth, I., Woweries, J., Eberhardt, P., Dolder, U., Herrmann, U., Schulte, W., Berthelier, J.J., Illiano, J.M., Hodges, R.R., & Hoffman, J.H. (1986) *In situ* gas and ion measurements at Comet Halley, *Nature* **321** 326–329

Krueger, F. (1987) Organic cometary grains, *Leiden Workshop on Fluffy Structures. Polycyclic Aromatic Hydrocarbons and Astrophysics* eds Leger, A., d'Hendecourt, L.B., & Boccara, N. (1987) D. Reidel Pub.

Larson, H.P., Weaver, H.A., Mumma, M.J. & Drapatz, S. (1989) Airborne infrared spectroscopy of comet Wilson (1986) and comparisons with comet Halley, *Astrophys. J.,* **378**, 1106–1114

Mathis, J.S., Rumpl, W., & Nordsieck, K.H. (1977) The size distribution of interstellar grains, *Astrophys. J.* **217** 425–433

Mazets, E.P., Sagdeev, R.Z., Aptekar, R.L., Golenetskii, S.V., Guryan, Yu.A., Dyachkov, A.V., Ilynskii, V.N., Panov, V.N., Petrov, G.G., Savvin, A.V., Sokolov, I.A., Khavenson, N.G., Shapiro, V.D., & Shevchenko, V.I. (1987) Dust in comet Halley from Vega observations, *Astron Astrophys.* **187**, 699–706

McDonnell, J.A.M., Alexander, W.M., Burton, W.M., Bussoletti, E., Clark, D.H., Grard, R.J.L., Grün, E., Hanner, M.S., Hughes, D.W., Igenbergs, E., Huczera, H., Lindblad, B.A., Mandeville, J.-C., Minafra, A., Schwehm, G.H., Sekanina, Z., Wallis, M.K., Zarnecki, J.C., Chakaveh, S.C., Evans, G.C., Evans, S.T., Firth, J.G., Littler, A.N., Massonne, L., Olearczyk, R.E., Pankiewicz, G.S., Stevenson, T.J., & Turner, R.F. (1986) Dust density and mass distribution near Comet Halley from Giotto observations, 1986, *Nature* **321** 338–341

McDonnell, J.A.M., Pankiewicz, G.S., Birchley, P.N.W., Green, S.F., & Perry, C.H. (1989) in *Workshop on Analysis of Returned Comet Nucleus Samples,* in the press.

Millman, P.M. (1976) Meteors and interplanetary dust, In: *Interplanetary dust and zodiacal light,* eds Elsässer, H., & Fechtig, H. Springer-Verlag, 359–372

Moore, M.H., & Donn, B. (1982) The infrared spectrum of a laboratory-synthesized residue—implications for the 3.4 micron interstellar absorption feature, *Astrophys. J. Lett.* **257** L47–L50

Mukai, T., & Fechtig, H. (1983) Packing effect of fluffy particles. *Planet Space Sci.* **31** 655

Nuth, J.A. (1985) *Nature* **318** 116

Observatory XX. The ultraviolet extinction hump, *Astrophys. J.* **199** 92–109

Olsson-Steel, D. (1987) The flux, structure and origin of radar meteors, *Leiden Workshop on Fluffy Structures*

Olsson-Steel, D. (1989) The flux, structure and origin of radar meteors, In: IAU Joint Discussion *The cosmic dust connection*

Rickman, H., Kamel, L., Festou, M., & Froeschlé, C. (1988) In: *Symposium on the Diversity and Similarity of Comets,* ESA SP-278, ESA Publications, ESTEC, Noordwijk, The Netherlands

Ruznaikina, T.V., & Maeva, S.V. (1988) Process of formation of the protoplanetary disk, In *COSPAR XXVII Helsinki proceedings.*

Ryan, M.P. Jr., & Draganic, I.G. (1986) An estimate of the contribution of high energy cosmic-ray protons to the absorbed dose inventory of a cometary nucleus, *Astrophys. Space Sci.* **125** 49–67

Sagdeev, R.Z., Blamont, J., Galeev, A.A., Moroz, V.I., Shapiro, V.D., Shevchenko, V.I., & Szegó, K. (1986) VeGa spacecraft encounters with Comet Halley, *Nature* **321** 259–262

Sagdeev, R.Z., Elyasberg, P.E., & Moroz, V.L. (1988) Is the nucleus of Comet Halley a low density body?, *Nature* **331** 240–242

Savage, B.D. (1975) Ultraviolet photometry from the Orbiting Astronomical Observatory XX. The ultraviolet extinction hump, *Astrophys. J.* **199** 92–109

Savage, R.D., & Mathis, J.S. (1979) Observed properties of interstellar dust, *Ann. Rev. Astron. Astrophys.* **17** 73–111

Schutte, W. (1988) The evolution of interstellar organic grain mantles, PhD thesis, University of Leiden

Schutte, W., & Greenberg, J.M. (1986) Formation of organic molecules on interstellar dust particles, In: *Light on dark matter,* ed. Israel, F.F., (Reidel), 229–232

Sekanina, Z. (1986) Nucleus studies of Comet Halley, *Adv. Space Res.* **5** 307–316

Sekanina, Z., & Larson, S.M. (1986a) Coma morphology and dust-emission pattern of periodic Comet Halley. IV. Spin vector refinement and map of discrete dust sources for 1910, *Astron. J.* **92** 462–482

Sekanina, Z., & Larson, S.M. (1986b) Dust jets in Comet Halley observed by Giotto and from the ground, *Nature* **321** 357–361

Sekanina, Z., & Yeomans, D.K. (1985) Orbital motion, nucleus precession, and splitting of periodic Comet Brooks 2, *Astron. J.* **90** 2335–2352

Simpson, J.E., Sagdeev, R.Z., Tuzzolino, A.J., Perkins, M.A., Ksanfomality, L.V., Rabinowitz, D., Lentz, G.A., Afonin, V.V., Ero, J., Keppler, E., Kosorokov, J., Petrova, E., Szabó, L., & Umlauft, G. (1986) Dust counters and mass analyser (DUCMA) measurements of Comet Halley's coma from VeGa spacecraft, *Nature* **321** 278–280

Snow, T.P., & Seab, C.G. (1980) An anomalous ultraviolet extinction curve in the Taurus dark cloud, *Astrophys. J.* **242** L83–L86

Stecher, T.P. (1965) Interstellar extinction in the ultraviolet, *Astrophys J.* **142** 1683–1684

Stecher, T.P., & Donn, B. (1965) On graphite and interstellar extinction, *Astrophys. J.* **142** 1681–1683

Stern, S.A., & Shull, J.M. (1988) The influence of supernovae and passing stars on comets in the Oort cloud, *Nature* **332** 407

Strazzulla, G. (1986) Primitive galactic dust in the solar system?, *Icarus* **67** 63–70

Strazzulla (1987) *Leiden Workshop on Fluffy Structures* van de Bult, C.E.P.M., Greenberg, J.M., & Whittet, D.C.B. (1984) Ice in the Taurus molecular cloud: modelling of the 3-μm profile, *Mon. Not. R. Astron. Soc.* **214** 289–305

Van de Hulst, H.C. (1957) *Scattering of light by small particles* (Wiley, N.Y.)

Van der Zwet, G.P., de Groot, M.S., Baas, F., & Greenberg, J.M. (1986) Molecular origin of the 216 nm interstellar hump, In: *Polycyclic aromatic hydrocarbons and astrophysics,* eds Leger, A., d'Hendecourt, L.B., & Boccara, N. (Reidel) 183–195

Verniani, F. (1969) Structure and fragmentation of meteorites, *Space Science Reviews* **10** 230–261

Verniani, F. (1973) Physical parameters of faint meteors *J. Geophys. Res.* **78** 8429–8462

Whipple, F. (1977) The constitution of cometary nuclei, In: *Comets, asteroids, meteorites: interrelations, evolution and origins,* ed. Delsemme, A., IAU Coll. No. 39, University of Toledo, Toledo, Ohio, 25–35

Whittet, D.C.B., Longmore, A.J., & McFadzean, A.D. (1985) Solid CO in the Taurus dark clouds *Mon. Not. R. Astron. Soc.* **216** 45P–50P

Wickramasinghe, D.T., & Allen, D.A. (1986) Discovery of organic grains in Comet Halley, *Nature* **323** 44–46

Willner, S.P., Russell, R.W., Puetter, R.C., Soifer, B.T., & Harvey, P.M. (1979) The 4 to 8 micron spectrum of the galactic center, *Astrophys. J.* **229** L65–L68

Willner, S.P., & Pipher, J.L. (1982) *Proc. Cal. Tech. Workshop on Galactic Center*

Wopenka, B. (1988) *Earth and Planet Sci. Lett.* **88** 221

Yamamoto, T., & Kozasa, T. (1987) Cometary nucleus as aggregate of planetesimals, *Inst. of Sp. and Astr. Science Research Note* 364

10

The structure and evolution of the Comet Halley meteor stream

Bruce A. McIntosh

1 INTRODUCTION

People who missed the once-in-a-lifetime opportunity to see Comet Halley can take consolation in being able to observe fragments of the comet twice a year, every year. In early May, and again in late October, the Earth's passage through a stream of particles from Comet Halley results in two meteor showers called the η-Aquarids and the Orionids respectively. The names result from the custom of naming meteor showers after the constellation from which the meteors appear to radiate.

Although there are two meteor showers, they result from a single stream of particles which has evolved from the debris of Halley's Comet. This chapter describes a model for the dynamical evolution of the particle stream. The model is based on the simple concept that particles released from the comet are to be found in orbits where the comet *was* early in its history and where it *will be* in the future. It can be shown, both intuitively and mathematically, that the stream evolves as a flat ribbon-like structure, which explains most of the observed, but heretofore puzzling, features of the meteor showers.

1.1 Observed properties of the meteor showers

Technically, a 'meteor' is the streak of light—a shooting star—seen on a clear night when a small solid particle burns up as it strikes the Earth's atmosphere at speeds ranging from 12 to 72 kilometres per second. As well as recording the light on photographic film or video tape, one can also detect the residual ionized trail by radar. The advantage of being able to record meteors by radar during daylight as well as at night is somewhat offset by the difficulties in interpreting the observations.

Orionid meteors are easily seen on nights between 16 October and 25 October. The strongest activity occurs on some five nights centred on October 21/22. (Because of the leap-year shift, times of occurrence are more easily specified in the measure 'solar longitude'. The centre is at 208° solar longitude.) The shower is not a strong one—about ten to twenty meteors per hour will be seen near the central peak—but the high velocity of the meteors, near 66 kilometres per second, does produce some spectacular individual events. The activity of the shower is quite variable from year to year—the peak rates may not occur at the usual 208°, or there may be more than one distinct maximum.

Observation of η-Aquarid meteors, in early May, is not so easy. The meteors appear to radiate from a point in the constellation Aquarius which rises above the horizon only a few hours ahead of the Sun. As a result, most of its activity occurs during the daytime. Radar measurements have contributed a good deal to our knowledge of the shower. Its activity pattern is quite similar to that of the Orionids: some activity over about fourteen days; a central core lasting about five days centred at solar longitude 45° (about 5/6 May), during which the rates are about 10 to 20 meteors per hour.

To a limited extent, the composition of a meteor particle may be deduced from the spectrum of the meteor. Halliday (1987) has recently examined a number of spectra of Orionid meteors. The observed meteors were quite bright, resulting from meteoroids with typical masses of several grams. Since the spectra were very similar to those of other fast shower meteors, Halliday concluded that the composition of Comet Halley particles was essentially similar to those from other comets having similar periods of revolution.

1.2 Historical

Just as returns of Comet Halley have been traced back in time in ancient records (Ho Peng Yoke 1964), so too have occurrences of these meteor showers (Imoto & Hasegawa 1958, Zhuang 1977). It is noteworthy that there is a higher incidence of strong η-Aquarid showers in the ancient records, in spite of the fact (noted above) that it is Orionid meteors which are more readily observable in the Northern Hemisphere. Among the strongest historical η-Aquarid showers were those of AD 443, 466 and 530. Zhuang (1977) translates the comments of ancient Chinese observers of these showers as: 'their number past counting', 'numbered in thousands'. A strong Orionid shower was recorded in AD 585. This era of unusually high activity in both meteor showers is consistent with the evolutionary model described below.

The association of meteor showers with comets began to be looked on with favour in the nineteenth century. A specific connection between the η-Aquarid shower and Halley's Comet was postulated about 1868; and the similarity of the orbits of the Orionids and η-Aquarids was recognized in 1911 (Olivier). But even as late as 1961, D.W.R. McKinley, one of the well-known authorities in meteor science, wrote (McKinley 1961), 'The relation between either the η-Aquarids or the Orionids and Halley's striking comet is tenuous and uncertain, though still an attractive

possibility.' The reasons for the uncertainty were twofold: (1) consistently accurate measurements of the meteor orbits were lacking and, (2) the exact nature of the differences between the dynamic evolution of the comet and of the meteor particles was not known.

2 THE GENERAL DEVELOPMENT OF A METEOR STREAM

2.1 Release of particles and dispersion around the orbit

Whipple's (1951) model of the structure and composition of comets is still a satisfactory starting point from which to develop the relationship between comets and meteor streams. The fragments of 'dirt' in his 'dirty-snowball' model are the particles which eventually become meteors. As the ices sublime, the gases carry off the solid particles. Whipple's formulation predicts ejection velocities of a few tens of metres per second, depending on particle size. Since the escape velocity from a comet-sized body is typically 1 m s^{-1}, the particles are thereafter free on their own orbits. Addition of the ejection velocity to the comet's orbital velocity changes the motion in a complex manner (see, for example, McIntosh 1973). Some particles will lag behind the comet, some will gradually get ahead of it. Eventually, leading particles meet lagging particles, closing the loop, and forming a continuous ring. The time required to form a continuous stream is about 15 to 20 revolutions for the Halley orbit (see Fig. 1).

2.2 Other forces

A small body in orbit around the Sun is subject to a range of perturbing forces. The gravitational attractions of the planets are common to both comet and particles. Forces such as light pressure and effects due to differential heating and energy loss are insignificant for the comet but important to the evolution of the particle stream. In contrast, the perturbations due to gas jet forces, so important to accurate prediction of comet orbits, are of no consequence to the gross statistical behaviour of particle streams.

The dust that is a component of comet tails is blown away by light pressure and does not remain in

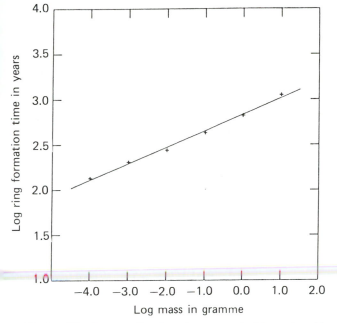

Fig. 1 — The time, as a function of mass, required for meteoroids ejected from Comet Halley to disperse into a ring. (From Jones & McIntosh 1986)

the vicinity of the compact meteoroid stream. Thus most meteor showers consist of larger particles, typically in the mass range: milligrammes to grammes. course there is a continuous gradation in this effect, and as a stream ages, more and more small particles are lost.

3 COMET HALLEY AND ITS METEOR SHOWERS

3.1 Description of present orbit

The major variables necessary to describe orbits are shown in Fig. 2. The comet travels up through the ecliptic plane at the ascending node well outside the Earth's orbit (1.8 AU) and down again at the descending node inside the Earth's orbit at 0.85 AU. The position of the line of nodes in the ecliptic plane is given by the angle Ω, the longitude measured from the vernal equinox. The position of perihelion is the angle ω measured from the ascending node.

For lack of a better model, a meteoroid stream is usually assumed to be a cylinder of particles with the

comet orbit at its core. For a meteor shower to occur, the position of the cylinder must be such that the Earth crosses through it. But clearly there is little guarantee that the Earth will go through the central core of the stream. Most meteor showers occur near one of the nodes of the parent comet orbit, the offset being usually less than 0.1 AU. This is the reason why many questioned the association between Comet Halley and the meteor streams since the nodes are 0.15 AU and 0.8 AU from the Earth's orbit.

To be more specific, Fig. 3 shows what one would see if one were situated on the comet orbit watching the Earth go by. At the time of the η-Aquarid shower the Earth is well away from the descending node but does pass within about 0.07 AU of the comet orbit. In October, the Earth passes Halley's orbit at a distance about 2.5 times greater than in May. For any reasonable cylindrical distribution of particles, the concentration encountered by the Earth should be significantly wider and stronger in the May shower than in the October shower. But the observational evidence is clearly that the showers are of approximately equal duration and equal strength.

A comparison of the orbital parameters of the comet and the showers is made in Table 1. The values for the showers are taken from Cook (1973) who based his selection of orbital elements mostly on the high-precision orbits calculated by Jacchia & Whipple (1961). These data comprise only a few orbits, so that some of the parameters are averages of limited accuracy. For example, the values of semimajor axis, a, for the Orionid shower range from 12 to 30. It will

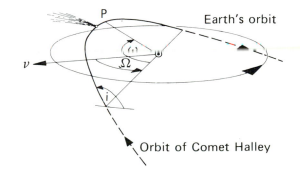

Fig. 2 — The motion of Comet Halley in the vicinity of the Earth's orbit. (From McIntosh & Hajduk 1983)

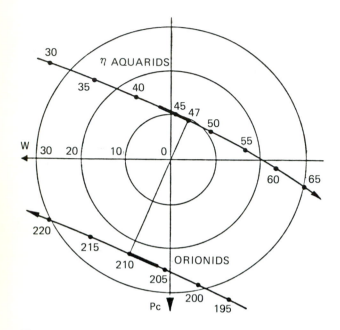

Fig. 3 — Motion of the Earth past the comet orbit. Direction of the comet is perpendicular to the paper at the origin. P_c points to the pole of the comet orbit. The thickened portions of the Earth-passage lines mark the positions of the meteor showers, and the numbers indicate the value of solar longitude of the Earth at positions along its motion. The circles are spaced at about 0.07 AU for the current orbit. (From McIntosh & Hajduk 1983)

be seen later that values of ω and Ω undergo rapid changes, and hence large differences in these parameters are to be expected. The important parameter is the difference between them, $\omega - \Omega$, which is indicative of the position of perihelion. This is quite similar among the three.

Table 1. Orbital parameters of Comet Halley and its meteor showers

	a	e	i	ω	Ω	$\omega - \Omega$
Halley's Comet	17.9	0.967	162.2	111.9	58.2	53.7
η Aquarids	13	0.958	163.5	95.2	42.4	52.8
Orionids	15	0.961	163.9	82.5	28.0	54.5

3.2 Fine structure in the stream

Another feature which must be explained by a model of the stream structure is the multipeaked activity which is frequently observed. A double peak is frequently seen in visual observations (Stohl & Porubcan

1978). Hajduk's analysis (1980) shows that up to five peaks must be accounted for. (See Fig. 4).

4 CLUES POINTING TO A BETTER MODEL

4.1 Clues from the evolution of the comet orbit

Yeomans & Kiang (1981) have determined the history of Comet Halley by numerical integration of the equations of motion. Errors which would normally occur when the comet made an unusually close approach to the Earth were taken care of by forcing the calculations to conform to the available ancient observations. With this method, they believe their calculations are reasonably accurate back to 1404 BC (46 revolutions).

Fig. 5 shows the behaviour of two significant parameters of the orbit over this period. It is clear that there has been a very rapid variation of the positions of the nodes. The ascending node started inside the Earth's orbit, crossed it, and moved out to its present position at 1.8 AU. The position of the descending

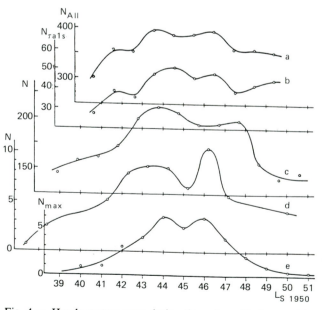

Fig. 4 — Hourly meteor rates during the η-Aquarid shower: (a) Ottawa radar data (1958–1967); (b) as (a) but long-duration echoes only; (c) New Zealand radar (1960, 63–65); (d) New Zealand visual data (1928–33); (e) number of peaks from visual and radar results (1910–1971). (From Hajduk 1980)

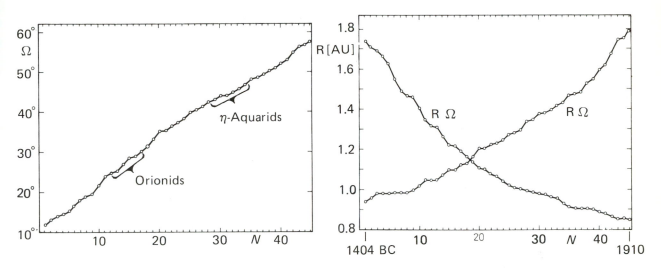

Fig. 5 — Left: Longitude of the ascending node of the comet orbit. Corresponding longitudes of the showers are marked. Right: Radius vectors at the ascending and descending nodes. N is the revolution number of the Yeomans & Kiang osculating orbits. (From McIntosh & Hajduk 1983)

node has moved in the opposite direction. Thus each of the nodes in turn has been at the Earth's orbit. At these times, meteoroids close to the comet orbit would have generated strong showers. It is not surprising, then, that most of the exceptional historical showers occurred in this era.

The rapid motion of the nodes is reflected also in the change of longitude of the ascending node, Ω, from 11° to 58° in this time span. The angle of perihelion, ω, increases at nearly the same rate as Ω, so that the location of perihelion remains nearly constant.

These changes in position of the nodes occur because, in accord with classical theory, the gravitational attraction of the major planets produces a rotation of the plane of the orbit about the major axis. Because of the small tilt of the axis to the ecliptic (18°) a small rotation engenders a large change in Ω and ω.

The overall variation in Ω shown in Fig. 5 appears to be a smoothly increasing trend. But what about the changes in Ω from one orbit to the next? This has been calculated analytically by Yabushita (1972), assuming Jupiter as the only perturbing force. Fig. 6 shows that the change in Ω over one revolution varies from practically zero to nearly two degrees, depending on the relative positions of the comet and Jupiter. If one

plots the values of $\Delta\Omega$ between successive Yeomans & Kiang orbits (right side of Fig. 6), these are also in accord with Yabushita's predictions. (Negative values in Yabushita's analysis result from his simplifying assumption of a parabolic orbit. For the real elliptical orbit the minimum value will be small but positive.)

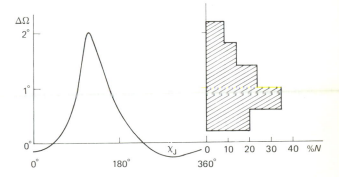

Fig. 6 — The change in Ω over one revolution, Left: as a function of X_J the longitude of Jupiter measured from the time of Comet Halley perihelion. (Adapted from Yabushita) Right: occurrence frequency of $\Delta\Omega$ values in the Yeomans & Kiang orbits. (From McIntosh & Hajduk 1983)

5 DEVELOPMENT OF THE SHELL MODEL

5.1 Perturbations to the particle stream

All of the planets contribute to the long-term perturbations of both the comet and the particle stream. The cumulative effect of the short-term perturbations—which are of an impulsive nature, as in the example of Fig. 5—on the evolution of the stream as a whole is different than for the comet. The greatest single perturbation is by Jupiter, and occurs when the body and Jupiter are simultaneously in the vicinity of minimum separation of their orbits. This is a coincidence which occurs rarely for the comet. Rather, it suffers a series of $\Delta\Omega$ perturbations which fluctuate about an average value resulting in the nearly linear trend of Fig. 5. But for a meteor *stream*, there are *always* some particles in the region of minimum separation. Thus some particles (a different group each time) suffer maximum perturbation *every* Jupiter revolution, and equally, some receive practically no perturbation. Since Jupiter makes about six revolutions to one revolution of the comet, the perturbations on the stream accumulate at a rate some six times faster than for the comet.

5.2 The shell model

5.2.1 Description of the development of a shell structure

The stream particles will effectively diffuse over a surface described by the rotation of the orbit about its major axis. While the particles with minimum perturbation will remain in orbits where the comet was centuries ago, the more highly perturbed particles will travel in orbits which have rotated faster than that of the comet, and will take up positions where the comet will be centuries hence.

Thus the structure which develops is a segment of a shell: a ribbon or belt. The ribbon thickens owing to gravitational perturbations and to a slight precession of the orbits (about one degree per millennium) (McIntosh & Hajduk 1983). Particles released at each revolution of the comet will not be found in orbital positions of earlier revolutions, and their maximum

rotational advance is dictated by the time since they were released.

There may be a fundamental limitation on the extent of the ribbon. Kozai (1979) has calculated that the motion of the comet librates. Specifically, the argument of perihelion, ω, librates between 47° and 133°. Assuming that the libration is sinusoidal, the rate of change of ω at $\omega = 90°$ from the Yeomans & Kiang orbits combined with Kozai's amplitude indicate that the period of this oscillation is about 300 revolutions. This places bounds on the shell segment such that it subtends an angle of 25° at the major axis of the ellipse as shown in Fig. 7.

Recall that a cylindrical model of the particle distribution as exemplified by Fig. 3 predicts shower duration and activity which should be considerably different for the Orionids than for the η-Aquarids. But with the ribbon model (Fig. 7), the Earth will encounter approximately the same thickness of ribbon at each crossing, accounting for the similar durations of the showers. Furthermore, the total flux of particles through a cross-section of the belt must be the same for both Earth-crossing regions, and therefore approximately equal activity is to be expected.

The schematic representation in Fig. 8 of the belt of particles at one of the Earth crossings shows in detail the structure in relation to the present orbit of Comet

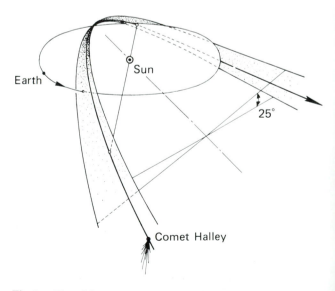

Fig. 7 — Pictorial representation of part of the ribbon of particles. (From McIntosh & Hajduk 1983)

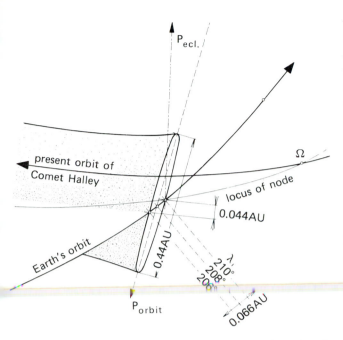

Fig. 8 — Detail of idealized particle belt structure in the region of the Earth's crossing at the time of the Orionid meteor shower. The locus of the ascending node is approximated by a smooth curve. (From McIntosh & Hajduk 1983)

Halley, to the Earth's orbit, and to the locus of the motion of the node.

McIntosh & Hajduk speculated that the libration of the orbits could account for the fine structure seen in the shower observations. The five major peaks observed seemed to require five librations, indicating a total age for the stream of 1500 revolutions, or about 100 000 years. However, the calculations to be described next show that the fine structure can develop much more rapidly.

5.2.2 Numerical calculation of the development of the shell

Attempts to model the dynamical evolution of meteoroid streams with a few tens of particles have met with limited success. Loss of particles or fortuitous groupings can make results questionable. Numerical integration of the motion of a single body over several millennia, taking account of all significant planetary perturbations, is a formidable task. To do so for a large number of particles is practicable only if simplifying assumptions are made. Jones & McIn-

tosh (1986) opted for a large number of particles— 500—but included only Jupiter perturbations. They estimate that Jupiter represents only about half of the total effect on Halley-like orbits. Neglect of perturbations from the other planets underestimates effects such as the long-term advance of the nodes. Also, light pressure enlarges the orbits of very small particles (McIntosh 1973). Nevertheless, Jupiter is the major influence on the dispersion of meteor-sized particles over the shell, and their simplified model illustrated the general evolution of the stream structure.

5.2.2.1 Initial conditions

It is desirable to examine the dynamic evolution of the particle stream from as early in time as possible. But it is obvious that if the starting orbit does not correspond to the actual orbit of Comet Halley, the results will be meaningless. One notes that the calculations of other workers (for example, Brady 1982) diverge considerably from Yeomans & Kiang at 1404 BC, and there is an element of doubt as to which are the most accurate.

Jones & McIntosh chose to start their calculations with 500 particles distributed uniformly around the Yeomans & Kiang orbit of 1404 BC. Some inaccuracy is introduced here because the particles must have been ejected from the comet earlier in order to have become distributed into a full ring by this time. As can be seen from Fig. 1, this dispersal time is only a few hundred years, which is short compared to the 3300 year time span since then.

To determine the effect of particle size, calculations were carried out for two values of mass: 1 gram and 10^{-4} gram. For each size, the magnitude of the ejection velocity was calculated from Whipple's formula (Whipple 1951). The angular component was selected randomly from an isotropic distribution, and the whole added vectorially to the velocity appropriate to the comet at that point. This procedure was similar to that described by Jones & Hawkes (1986).

5.2.2.2 Results and discussion

The motion of each of 500 particles has been numerically integrated from 1404 BC to the present taking into account the perturbations of both Jupiter and Saturn. The results are presented in Fig 9 by plotting the positions of the nodes of the final orbit for each particle. The coordinate system is centred at the sun and has the horizontal axis parallel to the vernal equinox. The general trend of the motion of the comet's nodes is shown by crosses. Each dot in the distribution along the crosses marks the position of the node of one of the 500 particles. The dots forming a circle should not be confused with those marking the particle nodes; the former indicate positions of the Earth at one-day intervals.

The relation between these plots and the belt of Figs 7 and 8 may be seen by visualizing a long cut slanting through the belt at an angle of 18°. It should be remembered also that the configuration of points represents only the particles from one return of the comet. The present belt is a superposition of particles from all past releases.

The figures show that, in only a few thousand years, the stream has developed into a configuration with considerable detailed structure and with an Earth-crossing width of several days. The fine structure is most marked for the larger particles. The greater ejection velocities of smaller particles cause the cross-section of the stream to become more diffuse, but bands of enhanced particle density are evident.

The very dense concentrations closest to the node of the starting orbit confirm that the motion of the nodes is always direct. But it is apparent that many particles do not advance very much and therefore crowd up. A release a few returns earlier than 1404 BC would have produced a concentration at the ascending node in the vicinity of the Earth resulting in a very strong meteor shower (Orionids). Particles released from the comet after this, would now have their nodes outside the Earth's orbit and will never be encountered as Orionid meteors.

Exactly which Halley return was the last one to supply particles to the current Orionid shower cannot be determined from these highly simplified calculations. The inclusion of the effect of radiation pressure and perturbations by all the planets expands the distribution of particle nodes further (see McIntosh & Jones 1988). It seems likely that the Orionid particles which we see today were expelled from the comet more than four or five thousand years ago.

The situation at the descending nodes is reversed; the nodes start well outside the Earth's orbit and move in. Figs 9 and 10 show that the η-Aquarid shower includes particles from more recent releases than does the Orionid shower.

6 TOTAL MASS OF THE STREAM; ITS AGE

It must be emphasized that Figs 9 and 10 represent the particle distribution from only one release. The entire stream is a superposition of similar but shifted distributions from many such releases. This is the context of the meaning of 'age' of the stream—essentially the time since the beginning of the orbital regime described earlier. There it was pointed out that, in the past, one of the nodes would have been near the orbit of Jupiter, raising the possibility of a close approach of the comet and planet. Yeomans (1986), by extrapolating his results, calculates that such an approach could have taken place 220 revolutions ago. The calculations of McIntosh & Hajduk, based on Kozai's results, would place this event at about 180 revolutions ago. Such a close approach supports two possibilities: one, that the comet was then captured into its current historical configuration; or, two, that that moment marked only the beginning of another Kozai libration. In the former case the age of the stream is low—only about 200 revolutions—while in the latter case it could extend to thousands of revolutions.

Given the total present mass of the stream and the amount of matter injected into it at each revolution, the age can be calculated. The mass of the stream can be determined if its cross-sectional area and mass flow per unit area are known.

Although the stream should not be expected to have the idealized form of Fig. 8, the theory of Kozai puts a reasonable bound on one dimension of the stream cross-section—that perpendicular to the orbit. The other dimension, the thickness of the belt in the radial direction, can be deduced from observation of the meteor showers.

The meteor mass flux can be sampled only along the Earth's crossing. Most values are based on radar meteor observations, the interpretation of which

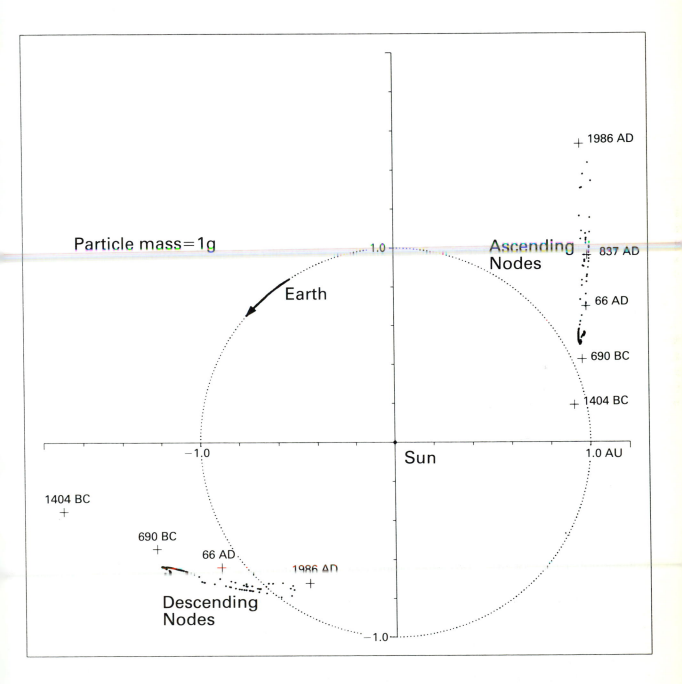

Fig. 9 — Distribution of ecliptic-plane crossings in 1986 of 500 particles released from the comet about 1404 BC. Particle mass 1 g. Positions of the Earth on its orbit are at one-day intervals. The crosses mark the locations of the nodes of the comet orbit for the returns indicated.

depends on rather poorly known observational selection factors. A recent determination (Hajduk 1982)
gives:

meteor mass flux $= 2 \times 10^{-16}$ kg m^{-2}s^{-1}.

Combining this value with the following dimensions
(see McIntosh & Hajduk 1983):

observed thickness of belt $= 0.044$ AU
estimated height of belt $= 0.44$ AU

gives:

total mass of the stream $= 2 \times 10^{14}$ kg.

On the other hand, Jones (1987 private communication) estimates the mass flux to be more than an order
of magnitude smaller:

meteor mass flux $= 7 \times 10^{-18}$ kg m^{-2}s^{-1}.

This gives:

total mass of the stream $= 7 \times 10^{12}$ kg.

The mass loss from Comet Halley for this appari-
tion is estimated at from 3 to 5×10^{11} kg. Since $\frac{1}{5}$ to $\frac{2}{3}$
of this is dust (Whipple 1987), then 2×10^{11} kg is a
reasonable estimate of the loss of solid particles.
Within this material are particles ranging in mass
from 10^{-17} g to 10^{+1}g. Much of the fine material is
blown out into the cometary tail region. The proportion of the total of solid particles retained in the
stream compared with that lost from the stream
depends critically on the particle number counts as a
function of particle mass. Provided the form of the
mass distribution is not significantly different from
that described by Hajduk (1987), the mass of the
particles retained in the stream will represent an
appreciable fraction of the total mass. Setting this
fraction at 50%, gives 1×10^{11} kg as the mass of
particles which will eventually be injected into the
stream from the current passage. The assumption that
the mass loss has been relatively constant for all
returns leads to an age range of:

4000 revolutions (Hajduk), to
140 revolutions (Jones)

These values for the age of the stream are very
uncertain because of the many uncertainties in the
values that entered into the calculations. This is true
also of other methods of calculating the age of the
comet (defined in the same way as age of the stream),
giving rise to a similarly wide range of values: from a
few tens of revolutions (Ferrin & Gil 1986) to a few
thousand revolutions (Hughes 1985).

Unfortunately, the range of ages encompasses both
values based on the postulate of many Kozai
librations, and on the postulate of capture at the last
close encounter with Jupiter. A more definitive study
of the age of the stream carried out by Jones et al.
(1989) puts it at 23,000 years.

A qualitative idea of the age of the comet may be
gained by comparing its current mass, i.e. the mass
remaining, with the mass of the stream, i.e. the mass
lost in the past. On the basis of the above numbers, the
total mass lost is from 6 to 10 times the mass of the
stream, for a range:

4×10^{13} to 2×10^{15} kg.

Although the present mass of Halley's Comet is still
uncertain, the central range of current estimates is
about 0.5 to 2×10^{14} kg (Whipple, 1988). Since these
values indicate that the comet has lost either as little as
17% of its mass or as much as 98% of its mass, they are
of little use in assessing the age of the comet.

7 THE VALUE OF CURRENT METEOR OBSERVATIONS

The flux of meteors in the two Halley showers,
observations of which have been collected under the
auspices of the International Halley Watch, was not
expected to be influenced by the current return of the
comet.† Rather, the accumulation and analysis of
these data should allow better determinations of
factors such as the mass in the stream and therefore
improve not only our understanding of the evolution
of the particle stream but also of the comet itself.

† This has been confirmed by the author's observations and
those of many others. See, for example, Hajdukova et al. (1987).

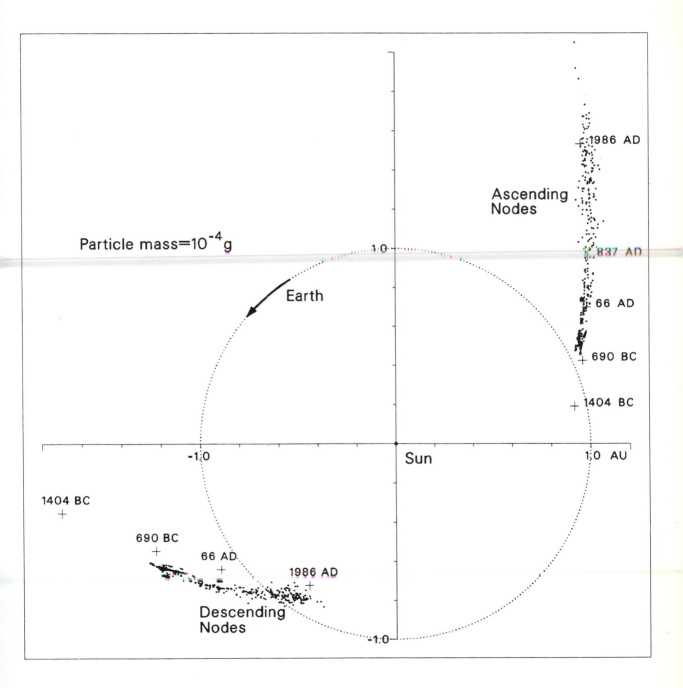

Fig. 10 — Distribution of ecliptic-plane crossings in 1986 of 500 particles released from the comet about 1404 BC. Particle mass 10⁻⁴ g. Positions of the Earth on its orbit are at one-day intervals. The crosses mark the locations of the nodes of the comet orbit for the returns indicated.

ACKNOWLEDGEMENTS

It is a pleasure to acknowledge the benefits of collaboration with my colleagues, Dr. A. Hajduk, Bratislava, Czechoslovakia, and Dr J. Jones, London, Canada.

REFERENCES

Brady, J.L. (1982) Halley's Comet: AD 1986 to 2647 BC, *J. Brit. astron. Assoc.* **92**(5) 209–215

Cook, A.F. (1973) A working list of meteor streams, *Evolutionary and Physical Properties of Meteoroids*, NASA SP-319, 183–191

Ferrin, I., & Gil, C. (1986) Sizes, aging rates and extinction dates of Comets Halley and Enke. *20TH ESLAB Symposium on the exploration of Halley's Comet*, ESA SP-250. **II** 427–432

Hajduk, A. (1980) The core of the meteor stream associated with Comet Halley, In: Halliday, I., & McIntosh, B.A. eds, *Solid particles in the solar system*, 149–152. D. Reidel Pub. Co., Dordrecht. Holland

Hajduk, A. (1982) The total mass and structure of the meteor stream associated with Comet Halley, In Fricke, W., & Teleki, G. eds, *Sun and planetary system*, 335–336. (*Astrophys. & Space Sci. Library* Vol. 96) D. Reidel Pub. Co., Dordrecht, Holland

Hajduk, A. (1987) Meteoroids from Comet P/Halley. The Comet's mass production and age, *Astron. Astrophys.* **187** 925–927

Hadjuková, M., Hajduk, A., Cevolani, G., & Formiggini, C. (1987) The Halley meteor showers in 1985–1986, *Astron. Astrophys.* **187** 919–920

Halliday, I. (1987) The spectra of meteors from Halley's Comet, *Astron. Astrophys.* **187** 921–924

Ho Peng Yoke (1964) Ancient and mediaeval observations of comets and novae in Chinese sources, *Vistas Astr.* **5** 127–230

Hughes, D.W. (1985) The size, mass, massloss and age of Halley's Comet, *Mon. Not. R. astron. Soc.* **213** 103–109

Imoto, S., & Hasegawa, I. (1958) Historical records of meteor showers in China, Korea, and Japan, *Smith. Contr. Astrophys.* **2** (No. 6) 131–144

Jacchia, L.G., & Whipple, F.L. (1961) Precision orbits of 413 photographic meteors, *Smithson. Contr. Astrophys.* **4** 97–129

Jones, J. (1982) The annual variation in the activity of the Geminid shower and the theory of the dispersal of meteoroid clusters, *Mon. Not. R. astron. Soc.* **198** 23–32

Jones, J. (1983) Radar observations of the Orionid meteor shower, *Mon. Not. R. astron. Soc.* **204** 765–776

Jones, J. (1985) The structure of the Geminid meteor stream. I: The effect of planetary perturbations, *Mon. Not. R. astron. Soc.* **217** 523–532

Jones, J., & Hawkes, R.L. (1986) The structure of the Geminid meteor stream. II: The combined action of the cometary ejection process and gravitational perturbations, *Mon. Not. R. astron. Soc.* **223** 479–486

Jones, J., & McIntosh, B.A. (1986) On the structure of the Halley Comet meteor stream, *20th ESLAB Symposium on the exploration of Halley's Comet*, ESA SP-250 **II** 233–237

Jones, J., & McIntosh, B.A., & Hawkes, R.L. (1989) The age of the Orionid meteor shower, *Mon. Not. R. astron. Soc.* **238**, 179–191

Kozai, Y. (1979) Secular perturbations of asteroids and comets, *Dynamics of the solar system.* ed. Duncombe, R.L., Dordrecht, D. Reidel, 231–237

McIntosh, B.A. (1973) The origin and evolution of recent Leonid meteor showers. In: *Evolutionary and physical properties of meteoroids*, IAU Colloquium 13, (NASA SP-319)

McIntosh, B.A., & Hajduk, A. (1983) Comet Halley meteor stream: a new model, *Mon. Not. R, astron. Soc.* **205** 931–943

McIntosh, B.A., & Jones, J. (1988) The Halley comet meteor stream: numerical modelling of its dynamic evolution. *Mon. Not. R. astron. Soc.* **235** 673–693

McKinley, D.W.R. (1961) *Meteor science and engineering*, McGraw-Hill Book Co. Inc., New York

Olivier, C.P. (1911) *Trans. Amer. Phil. Soc.* **22** (pt. i) 5–35

Stohl, J., & Porubcan, V. (1978) Orionid meteor shower: activity and magnitude distribution, *Contrib. astr. Obs. Skalnate Pleso* **10** 39–48

Whipple, F.L. (1951) A comet model. II: physical relations for comets and meteors, *Astrophys. J.* **113** 464–474

Whipple. F.L. (1987) The cometary nucleus: current concepts, *Astron. Astrophys.* **187** 852–858

Yabushita, S. (1972) The dependence on inclination of the planetary perturbations of the orbits of long-period comets, *Astron. & Astrophys.* **20** 205–214

Yeomans, D.K. (1986) Physical interpretations from the motions of Comets Halley and Giacobini–Zinner. *20TH ESLAB Symposium on the exploration of Halley's Comet*, ESA SP-250 **II** 419–425

Yeomans, D.K., & Kiang, T. (1981) The long-term motion of Comet Halley, *Mon. Not. R. astron. Soc.* **197** 633–644

Zhuang, T. (1977) Ancient Chinese records of meteor showers, *Chinese Astronomy* **1** 197–220

Part II
Nucleus

11

Surface features and activity of the nucleus of Comet Halley

H.U. Keller

INTRODUCTION

Our knowledge of the physical nature of cometary nuclei has made a major step forward as results from the imaging experiments became known after the flybys of Comet Halley by the Vega and Giotto spacecraft in March 1986.

It is not possible to observe cometary nuclei directly from the ground. The total reflected radiation of the supposedly bare nucleus at large heliocentric distances can be measured, determining the product of its cross-section times its geometric albedo. If the temperature of the nuclear surface can be determined by additional observations in the infrared wavelength range, these quantities can be separated into information on the size (cross-section) of the nucleus itself and its average albedo. Other contributions to our knowledge of the physical properties of cometary nuclei have come from rather indirect conclusions drawn from observations of cometary activity. This leads to a (minimum) size of the nucleus providing a surface large enough to produce the observed gas and dust production rates.

Observations made just before the Comet Halley encounters showed that several short periodic comets had low albedos, around 0.05, and were rather large (Hartmann *et al.* 1985). The reduced magnitude of Comet Halley determined at large heliocentric distances was $V(1,0) = 14^m$. If the observed surface area of 90 km² is introduced it yields a geometric albedo of 0.04 (Whipple 1986). This value is in good agreement with the albedo derived from the direct measurement of the reflectivity of the surface at small phase angles (peak value about 0.03 for a phase of about 20°) during the Vega-2 flyby (Sagdeev *et al.* 1986a). The reflectivity (brightness compared to that of a Lambertian surface with perpendicular insolation) measured by the Halley Multicolour Camera (Fig. 9) was much lower, only 0.005. However, the observation was made at a phase angle of 107°, and was consistent if the phase dependence were similar to that of the Moon.

The inhomogeneities (asymmetries, jets) of the dust production allow one to trace these back to the surface. Detailed investigations were carried out by observation of dust jets (e.g., Sekanina & Larson 1984). These and other observations of periodic phenomena such as expanding shells have led to estimates of the rotation properties of some nuclei (Whipple 1982) such as the periods and in some cases the orientations of the rotation axes, although the results could hardly be considered as definite; a fact fully illustrated in the case of Comet Halley. The values derived from the images taken by the spacecraft cameras seem to converge to a value of 2.2 days for the rotation period of the nucleus (Wilhelm 1986, Sagdeev *et al.* 1986.) However, various observers have presented compilations of results for the gas and dust production of Comet Halley that result in variations with a period of 7.3 or 7.4 days (Millis & Schleicher 1986, Festou *et al.* 1987 (suggesting 14.6 d), Stewart 1987). A combination of rotations (or nutation) around the axis of maximum moment of inertia and along the long axis was suggested by Sekanina (1987) and Julian (1987). Wilhelm (1987) demonstrated that free nutation (precession) cannot exceed a cone angle of a few degrees. Even small damping values would extinguish any nutation rather fast. Smith *et al.* (1987) claim that any rotation other than around the axis of maximum moment of inertia can be excluded from the imaging data of Vega-1 and -2 and the Halley Multicolour Camera (HMC) on board Giotto. For a more detailed discussion see other contributions in the present book. As yet, no clear solution to the problem of Comet Halley's rotation exists. The analysis of dust jets in high resolution ground-based observations seems to contradict the orientation of the rotation axis derived from the spacecraft images (Keller & Thomas 1988).

Many different observations and investigations have contributed to our present knowledge and perception of cometary nuclei, and much has still to be analyzed and observed. The major contribution has come from the huge effort mounted to observe Comet Halley, combining spacecraft and ground-based investigations. Only the imaging systems on board Vega-1 and -2 and Giotto could yield descriptions of a cometary nucleus surface. The images taken by the Halley Multicolour Camera (HMC) will be the prime source of information for this contribution, which describes surface features and activity of the nucleus of Comet Halley.

THE OBSERVATIONS

The HMC telescope had an effective speed of $f/7.7$ and a focal length of 998 mm. Its angular resolution was 22.4 μrad per pixel. It operated in line scan mode, taking advantage of the motion of the image caused by the rotation of the spacecraft. The exposure times varied from about 5 ms to 150 μs, decreasing with increasing offset angle of the object from the spin axis. The field of view (FOV) of a full CCD section with 390 × 292 pixels corresponded to 0.5° × 0.36°. A limited FOV of 0.1° × 0.1° (74 × 74 pixels) could be

transmitted during the final approach only every spin of 4 s (multidetector mode starting with image # 3436). Simultaneously, three images with half the resolution were taken through wideband filters in the red, blue, and clear. The camera has been described by Schmidt *et al.* (1986) and its operation by Keller *et al.* (1987a). The first image on encounter night was taken at 3 h 06 m before closest approach (CA) of the Giotto spacecraft on 14 March at 0003:01.9 h UT from a distance of 770 000 km. The last useful image (# 3504) was achieved from a slant distance of 1675 km a few seconds before CA when the communication with the spacecraft failed owing to dust impacts. The phase angle varied from 107° down to 90°.

The early calibration of the clear filter (300 to 1000 nm) images (Keller *et al.* 1987b) has been replaced by a complete calibration of all images taken during the fly-by (Thomas & Keller, 1990). The absolute calibration carries an estimated error of 5%. The point spread function (PSF) of the instrument was found to have an FWHM (full width half maximum) of about two pixels (Kramm *et al.* 1987). Images presented in this contribution have been processed by eliminating the dark current, the coherent noise, and in some cases by smoothing or filtering of high-frequency noise. The PSF was removed where indicated. The Vega spacecraft carried imaging systems of similar capabilities. Their encounters took place on 6 March at 0720:06 UT (Vega-1) and 9 March 1986 at 0720:00 UT (Vega-2) with minimum distances of 8889 and 8030 km, respectively (see Sagdeev & Szegö in this volume).

THE SHAPE AND ORIENTATION OF THE NUCLEUS

The nucleus was visible on several hundred HMC images from distances further out than 100 000 km. The unilluminated part of the surface was seen as a dark silhouette against the scattered light of the dust in the background (Fig. 1). The phase angle of 107° produced a rather small illuminated crescent. The nucleus was strongly elongated with visible dimensions of roughly 15 km by 8 km, much larger than anticipated. Based on the analysis of all the flyby images (Vega and Giotto, Sagdeev *et al.* 1986b, Wilhelm *et al.* 1986) the third dimension not visible on

Fig. 1 — The dust coma of Comet Halley with jets and the nucleus just resolved, seen from a distance of 69 350 km. The Sun is on the left, 4° below the horizontal and 17° behind the image plane. The image is radiometrically and geometrically corrected. The horizontal line of symmetry corresponds to 304 km.

Fig. 2 — Image #3461 is the last image to show almost the complete outline of the nucleus of Comet Halley. It is taken from a distance of 13 330 km (resolution 300 m/pixel). The Sun is to the left 28° above the horizontal and 15° behind the image plane. The image is radiometrically and geometrically corrected and slightly smoothed.

HMC images has also been determined to be about 8 km. The overall shape of the nucleus of Comet Halley can be approximated by an ellipsoid of 8 km × 8 km × 16 km. The long axis of the nucleus and the Comet–Sun direction both projected onto the image plane formed an angle of about 30° during the Giotto observations. The northern end of the nucleus pointed about 30° out of the image plane toward the approaching camera, causing a foreshortening of the longest dimension. The rotation axis is assumed to be perpendicular to the longest dimension of the nucleus (rotation around the axis of maximum moment of inertia). It pointed about 13° out of the image plane away from the observer (see Fig. 4 below). The rotation was prograde relative to Comet Halley's orbit; the northern tip was rotating away from the observer so that the morning terminator was visible.

THE LIMB

The aspect (and phase) angle changed by very small amounts from image to image for all but the last images. Images can therefore be combined. Later images with higher resolution, but therefore smaller

geometric FOV, can be implanted into earlier ones in a composite technique. In this way the high-resolution cut-outs can be placed in the right perspective of a larger part of the nucleus or of the total nucleus. One has to be aware of artificial effects of the varying resolution. Image # 3461 (Fig. 2) was taken from a distance of 13 330 km. It shows the whole nucleus with a resolution of 300 m per pixel. For comparison Fig. 3 displays a composite image based on image #3457 taken from a distance of 14 420 km with a resolution of 325 m. Implanted are images #3475, #3480, #3491, #3496, and #3500 with improving resolution of 210, 180, 120, 80, or 60 m per pixel, respectively. The offset angle changes from 2° to 11°, and the phase angle from 105° to 96°. The changes in the last two images (#3496 and #3500) could distort the overall aspect somewhat. The technique of image composition requires knowledge of the camera pointing with an accuracy of a few 10^{-3} degrees and subpixel registration. Fig. 4 displays a schematic drawing of the outline of the nucleus, and indicates some of the more prominent features on its surface.

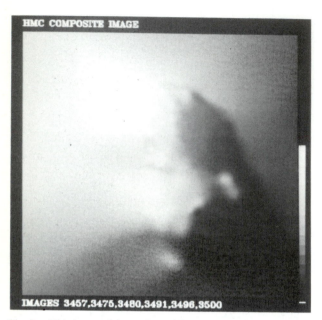

Fig. 3 — This image of the nucleus of Comet Halley is composed of 6 images taken by the Halley Multicolour Camera during the encounter of the Giotto spacecraft on 13 March 1986. Images #3475, #3480, #3491, #3496, and #3500 replace parts of the outermost image #3457. The images were taken at 14 420, 9510, 8150, 5150, 3800, and 2730 km from the nucleus. The resolution improves from 325 to 60 m per pixel while the brightest part of the image is approached. The Sun is 30° above the horizontal on the left and between 15° and 12° below the image plane. Images are radiometrically and geometrically corrected. Noise is removed and the PSF is deconvolved. The image is displayed in pseudo colours.

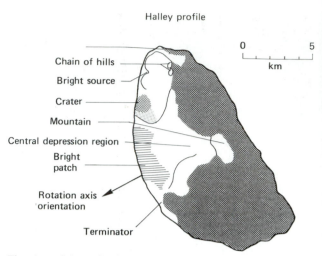

Fig. 4 — Schematic drawing of major nucleus features. The orientation is the same as that of the image of Fig. 3. Adapted from Keller *et al.* (1987b).

The maximum length of the visible cometary outline on HMC images was 14.9 km, the maximum width was 8.2 km (Keller *et al.* 1987a) measured perpendicular to the apparent long axis. The uncertainty of the limb on the night side is determined by the resolution at the southern end, and in particular by the limited signal to noise. Detailed comparison of various images is being performed to quantify the error limits. At present the uncertainty of limb details is of the order of two pixels or about 500 m at the southern dark end. The illuminated limb was partly obscured by scattered light from the emitted dust. At various places the contrast in the image was so low that it could not be resolved in spite of the relatively high signal. The limb of the northern tip was visible in between jets. The total visible surface area was 90 km². Major features are marked in the sketch outlining the nuclear surface (Fig. 4).

SURFACE

A large unilluminated part (75%) of the visible surface was separated from the insolated sunward crescent (25%) by the morning terminator. This dividing line was inclined toward the Sun on the northern end. This is a result of the elongated shape of the nucleus, the long axis of which formed an angle between 45° and 50° with the projected Sun–Comet direction. The terminator touched the Sunward limb at the lower end. This end was bulkier than the northern tip. The terminator receded from the Sun direction at the centre. Here the course of the terminator was not so well defined; the transition from the illuminated surface to the dark part was more gradual, indicating a slow change of solar inclination angle. The 'effective curvature' of the body seems to be larger than, for example, at the northern end. This central depression tapered toward a bright spot located almost in the centre of the visible outline of the nucleus. This peculiar feature—the mountain—displayed a very sharp intensity gradient on its northern tip (see Fig. 5*). Isophotes show that the mountain is separated from the central depression by a dark stripe as the illuminated top of a mountain would be. Images with high resolution show 4 almost periodic structures along the northern terminator of a typical size between 0.5 and 1.0 km, the chain of hills (see Fig. 6).

* to be found in the colour illustration section in the centre of the book.

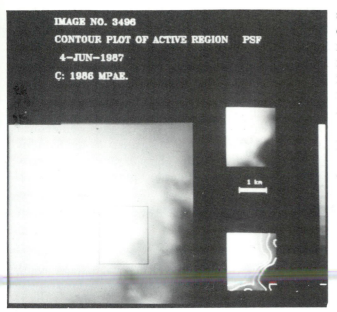

Fig. 6 — Image #3496 was taken from a distance of 3800 km yielding a resolution of 85 m per pixel. The Sun is 30° above the horizontal on the left and 6° behind the image plane. Processing level, see Fig. 3. Parts of the active area are shown in the cut-out (upper right) together with isophotes (lower right). The upper bright spot is connected to a dust filament (cf. Fig. 12).

The southern end of the terminator seemed to be smoother, although this may be due to the inferior resolution (see also Fig. 2).

A rather narrow, dark (unilluminated) band formed the northern tip of the nucleus, reminding one of a bent 'forefinger'. It was about 1.5 km long. This band was separated from the bulk of the nuclear surface on the night side by a brightly illuminated feature (a ridge?) about 400 m across.

The insolated part of the surface just south of this feature was the area of strongest activity. Several bright spots of a size less than 1 km could be detected (see Fig. 6). Some of them displayed rather well-defined boundaries on high-resolution close-up images excluding concentrated dust clouds. Brightness variations indicated a certain variability of the reflectivity of the surface (up to a factor of two), although some of this variation was almost certainly caused by topography or a combination of both. Several of these bright spots seemed to be starting points of small jets (jet N3 and N4 on Fig. 13*). The typical size of these features and of the also visible

somewhat darker spots corresponded to the angularity of the terminator (chain of hills, see Fig. 6). A more reliable determination of the surface properties requires the knowledge of the signal contributions from the dust above the surface. This may be possible in the future, using the information of the filter images and/or by modelling the dust distribution.

Just south of the active northern area the most prominent single feature on the surface was found, the 'crater' (see Fig. 8). It seemed to have a rather regular oval shape on early images with lower resolution (Fig. 2). The use of the word crater does not imply that this feature is a physical crater (certainly no volcano and very probably not caused by impact). The expression is simply used descriptively. The Sunward part of the crater bottom appeared dark. This was caused either by almost grazing incidence of the solar radiation or by the shadow of the northern Sunward rim. However, images with the highest available resolution (Fig. 8, 120 m per pixel) show that the rim and bottom were structured. The Sunward rim coincided with the limb of the nucleus. One has to take into account that the crater appeared considerably distorted owing to

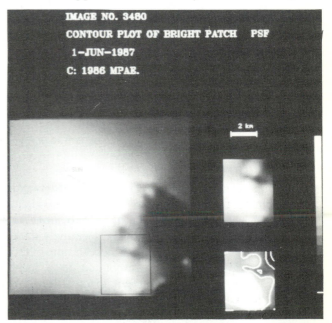

Fig. 7 — Image #3480 was taken from a distance of 8150 km (resolution 180 m per pixel). The Sun is 13.5° behind the image plane. Processing level, see Fig. 3. The cut-out shows the area of the bright patch (isophotes on the lower right). Some horizontal structure within the bright area at the limb is visible.

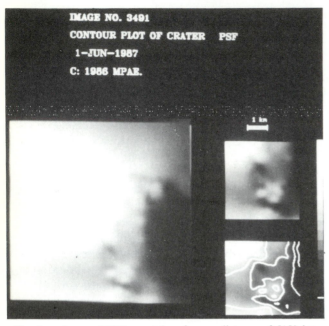

Fig. 8 — Image #3491 was taken from a distance of 5150 km (resolution 116 m per pixel). The Sun is 11° behind the image plane. Processing level, see Fig. 3. The cut-outs show the crater and corresponding isophotes. The walls and the bottom of the crater clearly show resolved structures. A limit toward the north is not obvious. The left rim probably coincides with the limb of the nucleus.

the perspective. Application of photoclinometric methods (Schwarz *et al.* 1987) yields a diameter of about 2 km and a depth of only 100 to 200 m. The latter value (in particular) depends on the amount of light scattered by the dust above the surface. South of this prominent crater there may have been one or two more, shallow, features just barely visible in the calibrated non-enhanced data.

The large central depression ran into the Sunward limb forming the bright patch (Fig. 7). The whole region between the terminator and the limb appeared to have a rather smooth surface. A seemingly dark ledge ran parallel to the terminator on both sides of the depression. This feature corresponded to a local decline of the brightness gradient toward the limb rather than to a decrease of brightness, as can be seen in the intensity profile of Fig. 9. It seemed to separate the bright patch from the inner part of the central depression. From the ledge toward the limb the brightness increased rather strongly. Images of this area with the highest available resolution (210 m per

pixel) showed a weak, essentially horizontal structure (Fig. 3). A somewhat darker streak separated the lowest third.

A ledge (running perpendicular to the Comet–Sun direction) just north of the outcropping dark part of the nucleus seemed to be enhanced or outlined by a brighter seam. It ran parallel to a more pronounced, similar feature coinciding with the well-defined terminator. This again indicated a topographic feature, the Sunward edge being brighter than the side facing away from the Sun. The limb of the nucleus ran from the Sunward rim of the crater to the notch in the dark area where the terminator extended toward the Sunward side. A narrow, bright line just at the resolution limit of the camera indicated the limb at this point.

South of the outcropping of the terminator a bright spot was visible. Dust was spewing out. The rather sharp separation of the dust from the dark feature in an almost straight line indicates that this dark part of the nucleus was in the foreground, obscuring the dust jet at its northern extension.

The overall outline of the nucleus was rather smooth and roundish. Protrusions and features were small compared to the radius of curvature of the main body. A marked exception seemed to be the crater with its diameter of 2 km. It is conceivable that this was a coherent feature formed by erosion of the subliming ice. Further investigation may reveal the

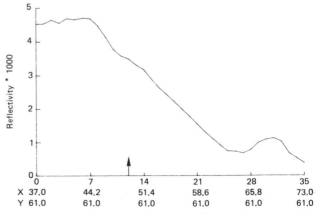

Fig. 9 — A scan across the nucleus (along line 61 of image #3477, distance 8970 km) from the mountain (right hand peak) toward the bright limb (left hand peak). The arrow indicates the ledge. The length of the scan is 7.3 km. The reflectivity relative to a Lambertian surface with inclination angle zero is given as ordinate.

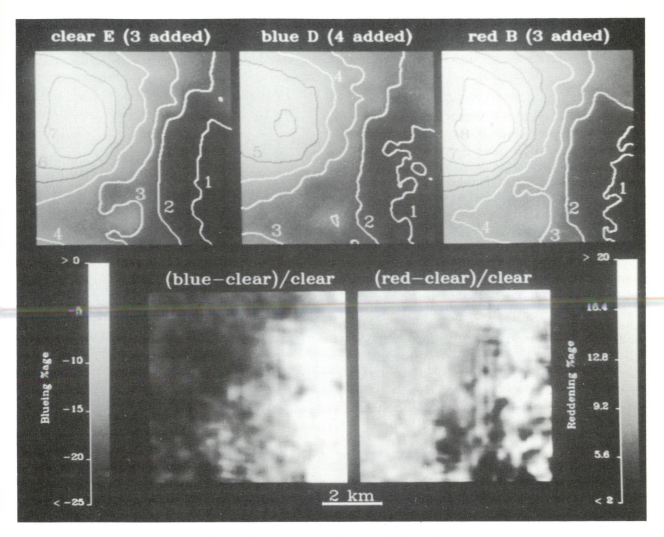

Fig. 10 — Reflectivity maps through 3 different filters (a: clear, b: blue, c: red). The scaling is such that an object reflecting the solar spectrum would give identical reflectivity values for all 3 maps. The 0.001 contours for the clear and blue images are almost identical. The peak reflectivity in clear is 0.0086 while in blue it is 0.0070. Differences between clear and red are also evident. These differences are emphasized in d) and e) where the [colour-clear]/clear images are displayed (units percentage numbers). The nuclear surface is clearly of different colour compared to the dust; both are redder than the solar spectrum.

origin of this feature. Its Sunward rim coincided with the limb and therefore formed an outcropping. There was relatively little (if any at all) dust activity directed toward the observer from the rim and surface of the crater. A dark wedge of lower brightness extended from this rim away from the nucleus. It reflected the brightness of dust in the background framed by the crossing dust jets emanating from the surface north and south of the crater. A recent compilation of images of surface details can be found in Keller *et al.* (1988).

Images taken through the broad band filters in red (effective wavelength 813 nm) and blue (440 nm) have been used to determine the colour of the nucleus surface and of the dust. Figure 10 shows the northern active area. An excess in the red and a deficite in the blue band is visible by comparing the isophotes. The nucleus was reddened compared to solar light by 6.3% ± 3% per 100 nm between 440 and 810 nm, but less reddened than the dust for which the corresponding figure was found to 7.3% (Thomas & Keller 1989).

GLOBAL ACTIVITY

Images of the cometary coma taken by the cameras of the spacecraft days before their encounters already showed the asymmetry of the inner coma (Sagdeev *et al.* 1986a, HMC images not yet published), also visible on high-resolution observations from the ground (Cosmovici *et al.* 1986).

The first images taken during the Giotto encounter from a distance of 770 000 km showed that by far the greater part of the dust production of Comet Halley at that time was confined to a cone with an angle of 70° projected on the sky (Keller *et al.* 1986a). This cone was not centred on the Comet–Sun direction as one might have expected. The strongest dust emission pointed about 50° to the south (Fig. 11) nearly in the direction of the rotation axis of the nucleus. Coming closer to the nucleus the direction of maximum emission changed toward smaller angles (35°) relative to the Comet–Sun line (Keller *et al.* 1986c). Close to the nucleus at distances smaller than its length the dust emission of the then clearly visible jet north (13°) of the Comet–Sun line dominated (compare image #3461, Figs. 2 and 14*).

This global emission pattern has been interpreted in the following way (Keller *et al.* 1986c, 1987b). Three major areas of emission have to be accounted for: (1) The area with strong visible activity on the northern Sunward limb that also showed the brightest emission in all HMC images; (2) The area at the bright patch at the centre of the nucleus, showing less bright ($\frac{2}{3}$ of the peak brightness) activity; and (3) A source area on the other side of the nucleus not visible to HMC. It is inferred that this source was located close to the rotation pole pointing south (actually the north pole of the nucleus because of its prograde rotation relative to its retrograde orbital motion, following IAU conventions). This source was the most active, and it supposedly produced the strong dust emission dominating at larger distances from the nucleus. The general emission pattern seemed to follow the respective directions of the mean normal of the insolated productive area. Model calculations assuming a uniform (independent of the inclination angle) dust production of the insolated surface reproduce the direction of the peak of dust 50° south of the Sun–Comet line (Thomas & Keller, 1988). The global appearance of the inner dust coma was surprisingly similar during all three spacecraft encounters (Keller *et al.* 1986b).

ACTIVE AREAS

The extent of the northern active area was 3 km along the visible limb. Assuming that this region extended about as far on the invisible hemisphere as it did on the visible hemisphere, its size corresponds to a circle with a diameter of about 3 km. The total gas production rate from such an area can then be estimated to be 1.3×10^{29} molecule s^{-1} (Huebner *et al.* 1986), assuming that the sublimation was controlled by water (80% with 20% more volatile admixture, Krankowsky *et al.* 1986). The other major visible source area at the centre of the nucleus was of similar dimensions. Both areas together produced about 2.5×10^{29} molecule s^{-1} into the general direction toward the Sun. As mentioned above, the hidden source region produced about 1.6 times more into the southerly direction. Its size should have been by the same amount larger than the combined areas of the visible sources. This adds up to a total active area of less than 40 km^2 or about 10% of the total surface area of the nucleus—a very surprising result indeed.

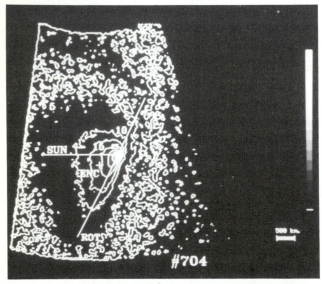

Fig. 11 — Image #704 was taken from a distance of 767 000 km. The length of the central vertical line corresponds to 6530 km. The direction toward the Sun is indicated. The Sun is 17° behind the image plane. Also indicated are the direction of the rotation axis (ROT) and the point of closest approach (ENC).

* to be found in the colour illustration section in the centre of the book.

Only a minor part of the illuminated surface ($\sim 20\%$) reveals strong activity. The water ice is not exposed to the solar radiation, with the possible exception of the active areas. Almost all of the surface of the nucleus of Comet Halley is covered by a non-volatile crust or mantle.

The total gas production rate from the active areas becomes 6.5×10^{29} molecule s^{-1}, a value very similar to the *in situ* observations. This cross-correlation with the gas production is introduced because the total dust production is very much less well known. *In situ* measurements of the dust confirm the strongly structured nature of the emission visible as dust jets (McDonnell *et al*. 1986, Simpson *et al*. 1986, Vaisberg *et al*. 1986, Edenhofer *et al*. 1986). The generalization of measurements along the spacecraft trajectory will necessarily lead to wrong estimates. The accurate determination of the total dust production rate from the optical observations (from ground and from the spacecraft) remains a major task for the future. The essential uncertainty in this procedure will be the scattering properties of the dust grains. They are also required for the determination of the optical thickness of the dust emission visible in the HMC images. The optical thickness derived from the strongest emission at the northern tip is estimated to be 0.3, assuming particles with a representative radius of 1μm and a scattering albedo of 0.03 at a phase angle of $107°$. The scattering of dust is optically thin. The obscuration of the limb and part of the surface was caused by the fact that the albedo of the dust was still appreciably higher than the reflectivity of the surface at a phase angle of $107°$ (about 0.005). The scattered light from the dust could therefore dominate in the combined signal (Keller *et al*. 1987b).

DETAILED ACTIVITY

A closer look at the isophotes on carefully reduced HMC images reveals a wealth of fine structure in the dust emission. Fine rays—dust filaments—were visible. So far, 17 filaments have been identified and classified according to their place of origin (Thomas & Keller 1987a). Most of them emanated from the northern area of activity (Fig. 12). These filaments were rather weak, but well-defined and very narrow (less than $10°$ opening angle). They could be observed

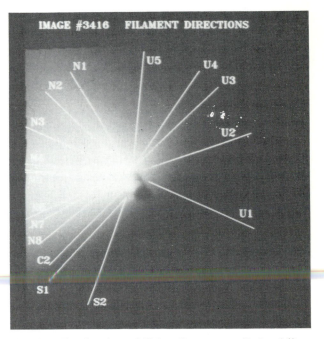

Fig. 12 — The directions of 17 dust filaments are displayed (from Thomas & Keller 1987a). The filaments originate from three active areas N, C, and S. Some are of unidentified origin, U. Image #3416 was taken from 25 620 km.

only from the spacecraft (see also Smith *et al*. 1986). They should not be confused with the jets visible from ground-based observations. The full southern emission corresponded to a jet in the classical sense (Cosmovici *et al*. 1986). It could be identified on ground-based images also, confirming that it is directed away from HMC during the Giotto encounter (Keller & Thomas, 1988). The dust filaments were fine structures within the jets. They were small compared to the jet itself; in fact they contributed so little to the total signal that in many cases it was difficult to detect them in the images directly. They could more easily be picked up by using gradient and shifted images. Five filaments could be found emanating from the northern area (Fig. 13*). Some of them were discernible over only a short range from the nucleus. One of these filaments (N3 in Fig. 13*) could be directly linked to a bright spot on the surface interpreted as the starting point or source of this filament. The source had a diameter of about 500 m, and its reflectivity was enhanced by about 50%. The filaments can be interpreted as local enhancements of

* to be found in the colour illustration section in the centre of the book.

the dust production (30–50% derived from the brightness differences, Thomas & Keller 1987b) within the large active areas. This could be caused by either an increased dust-to-gas ratio or a local variation of the sublimation, not necessarily requiring chemical inhomogeneities. However, enrichments with more volatile gas cannot be excluded.

The intensity distribution as a function of distance from the nucleus follows a $1/R$ law at distances larger than the size of the nucleus (>10 km). A region of acceleration closer up could not be detected; it should steepen the gradient beyond that of the $1/R$ law. In contrast, the intensity increase became shallower than $1/R$ closer to the nuclear surface in all cases that have been investigated (Thomas & Keller 1987b). As already pointed out, the contribution from a filament to the total signal was so small that it is almost impossible to separate it with sufficient accuracy. The determination of a global region of acceleration of the dust in the whole jet is difficult because near the surface the intensity distribution is dominated by geometrical effects of the extended source. Peripheral parts of the source contribute less and less to the column density near the surface. This 'dilution' flattens the intensity profile ($I \cdot R \neq$ const., it decreases for small R, see Thomas *et al.* 1988). The region of acceleration of dust particles by hydrodynamic drag of the gas molecules is expected to be rather short for particles of the typical size of $1\mu m$. 50% of the particles reach terminal velocity within a fraction of a kilometre (Hellmich 1981, Gombosi *et al.* 1985), although the calculations of Kitamura (1986) indicate somewhat larger scale-lengths. The rather flat intensity distribution near the source and its steepening toward the $1/R$ law at larger distances could of course be explained by fragmentation of the dust particles, in particular during their acceleration phase and their initial heating up as they separate from the surface and later from the gas component. If particle fragmentation alone is assumed, a lower limit for the increase in particle numbers of 25% is required (Thomas & Keller 1987b). Considering that the acceleration of the dust particles is not visible even on the smallest scale (100 m), initial break-ups could be much more frequent.

At least 5 dust filaments seemed to point away from the Sun on the dark side of the nucleus (Fig. 12). These were in general weaker than those extending toward the Sun. The apparent directions of emission do not necessarily require activity centres on the night side of the nucleus.

All the filaments could have emanated from the small (17° wide) strip near the dusk terminator on the other side of the nucleus. The projection of the filaments running nearly parallel to the line of sight can produce the apparent large angles away from the Sun direction (Keller *et al.* 1986a). However, the multitude of filaments pointing into the night side and the detection of a jet originating from the dark surface during the Vega-2 encounter (Smith *et al.* 1986) makes it difficult to reject the possibility of there being weak emissions from the night side of the nucleus.

These filaments may indicate decaying activity centres after their transit onto the night side of the nucleus close to the evening terminator. If this is the case, some of the solar input energy had to be stored in the surface of the nucleus. This requires a certain heat capacity of the dust–ice mixture, or energy sources by chemical fractionation leading to pockets of more volatile material.

The brightness variation along the dark side of the nucleus increasing from the southern end toward the northern tip (Fig. 14*) was caused by scattered light from dust between the surface and the observer. This brightness variation corresponds to a similar variation of the dust column density that had to increase by about a factor of three over a lateral distance of about 10 km; a very small number compared to the distance of the camera from the surface (13 300 km in the case of image #3461, Fig. 14*). It has to be inferred that the dust distribution changed immediately above the surface. This implies that dust was swept across the nucleus surface from the active regions. The dust was entrained in the breeze of the laterally expanding gas, requiring that the gas pressure is lower above non-active areas. This is corroborated by the fact that there was already an enhanced dust particle flux before the Giotto spacecraft reached the angular direction of the terminator (see, e.g. McDonnell *et al.* 1986). Further investigations will provide more insight into the hydrodynamic acceleration processes of the dust.

* to be found in the colour illustration section in the centre of the book.

CONCLUSIONS FROM THE OBSERVATIONS

Immediately after the Giotto flyby, it was obvious that only a minor part of the illuminated surface of the nucleus was active in the sense of unrestricted sublimation of water ice. Large parts of the surface had to be covered by a non-volatile crust or mantle. The dark side of the nucleus was essentially inactive at least so far as the production of dust is concerned. It is almost impossible to rule out night side gas production based on the so far analyzed HMC images. This caused an asymmetry of the inner dust coma (ratio of the Sunward to anti-Sunward hemisphere 3:1 (Thomas & Keller, 1988), leading eventually to the determined shift of the centre of brightness toward the Sun on ground-based images with their limited resolution. The dust emission was structured in large jets (by the standard of the high-resolution images of HMC). Their opening angles were of the order of 30° to 40°. Fine structures within these jets were visible. These dust filaments detected by HMC indicate variable output within the areas of activity. These filaments were very well collimated, and some of them can be followed out to several hundreds of kilometres, although their initial cross-sections (starting points) are only about 500 m in diameter. It is not clear by what process(es) these filaments are focused so strongly. Available model calculations (Kitamura 1986) suggest larger opening angles. One possibility could be that these filaments are not dust jets by themselves but rather only inhomogeneities of inner parts of the large dust jet emanating from the whole active area. In the centre of a large active source area the divergence of the gas and therefore the dust particles should be rather small. The overall picture of the dust emission resembled more the acceleration of dust from an extended (flat) surface than that of a concentrated relatively small source. There were no indications of the 'tulip' shaped density distributions with enhancements of dust densities along the mantle of the cone confining the emission. However, it is difficult to separate single jets in the observations. The extremely low reflectivity of the nuclear surface is probably caused by its porosity. The albedo of the dust grains, although small, was larger than that of the surface (Divine et al. 1986). The mineral and organic particles (Kissel et al. 1986) seem to be very fluffy,

considering their low observed densities. They could cause a small-scale roughness by forming cavities on the surface of the nucleus, thus trapping the light. It is unclear whether the mantle on the surface was coherent or merely the leftover regolith from the sublimation process. Nevertheless, the interior of the nucleus did not look, or at least did not reflect the light, differently when compared to the surface, otherwise the reflectivity variations across active areas should have been stronger. The Vega-2 images looking onto the illuminated surface with low phase angles did not show any variations exceeding about 50% either (Sagdeev et al. 1986a).

The suggested low density of the whole nucleus (0.2 g cm^{-3} Rickman 1986, Whipple 1986 for a discussion) could be explained by a matrix of fluffy dust particles filled partly with volatile ice rather than by a conglomerate of ice grains or snow flakes (Whipple 1987) mixed with dust grains as impurities. It will be important to determine the dust-to-gas mass ratio accurately. The range of uncertainty (0.2 to 6, McDonnell et al. 1986, Mazets et al. 1986, Edenhofer et al. 1986) is still large in spite of the in situ measurements. Another open question is the age of the observed topographic features. The layer lost during one revolution of the comet around the Sun can be estimated to be about 6 m (Huebner et al. 1986). The crater should have been visible over the whole time of the recorded appearances of Comet Halley (about 30 apparitions). Topographic features could then date back to the formation of the comet. On the other hand the distribution of the meteor showers correlated with Comet Halley indicates a much longer stay of Comet Halley in its present orbit (Hajduk 1986). The elongated shape (2:1) of the nucleus cannot have formed through erosion by sublimation. It has to be traced back to the creation process of the nucleus, or it is the result of a break-up of a much larger original nucleus. The observed inhomogeneities and the elongated shape suggest that the nucleus is composed of smaller bodies rather than being formed by accretion of small grains.

QUESTIONS FOR THE FUTURE

The shape, appearance, activity, distribution, and extreme darkness of the nucleus of Comet Halley give it an individual character. How typical is this nucleus for the species of cometary nuclei? Which of its properties can be generalized? Part of the answers could be gained by the extended Giotto mission to Comet Grigg-Skjellerup if the spacecraft and camera are still operational. Further important questions concern the physical nature of the nucleus. How much dust is there, what is the average dust (non-volatile)-to-ice (volatile) ratio, is it homogeneous throughout the nucleus? Does the nucleus consist of ice with imbedded dust particles, or do fluffy particles form a matrix with ice within their cavities, or are the grains covered with ice?

Connected to this are the questions of whether the observed mantle is really regolith, and how thick is the crust? What is the density of the nucleus; is it really as low as 0.2 g cm^{-3}?

ACKNOWLEDGEMENT

I am indebted to my colleagues participating in the Halley Multicolour Camera effort, and acknowledge our many discussions and fruitful collaboration leading to the results and their interpretation. R. Kramm and N. Thomas produced most of the images.

REFERENCES

Cosmovici, C.B., Green, Simon, F., Hughes, David, W., Keller, Horst Uwe, Mack, P., Moreno-Insertis, F., & Schmidt, Hermann, U. (1986) Groundbased CCD-Observations of Comet Halley with the Giotto-HMC-filters. *Symposium on the Exploration of Halley's Comet,* ESA-SP 250, 375–379

Divine, Neil, Fechtig, H., Gombosi, T.I., Hanner, M.S., Keller, Horst Uwe, Larson, S.M., Mendis, D.A., Newburn jr., Ray L., Reinhard, Rüdeger, Sekanina, Z., & Yeomans, D.K. (1986) The Comet Halley dust and gas environment. *Space Science Reviews* 43 1–104

Edenhofer, P., Buschert, H., Bird, M.K., Volland, H., Porsche, H., Brenkle, J.P., Kursinsky, E.R., Mottinger, N.A., & Stelzried, C.T. (1986) Dust distribution of Comet Halley from the Giotto radio science experiment. *Symposium on the Exploration of Halley's Comet,* ESA-SP 250, 215–218

Festou, M.C., Drossart, P., Lecacheux, J., Encrenaz, T., Puel, F., & Kohl-Moreira, J.L. (1987) Periodicities in the light curve of P/Halley and the rotation of its nucleus. *Astron. Astrophys.* 187 575–580

Gombosi, T.I., Cravens, T.E., & Nagy, A.F. (1985) Time-dependent dusty gasdynamical flow near cometary nuclei. *Astrophys. J.* 293 328–341

Hajduk, A. (1986) Meteoids from Comet Halley and the comet's mass production and age. *Symposium on the Exploration of Halley's Comet,* ESA-SP 250, 239–243

Hartmann, W.K., Cruikshank, D.P., & Tholen, D.J. (1985) Outer solar system materials: ices and color systematics. In: *Ices in the solar system,* (ed. J. Klinger *et al.*) 169–181, Dordrecht

Hellmich, R. (1981) The influence of the radiation transfer in cometary dust halos on the production rates of gas and dust. *Astron. Astrophys.* 93 341–346

Huebner, Walter F., Delamere, W. Alan, Reitsema, Harold, Keller, Horst Uwe, Wilhelm, Klaus, Whipple, Fred L., & Schmidt, Hermann U. (1986) Dust-gas interaction deduced from multicolour camera observations. *Symposium on the Exploration of Halley's Comet,* ESA-SP 250, 363–364

Julian, W.M. (1987) Free precession of the Comet Halley nucleus. *Nature* 326 57–58

Keller, Horst Uwe, Kramm, Rainer, & Thomas, Nicolas. (1988) Surface features on the nucleus of Comet Halley. *Nature* 331 227–231

Keller, Horst Uwe & Thomas, Nicolas. (1988) On the rotation axis of Comet Halley. *Nature* 333 146–148

Keller, Horst Uwe, Arpigny, Claude, Barbieri, C., Bonnet, C.M., Cazes, S., Coradini, M., Cosmovici, C.B., Delamere, W. Alan, Huebner, Walter F., Hughes, David, W., Jamar, C., Malaise, D., Reitsema, Harold, Schmidt, Hermann U., Schmidt, Wolfgang, K.H., Seige, P., Whipple, Fred L., & Wilhelm, Klaus. (1986a) First Halley multicolour camera imaging results from Giotto. *Nature* 321 320–326

Keller, Horst Uwe, Arpigny, Claude, Barbieri, C., Bonnet, R.M., Cazes, S., Coradini, M., Cosmovici, C.B., Curdt, Werner, Delamere, W. Alan, Huebner, Walter, F., Hughes, David H., Jamar, C., Kramm, Rainer, Malaise, D., Reitsema, Harold, Schmidt, Hermann U., Schmidt, Kurt, Schmidt, Wolfgang K.H., Seige, P., Whipple, F.L., & Wilhelm, Klaus. (1986b) Observations by the Halley multicolour camera. *Symposium on the Exploration of Halley's Comet,* ESA-SP 250, 347–350

Keller, Horst Uwe, Delamere, W. Alan, Huebner, Walter F., Reitsema, Harold, Schmidt, Hermann U., Schmidt, Wolfgang, K.H., Whipple, Fred L., & Wilhelm, Klaus. (1986c) Dust activity of Comet Halley's nucleus. *Symposium on the Exploration of Halley's Comet,* ESA-SP 250, 359–362

Keller, Horst Uwe, Schmidt, Wolfgang K.H., Becker, Christian, Curdt, Werner, Engelhardt, W., Hartwig, H., Kramm, J. Rainer, Meyer, H.J., Schmidt, R., Gliem, F., Krahn, E., Schmidt, H.P., Schwarz, Gottfried, Turner, J.J., Boyries, P., Cazes, S., Angrilli, F., Bianchini G., Fanti, G., Brunnello, P., Delamere, W. Alan, Reitsema, Harold, Jamar, C., & Cucciaro, C. (1987a) The Halley multicolour camera. *J. Phys. E.: Sci. Instrum.* 20 807–820

Keller, Horst Uwe, Delamere, W., Alan, Huebner, Walter F., Reitsema, Harold J., Schmidt, Hermann U., Whipple, Fred L., Wilhelm, Klaus, Curdt, Werner, Kramm, J. Rainer, Thomas, Nicolas, Arpigny, Claude, Barbieri, C., Bonnet, R.M., Cazes, S., Coradini, M., Cosmovici, C.B., Hughes, David W., Jamar, C., Malaise, D., Schmidt, Kurt, Schmidt, Wolfgang K.H., & Seige, P. (1987b) Comet P/Halley's nucleus and its activity. *Astron. Astrophys.* 187 807–823

Kissel, J., Sagdeev, R.Z, Bertaux, J.L., Angarov, V.N., Audouze, J., Blamont, J.E., Büchler, K., Evlanov, E.N., Fechtig, H., Fomenkova, M.N., von Hoerner, H., Inogamov, N.A., Khromov, V.N., Knabe, W., Krueger, F.R., Langevin, Y., Leonas, V.B., Levasseur-Regourd, A.C., Managadze, G.G., Podkolzin, S.N., Shapiro, V.D., Tabaldyev, S.R., & Zubkov, B.V. (1986) Composition of Comet Halley dust particles from VeGa observations. *Nature* 321 280–282

Kitamura, Yoshimi. (1986) Axisymmetric dusty gas jet in the inner coma of a comet. *Icarus* 66 241–257

Kramm, J. Rainer, Möhring, W., & Keller, Horst Uwe. (1987) Image restoration using the point spread function of the Halley multicolour camera. *Geophys. Res. Lett.* **14** 677–680

Krankowsky, D., Lämmerzahl, P., Herrwerth, I., Woweries, J., Eberhard, P., Dolder, U., Herrmann, U., Schulte, W., Berthelier, J.J., Illiano, J.M., Hodges, R.R., & Hoffman, J.H. (1986) *In situ* gas and ion measurements at Comet Halley. *Nature* **321** 326–329

Mazets, E.P., Sagdeev, R.Z., Aptekar, R.L., Golenetskii, S.V., Guryan, Yu.A., Dyachkov, A.V., Ilyinskii, V.N., Panov, V.N., Petrov, V.N., Savvin, A.V., Sokolov, I.A., Frederiks, D.D., Khavenson, N.G., Shapiro, V.D., & Shevchenko, V.I. (1986) Dust in Comet Halley from VeGa observations. *Symposium on the Exploration of Halley's Comet,* ESA-SP 250, 3–10

McDonnell, J.A.M., Kissel, J., Grün, E., Grard, R.J.L., Langevin, Y., Olearczyk, R.E., Perry, C.H., & Zarnecki, J.C. (1986) Giotto's dust impact detection system DIDSY and particulate impact analyser PIA: interim assessment of the dust distribution and properties within the coma. *Symposium on the Exploration of Halley's Comet,* ESA-SP 250, 25–38

Millis, R.L., & Schleicher, D.G. (1986) Rotational period of Comet Halley. *Nature* **324** 646–649

Reitsema, Harold, Delamere, W. Alan, Huebner, Walter F., Keller, Horst Uwe, Schmidt, Wolfgang K.H., & Whipple, Fred L. (1986) Nucleus morphology of Comet Halley. *Symposium on the Exploration of Halley's Comet,* ESA-SP 250, 351–354

Rickman, H. (1986) Masses and densities of Comets Halley and Kopff. *ESA Workshop on Comet Nucleus Sample Return Mission,* ESA SP-249, 195–205

Sagdeev, R.Z., & Szegö, K. (1990: the present volume) Near nuclear region of Comet Halley based on the imaging results of the VeGa mission. Comet Halley 1986—World-Wide Investigations, Results and Interpretations

Sagdeev, R.Z., Szabó, F., Avanesov, G.A., Cruvellier, P., Szabó, L., Szegö, K., Abergel, A., Balazs, A., Barinov, I.V., Bertaux, J.-L., Blamont, J., Detaille, M., Demarelis, E., Dul'nev, G.N., Endröczy, G., Gardos, M., Kanyo, M., Kostenko, V.I., Krasikov, V.A., Nguyen-Trong, T., Nyitrai Z., Reny, I., Rusznyak, P., Shamis, V.A., Smith, B., Sukhanov, K.G., Szabó, F., Szalai, S., Tarnopolsky, V.I., Toth, I., Tsukanova, G., Valnicek, B.I., Varhalmi, L., Zaiko, Yu.K., Zatsepin, S.I., Ziman, Ya.L., Zsenei, M., & Zhukov, B.S. (1986a) Television observations of Comet Halley from VeGa spacecraft. *Nature* **321** 262–266

Sagdeev, R.Z., Krasikov, V.A., Shamis, V.A., Tarnopolski, V.I., Szegö, K., Tóth, I., Smith, B., Larson, S., & Merényi, E. (1986b) Rotation period and spin axis of Comet Halley. *Symposium on the Exploration of Halley's Comet,* ESA-SP 250, 335–338

Schmidt, Wolfgang K.H., Keller, Horst Uwe, Wilhelm, Klaus, Arpigny, Claude, Barbieri, C., Biermann, L., Bonnet, R.M., Cazes, S., Cosmovici, C.B., Delamere, W. Alan, Huebner, Walter F., Hughes, David W., Jamar, C., Malaise, C., Reitsema, Harold, Seige, P., & Whipple, Fred L. (1986) The Giotto Halley multicolour camera. *The Giotto Mission—Its Scientific Investigations,* ESA SP-1077, 149–172

Schwarz, Gottfried, Craubner, Heidelotte, Delamere, W. Alan, Goebel, M., Gonano, M., Huebner, Walter F., Keller, Horst Uwe, Kramm, J., Rainer, Mikusch, E., Reitsema, Harold, Whipple, Fred L., & Wilhelm, Klaus. (1987) Detailed analysis of a surface feature on Comet P/Halley. *Astron. Astrophys.* **187** 847–851

Sekanina, Z. (1987) Nucleus of Comet Halley as a torque-free rigid rotator. *Nature* **325** 326–328

Sekanina, Z., & Larson, S.M. (1984) Coma morphology and dust-emission pattern of periodic Comet Halley. II Nucleus spin vector and modeling of major dust features in 1910. *Astron. J.* **89** 1408–1425

Simpson, J.A., Rabinowitz, D., Tuzzolino, A.J., Ksanfomality, L.V., & Sagdeev, R.Z. (1986) Halley's comet coma dust particle mass spectra, flux distributions and jet structures derived from measurements on the VeGa-1 and VeGa-2 spacecraft. *Symposium on the Exploration of Halley's Comet,* ESA-SP 250, 11–16

Smith, B., Szegö, K., Larson, S., Merényi, E., Tóth, I., Sagdeev, R.Z., Avanesov, G.A., Krasikov, V.A., Shamis, V.A., & Tarnapolsky, V.I. (1986) The spatial distribution of dust jets seen at VeGa-2 flyby. *Symposium on the Exploration of Halley's Comet,* ESA-SP 250, 327–332

Smith, B.A., Larson, S.M., Szegö, K., & Sagdeev, R.Z. (1987) Rejection of a proposed 7.4-day rotation period of the Comet Halley nucleus. *Nature* **326**, 573–574

Stewart, A.I.F. (1987) Pioneer Venus measurements of H, O, and C production in Comet P/Halley near perihelion. *Astron. Astrophys.* **187** 369–374

Thomas, Nicolas, & Keller, Horst Uwe. (1987a) Fine dust structures in the emission of Comet P/Halley observed by the Halley multicolour camera on board Giotto. *Astron. Astrophys.* **187** 843–846

Thomas, Nicolas & Keller, Horst Uwe. (1987b) Comet P/Halley's near-nucleus jet activity. *Symposium on the Diversity and Similarity of Comets,* ESA SP-278, 337–342

Thomas, Nicolas & Keller, Horst Uwe. (1988): Global distribution of dust in the inner coma of comet P/Halley observed by the Halley Multicolour Camera. Proceedings of the Symposium Dust in the Universe, 540–541

Thomas, Nicolas & Keller, Horst Uwe. (1989) The colour of comet P/Halley's nucleus. *Astron. Astrophys.* **213** 487–494

Thomas, Nicolas & Keller, Horst Uwe. (1990) Photometric calibration of the Halley multicolour camera, submitted to Applied Optics

Thomas, Nicolas, Boice, Daniel, Huebner, Walter F., & Keller, Horst Uwe. (1988) Intensity profiles of dust near extended sources on Comet Halley. *Nature* **332** 51–52

Vaisberg, O., Smirnov, V., & Omelchenko, A. (1986) Spatial distribution of low-mass dust particles (m < 10^{-10}g) in Comet Halley coma. *Symposium on the Exploration of Halley's Comet,* ESA-SP 250, 17–23

Whipple, Fred L. (1982) The rotation of comet nuclei. In: *Comets* (ed. L.L. Wilkening,) 227–250, Tucson

Whipple, Fred L. (1986) The cometary nucleus current concepts. *Symposium on the Exploration of Halley's Comet.* ESA-SP 250, 281–288

Whipple, Fred L. (1987) The black heart of Comet Halley. *Sky & Telescope* **73** 242–245

Wilhelm, Klaus. (1986) *HMC in-flight calibrations. Some geometric and radiometric aspects.* Internal Report of MPAE, MPAE–W–55–86–06

Wilhelm, Klaus. (1987) Rotation and precession of Comet Halley. *Nature* **327** 27–30

Wilhelm, Klaus, Cosmovici, C.B., Delamere, Alan W., Huebner, Walter F., Keller, Horst Uwe, Reitsema, Harold, Schmidt, Hermann U., & Whipple, Fred L. (1986) A three-dimensional model of the nucleus of Comet Halley. *Symposium on the Exploration of Halley's Comet,* ESA-SP 250, 367–369

12

The near-nuclear region of Comet Halley, based on the imaging results of the Vega mission

R.Z. Sagdeev and K. Szegö

INTRODUCTION

The 1982 edition of the catalogue of cometary orbits (Marsden 1982) refers to 1109 cometary apparitions of 710 individual comets. Of those, 121 are classified as 'periodic', having periods less than 200 years. The other 589 comets were observed either between 146 BC–AD1758 (104 apparition + 2 apparition of Tempel-Tuttle + several of Halley) or after that time. The number of observed apparitions per year is naturally increasing; it was 1–4/y before 1850, 3–8/y around 1880–1890, and about 15/y nowadays.

Not counting P/Halley and P/Tempel-Tuttle, periodic comets have been registered only in the last 200 years. We know only 78 comets with more than one appearance; but only 11 were observed more than 10 times. At the top of the list is P/Tempel-Tuttle with 52 registered apparitions, and second is P/Halley, with 30 registered apparitions. This exceptional comet was the target of many space missions in March 1986.

One of the most ambitious goals was to image the nucleus itself, never seen before by the human eye. For this reason a camera system (Sagdeev *et al.* 1986a) was put aboard the Vega spacecraft to find, identify and track the nucleus autonomously. Images were taken consecutively with the onboard Ritchey-Chretien telescope of 1200 mm focal length, with various filters. The spectral characteristics of the filters are shown in Fig. 1. The view angle was $15 \times 20 \ 10^{-6}$ rad per pixel. Imaging started two days before the encounter, from a distance of 14 million km, and continued until two days after the encounter; the imaging systems on both Vega spacecraft worked for about 9–10 hours, producing about 1500 images as planned. In this paper we concentrate on about a 20 min period, when the nucleus was resolved.

In the case of Vega-1, there is a set of 63 images 128×128 pixels taken with NIR, VIS, and RED filters. The first image of the set was taken from a distance of 51 290 km, 638 s before closest approach (CA) which took place on 6 March 1986 at 0720:06 UT with a relative velocity of 79.177 km s^{-1}. The relative velocity vector was pointing toward $b = 6.3359$, $l = 355.71$ in our cometocentric coordinate system. (This is the system we shall use in this paper:

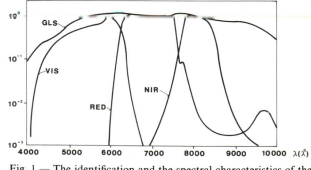

Fig. 1 — The identification and the spectral characteristics of the filters.

its x–y plane is parallel to the ecliptic plane, and the z axis points toward ecliptic north. It is centred on the centre of gravity of the nucleus. The x axis is parallel to the vernal equinox line; the angles b and l are defined in Fig. 2.) The distance between the nucleus and the spacecraft was 8889 km at CA in the $b = 14.6$, $l = 87.4$ direction. The last image was taken from a distance of 42 585 km, 526 s after CA. The exposure time was in the 80 ms–1.28 s range, so that the blurr never exceeded 1 pixel. The Sun was seen from the nucleus in the $b = -1.9556$ $l = 64.242$ direction at CA.

From the data set of Vega-2 images we analyze in this paper 11 images of 512×512 pixel. The first one was taken 370 s before CA, at a distance of 29 540 km from the nucleus, the last one 558 s after CA, from 43 570 km. The CA was at 9 March 1986, 0720:00 UT, when the spacecraft was 8030 km from the nucleus in $b = 1\overline{1}.8$, $l = 84.8$ direction. The relative velocity of the spacecraft was 76.725 km s^{-1} pointing in the $b = 7.1875$, $l = 353.25$ direction. When taking this set of images, the GLS, NIR, VIS, and RED filters were used. Images taken by the GLS filter were intentionally overexposed by a factor of two in comparison to the exposure time defined by the onboard system. As this

set was exposed using a back-up onboard system, the near-nucleus region is well exposed, but there are only two images of this set in which the nucleus, and one more where at least its limb, is well exposed. The Sun was seen from the nucleus in the $b = -0.4874$ $l = 59.657$ direction at CA.

For convenience, we give the CA data for Giotto also. The CA took place on 14 March 1986 at 0003:00 UT when the spacecraft was 605 km from the nucleus; the relative velocity, pointing in the $b = 10.179$, $l = 340.56$ direction, was 68.373 km s^{-1}. The Sun position as seen from the nucleus was $b = 1.5279$, $l = 53.36$ (Keller *et al.* 1986a).

DATA PROCESSING

The data processing procedure has been described in detail by Sagdeev *et al.* (1986b). It made use of the ground calibration data as well as in-flight calibration data obtained with built-in lamps and by the observation of Jupiter and Saturn. Jupiter was observed after the encounter to check also whether the imaging system suffered any damage or not. The results were reassuring. For an evaluation of the ground calibration see Abergel *et al.* (1987) and Sagdeev *et al.* (1986c). The raw data on both spacecraft were contaminated by some coherent noise of unknown origin, which was removed by Fourier filtering. Noise was also removed by a special matrix filtering procedure, see Sagdeev *et al.* (1986b).

The Vega-1 images needed further processing, because of two problems. One is that the offset value of the linear relationship between the incoming light and the outgoing dn signal was shifted from zero to -64 ± 15 dn value (Abergel *et al.*, 1987). The second is that the images are slightly out of focus. The restoration procedure, based on the point spread function derived from Jupiter data, has been completed (Merenyi *et al.*, 1989).

Based on the calibration data, the dn values of the image can be converted to absolute intensity values.

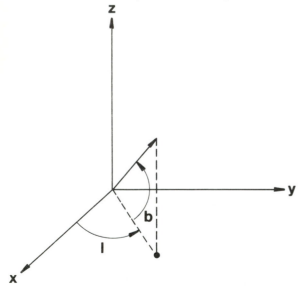

Fig. 2 — The definition of the angles l and b in our cometocentric coordinate system. The x axis is parallel to the vernal equinox line, the x–y plane is parallel to the ecliptic plane, and z points to ecliptic north. The angle b is measured from the x–y plane, the angle l from the x axis, counterclockwise.

RESULTS

The scientific objectives of the imaging experiment concerning the near nuclear region were to find out the existence, visibility, size, surface structure, albedo, photometry, activity pattern, spatial orientation, and rotation of the nucleus, and to obtain all possible information from the images, combined with other results where appropriate. These objectives are interrelated; an answer to one problem helps one to find an answer to another.

The results summarized below will certainly be improved later. However, the general character proposed for the nucleus is not likely to change much from the conclusion announced on 11 March (5 days after Vega-1, 2 days after Vega-2 flyby): that P/Halley is a consolidated body of irregular shape, with very low albedo. The main *sizes* are shown in Fig. 3.

The *albedo* was obtained by using the Jupiter data as baseline. The behaviour of the scattering indicatrix of the brightest spot on Vega-1 images was similar to the scattering indicatrix of the Moon. Since the mean brightness factor of the Vega-1 images at $32°$ phase angle was 0.02, the geometric albedo is 0.04 ($+0.02$, -0.01). No differences were seen between images taken with different filters.

The *visibility* of the nucleus is certainly different for a camera inside the coma than from the ground,

A - A [km] = 7.5 ± 0.8
B - B [km] = 8.2 ≏ 0.8
C - C [km] = 16.0 ± 1.8

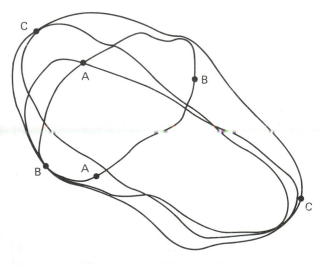

Fig. 3 — The main sizes of the nucleus, in three perpendicular directions. On top right the distances are given between the points of the drawing denoted by the same letter.

because of the inhomogeneity of the coma. Close to the nucleus cone-like and sheet-like dusty jets and fairly dust-free surface regions can be distinguished, hence the amount of dust in the radial direction is different from what would follow from the $1/r^2$ law. From the ground the whole dust envelope containing the curved and dispersed jets is seen; so the conditions for ground-based observation are much less favourable even if distance would allow us to resolve the nucleus. As can be seen, for example from Fig. 6, strong dust jets screen both some part of the surface and of the limb for the onboard camera also.

Our first concern was to find out the optical thickness of the dust over the jet-free regions. We found (Sagdeev *et al.* 1985) that at these places the dust layer over the surface is optically thin; its optical thickness is less than 0.1. The reasoning is as follows. It was assumed that the nucleus scatters light in the same way as the Moon does (but with its own albedo), and the dust above it forms a flat, homogeneous layer. In this layer the dust single-scattering albedo was chosen to be $a = 0.5$, which is the minimal value for large ($2\pi r/\lambda \gg 1$) particles contributing most to the brightness measured. Multiple scattering in this layer was also taken into account. Then a cross-section of the surface was selected, where the cross-section is almost a semicircle. The photometric profile of this configuration with different optical thickness values was calculated and compared with the data. The phase dependence of the brightest spot of this configuration was also checked against the data. Only small, <0.1 optical thicknesses could be fitted. Though measurements later revealed that the scattering properties of the surface are somewhere between a Moonlike and a Lambertian surface (B.A. Smith; personal communication), the above results are upheld.

One of the most challenging tasks has been the *3-dimensional reconstruction* of the nucleus shape, its orientation in space, and the 3-d structures of the surrounding jets. The Vega results are unique in this sense, since, owing to the different strategy, the Giotto data gave two-dimensional imaging information, with better resolution. The two missions were complementary.

The most straightforward task was the reconstruction of the *3-d jet structure*, for the Vega-2 flyby

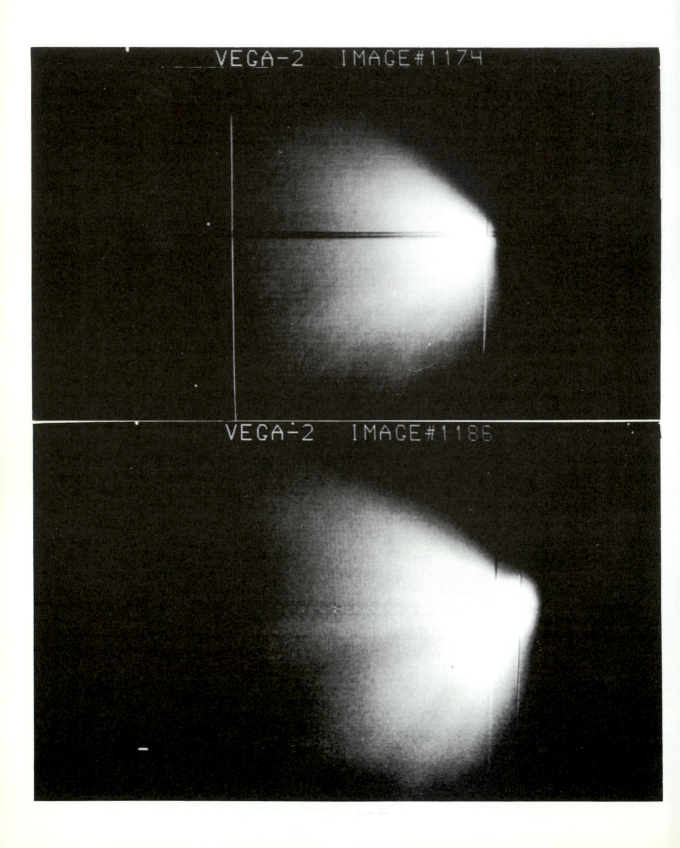

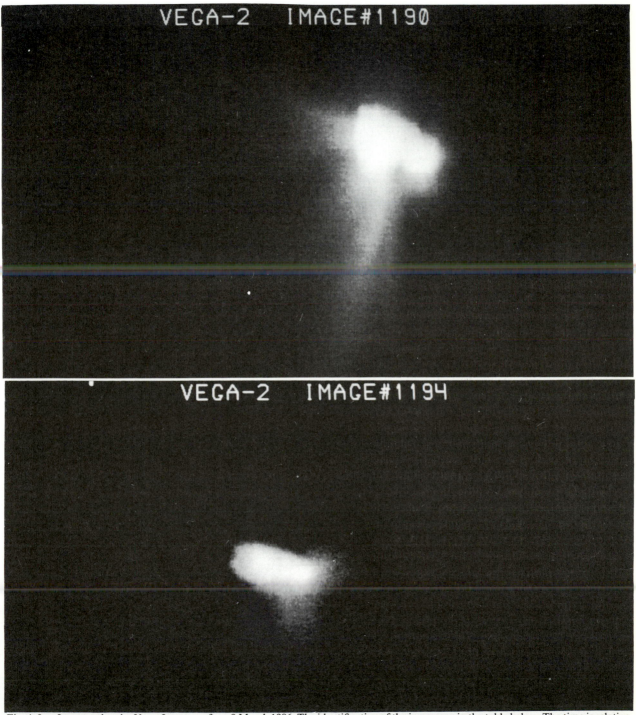

Fig. 4–9 — Images taken by Vega-2 spacecraft on 9 March 1986. The identification of the images are in the table below. The time is relative to CA in seconds. The direction is the camera position as seen from the nucleus. The phase angle is the angle between the Sun–comet–spacecraft. The Sun angle means the angle measured counterclockwise between the vertical direction and the projection of the nucleus–Sun direction onto the image plane.

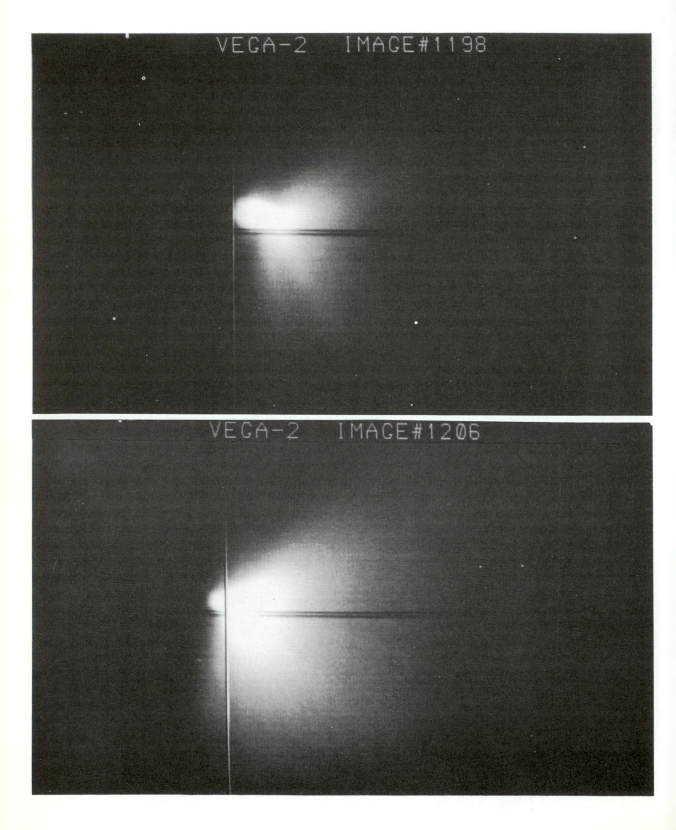

employed 11 full-format images. Six of these are shown in Figs 4–9. The concept of the reconstruction is based on the fact that each jet or jet complex can be thought of as an organized set of spatial vectors. On any given image, the vector sets are seen projected onto the image plane, which is always normal to the spacecraft-comet vector. Owing to the mechanical construction of the spacecraft and its three-axis stabilization, the horizontal (row) direction of the images is aligned parallel to the ecliptic plane. Thus, measuring the position of the same feature on different images, and knowing the changing spatial relationship between the spacecraft and the comet, the absolute spatial orientation of the feature can be derived. So the problem lies in identifying the same features on different images. To enhance the core and boundaries of the jet features, a shift differentiation algorithm (Larson and Sekanina 1984) was used. We also iterated between features to reach unambiguous identification. The most viable solutions are shown in Fig. 12, as they would appear on a sphere centred around the nucleus. For convenience, the spacecraft trajectory and the recording of the images are also exhibited. Most of the sources form a large linear feature crossing local noon. This linear feature on the nucleus surface is about the 'waist' of image 1190, cf. Fig. 6, or somewhat to the left. This follows, for example, from the comparison of Figs 6 and 7, in which the jet activity is seen on different sides of the images. There is, however, one source which is on the night side, unless the irregular shape of the nucleus does not allow it to be lit. The results show that the spacecraft crossed one of the jets; this was confirmed by the SP-1 dust experiment (Vaisberg *et al.* 1986) and by the general appearance of Fig. 7, which was taken about that time. Individual jet sources at this stage have not been identified unambiguously on the Vega-1 images. This is partly due to their small length on the images.

If we are able to identify the projection of the long axis of the nucleus on the images, the algorithm yields the orientation of the long axis. The procedure is difficult because the nucleus is only partly lit by the Sun. The dark side over the terminator can be identified only if we make some assumption about the shape of the nucleus. In the first approximation it was considered to be a cone, capped at each end by hemispheres (avocado-like shape). Then we iterated between the data so long as the solution obtained met all the constraints. The result is that the long axis (if it is oriented from the small end toward the big end) at the Vega-1 CA pointed in the direction $b = 15°$, $l = 79°$; at the Vega-2 CA in the $b = 9°$, $l = 310°$ direction. The error is about 5 degrees for the l values and 2 degrees for the b values.

The next step toward the 3-d reconstruction of the nucleus was the 3-d reconstruction of the *light intensity distribution* during the Vega-1 flyby (Sagdeev *et al.* 1986d). Some of the 63 flyby images are shown in Figs 10–11. These are noise filtered, but not sharpened images. The scattered light comes partly from the nucleus, partly from the surrounding dust. The gas can be neglected in this respect. Though the principle is similar to the reconstruction used in medical tomography, there are non-negligible differences. One is that the scattering angle changes very much from frame to frame, but our knowledge of the dust scattering indicatrix is poor. The second is that the coma does not accommodate well with the intersections of the view angle of the different images; the coma's dust extends beyond the intersection. The results—the isodense surfaces in the inner coma within a $64 \times 64 \times 75$ km^3 cube—are shown in Fig. 13. The intensity levels are also indicated. Each level contains contributions from the nucleus and from the jets; the part of the nucleus beyond the terminator may be underestimated. On the sharpened images the illuminated limb and terminator can be identified

Image no.	Time	Direction		Distance from nucl.	Phase angle	Sun angle	Filter
		l	b				
1174	−370	157.8	−3.7	29 540	98.4	89.0	NIR
1186	−102	128.6	3.4	11 200	69.3	91.8	GLS
1190	−1.5	85.6	11.7	8 030	8.7	113.3	NIR
1194	98.7	40.4	13.6	11 040	23.5	234.7	VIS
1198	187	22.5	12.1	16 460	38.0	253.7	RED
1206	373	8.8	10.1	29 700	51.0	261.1	NIR

Table 1 — Technical data for images shown in Figs. 4-9.

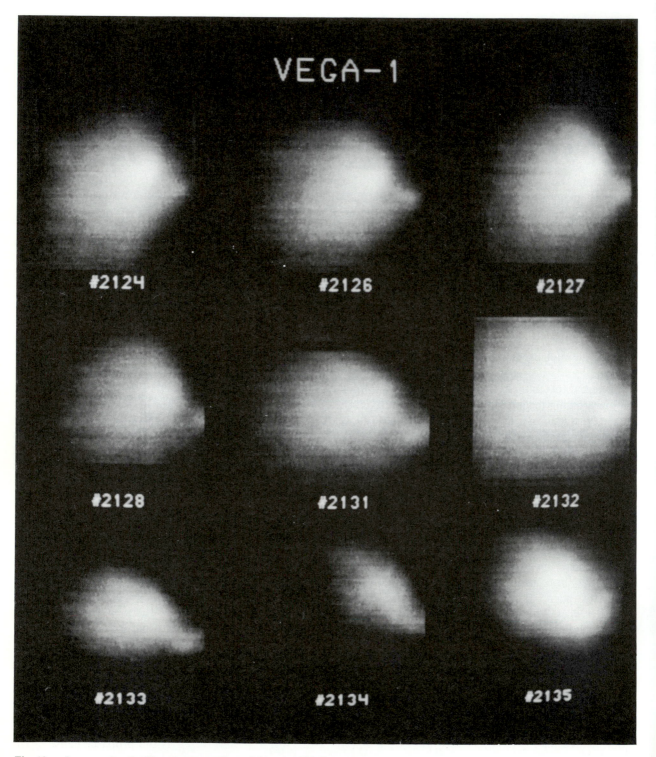

Fig. 10 — Images taken by Vega-1 spacecraft on 6 March 1986. Data as for Fig. 4–9.

Table 2 — Technical data for images shown in Fig. 10.

Image no.	Time	Direction		Distance from nucl.	Phase angle	Sun angle	Filter
		l	b				
2124	−215.6	149.2	1.1	19246	85.2	92.0	VIS
2126	−171.6	143.7	2.6	16 236	79.7	92.4	NIR
2127	−158.6	141.7	3.2	15 385	77.7	92.7	VIS
2128	−134.6	137.3	4.4	13 878	73.3	93.3	RED
2131	−83.7	124.2	7.8	11 088	60.3	96.6	RED
2132	−70.6	120.0	8.9	10 501	56.0	98.3	NIR
2133	−53.6	113.2	10.4	9 850	49.7	101.3	VIS
2134	−23.6	99.5	13.0	9 083	36.9	110.3	RED
2135	−6.6	90.8	14.2	8 904	29.3	119.1	NIR

within 2 pixel error.

In the first approximation the *shape* of the nucleus has also been reconstructed (Sagdeev *et al*. 1986d). Here, however, we are facing real difficulties. First, the limb and the terminator (the whole nucleus and its lighted part) should be identified in each image. As all of our images were exposed with a phase angle smaller than 90° (in Giotto the phase angle was 107°) the dust coma does not help us to identify dark limbs. The offset problem of the Vega-1 camera mentioned above also works against us. The definition of the limb in such a situation cannot be reduced to a simple mathematical algorithm; we have to combine different methods with the imagination of the experimenter.

One mathematical method of determining the limb is the algorithm of gradients (Sagdeev *et al*. 1986b) which fits a curve through those points of the image where the change of the light intensity is maximal. A second one is the algorithm of texture (Sagdeev *et al*. 1986b). This enhances the difference between small-scale variation of the signals on the surface relative to the large-scale background variation which is associated with the nucleus as a whole. The combination of these two methods (sometimes applied together with the algorithm of second derivatives to find the maximal curvature of the light intensity change) gives a fairly good estimate for the bright limb and for the terminator, provided that the dust jets do not screen them too much. The position of the dark limb, or that part of the limb which is screened by jets on certain images can, however, be estimated only by understanding the flyby geometry and by iterating among the images. This procedure at present gives about 1

km error in limb determination.

Another method which is less sensitive to the limb definition (Szegö *et al*. 1987) uses a zero order shape of the nucleus (in our case the avocado shape mentioned before) and a polyhedron is built around it. The polyhedron is also lit by the Sun, and the brightness of the polyhedron is compared to the observed brightness of the images; the shape of the polyhedron (and also the computed orientation of the long axis) is corrected accordingly. The results of this analysis (Merenyi *et al*., 1989) confirmed that the relative position of the small and large end of the nucleus is correctly understood.

The figure of the nucleus is shown in Fig. 14. The volume of this figure is about 365 km³ and its surface about 350 km².

The near-nucleus *coma photometry* is not only interesting in itself, but it helps to identify the limbs screened by dust. Therefore, light distribution was measured along jets and along jet-free regions. To analyze the results, the modification of the Haser model was necessary, because the viewing geometry is very different than from the ground. Constant outflow velocity, also, cannot be supposed. Though an overall $1/R$ behaviour canbe fitted both to the jets and to the jet-free coma, the analysis (Toth *et al*. 1987) revealed two distinct coma regions. The boundary is about 30–40 km from the nucleus. Though several explanations are possible, at this stage we assume that dust grain disintegration takes place in the close regions. It is very likely that jet sources are pristine ice; hence dust particles receive a heat shock after leaving the surface. The heat shock is the probable cause of the disintegration (Szegö *et al*. 1989a).

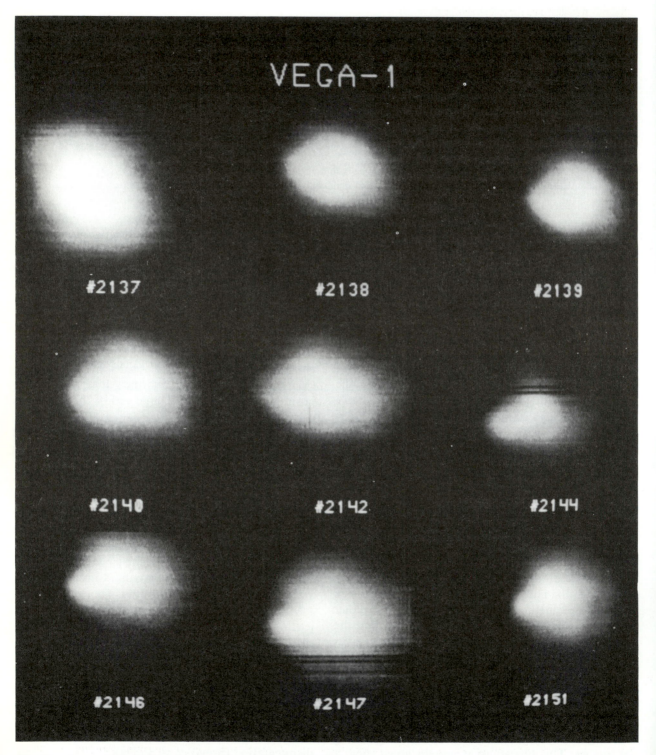

Fig. 11 — Images taken by Vega-1 spacecraft on 6 March 1986. Data as for Figs 4–9.

Table 3 — Technical data for images shown in Fig. 11.

Image no.	Time	Direction l	b	Distance from nucl.	Phase angle	Sun angle	Filter
2137	34.4	69.8	15.9	9 297	15.2	161.5	RED
2138	58.4	58.9	16.0	10 020	15.0	195.9	NIR
2139	66.4	55.6	15.9	10 327	16.1	205.3	VIS
2140	90.4	47.1	15.4	11 413	21.4	225.2	RED
2142	125.4	37.5	14.5	13 237	29.0	240.0	VIS
2144	170.4	28.9	13.4	16 157	36.5	248.8	NIR
2146	211.3	23.4	12.5	18 945	41.5	253.1	RED
2147	228.4	21.6	12.1	20 151	43.2	254.3	NIR
2151	295.4	16.2	11.1	25 021	48.2	257.6	VIS

Preliminary steps have been taken to identify *surface structure* on the nucleus (Sagdeev *et al.* 1986b, Mohlmann *et al.* 1986, Szegö *et al.*, 1989b), but much work needs to be done before conclusions can be reached. The *rotation* of the nucleus is discussed in the next section.

DISCUSSION

The measurements aboard the space missions to Comet Halley and the coordinated ground-based observations dramatically enhanced and in many respects drastically changed our knowledge of comets. In this part of the paper we discuss how the imaging results contributed to a modified (or new) model of the cometary nucleus and its relation to other data. A section will also be devoted to the problem of the rotation period.

One surprise was the very low albedo value, also confirmed by the Giotto measurements (Keller *et al.* 1986b). There are very few objects in the solar system

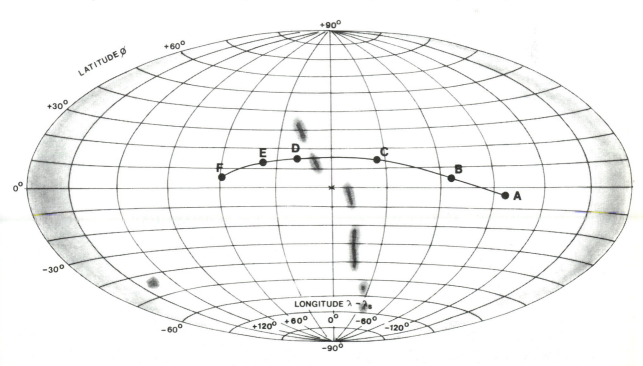

Fig. 12 — Jet sources, as reconstructed from Vega-2 images on a sphere centred around the nucleus. The spacecraft path projected onto this sphere, and the recording of the images of Figs 4–9 is also denoted by the dots A–F. The identified jet sources are shaded on the sphere.

with such a low albedo, e.g. the Uranus ring particles, Iapetus (satellite of Saturn), Amalthea (satellite of Jupiter); though this albedo value is not exceptional for comets (Hartmann *et al.* 1987). Though the amount of carbon found in the dust was higher than expected (Kissel *et al.* 1986), this alone can hardly explain the surface darkness. The fluffy structure of the surface material, as suggested by Greenberg & Grim (1986) should contribute considerably to the low reflectiveness. This surface structure also explains the low heat conductivity which follows from the general activity pattern. In a surprising analysis Rickman (1986) claimed that the whole density of the nucleus is 0.1–0.3 g/cm³, assuming ~500 km² for the volume; far less than it was thought. In a recent analysis based also on the PUMA experiment data a more realistic 0.6 g/cm³ was obtained (Sagdeev *et al.* 1987a). These are consistent with a recent idea of Whipple (1987) that the nucleus resembles a dirty snowdrift that somehow has not become compressed.

The imaging experiments revealed that surface activity with respect to gas and dust production is very non-uniform. Model calculations (Horanyi *et al.* 1984) for the thermodynamics of a nucleus covered by dust, repeated with the current parameters learnt after the missions, concluded that the surface temperature of

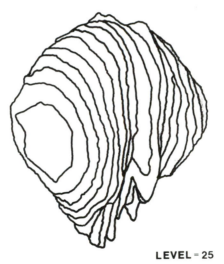

LEVEL = 25

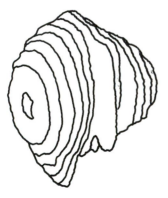

LEVEL = 50

LEVEL = 150

Fig. 13 — Light intensity distribution during the Vega-1 flyby around the nucleus. The levels are dn values as indicated.

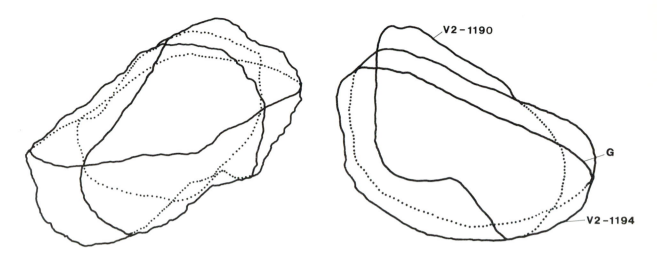

Fig. 14 — The three dimensional model of Halley's nucleus from three different viewing angles.

the inactive layers should be about 400 K for a dust mantle thicker than 1 cm. This model predicts considerably lower temperatures for the active parts to account for the observed gas production. The surface temperature was measured by IKS on Vega-1 (Emerich *et al.* 1986). It was found that the surface temperature is between 320 and 400 K for a surface smaller than the visible cross-section of the nucleus. We take this result literally, i.e. we assume that IKS measured the temperature of the inactive surface which is smaller than the visible one. The question arises whether there are two types of surfaces (Whipple 1986) or a few-centimetre thin mantle.

Moore *et al.* (1983) considered whether cosmic ray irradiated cometary type ice mixtures can form such an inactive layer. Various ice mixtures containing H_2O, NH_3, CH_4, N_2, CO, CO_2 etc. were irradiated around 20 K. All experiments confirmed the synthesis of new species, and approximately 1% of the mixture was converted into non-volatile residue. As 100 MeV protons penetrate a few centimetres into water ice, high-energy cosmic ray irradiation may lead to the formation of a metre or more thick inactive layer after the first few apparitions, unless thermal stresses destroy it. We need to study further the surface structure on the Vega images to reach a conclusion; but we tend to assume the two-type surface model for the nucleus, since it is difficult to conceive that an extended dust layer about one or few cm thick can stabilize itself to such a degree that surface structures

can form. In this context we also note that the centrifugal escape velocity at the long ends is of the same order of magnitude as that which follows from the actual angular velocity. (Better estimates are difficult, but it is not a bad approximation that the nucleus is as long as it can be.)

The very marked variability of the near-nuclear coma region was also a new observation. The analysis of the 1910 apparition of Comet Halley (Sekanina & Larson 1986) had already called attention to the variability of dust jets on a time scale of days. The present observations added many new facts. Sunward-tailward asymmetry was generally observed in brightness features.

Vega-1 data analysis (Sagdeev *et al.* 1986a) showed doubled coma radiance in a 24 h interval. Using a high-speed and very sensitive data acquisition system (Baumbaugh *et al.* 1986), very short term ($\frac{1}{4}$ to 40 s), drastic changes and some others varying by hours were reported. No quantitative data were given to characterize the intensity of the changes. Sterken *et al.* (1986) found large-amplitude night-by-night changes in the brightness of the comet in a range of about 2 magnitudes. It was noted that when these bursts occurred, they showed up in all emission lines. Short-term fluctuations more than 0.1 mag in the time interval of 6 h were also reported by them. The same erratic trends were also observed by Larson *et al.* (1986) in the velocity of water ions. By analyzing the shift of the spectral lines preperihelion and postperihelion, they deduced -0.21 ± 0.12 km s^{-1} postperihelion and -0.18 ± 0.15 km s^{-1} preperihelion velocities. From the spread of the line profiles a different, 0.9 ± 0.2 kms^{-1} and 1.4 ± 0.2 kms^{-1} outflow velocity was deduced for the preperihelion and postperihelion, respectively. They found that the velocity distribution function in the coma varies.

The dust composition changed as a function of distance around CA (Kissel 1986). CN jets were observed during March and April without seeing average-type accompanying dust jets. Though the Lyman-alpha observation aboard Suisei proved the periodicity of the jet activity, the individual activity pattern is quite irregular (Kaneda *et al.* 1986). This list could have been easily continued. The conclusion is that even the active surfaces are different, probably both in composition and activity pattern. Jet sources can be silent and be reactivated later on. These phenomena can qualitatively be understood as the result of a quickly changing thin dust layer over the active surface; however, much work on modelling is necessary for quantitative results.

To account for the observed gas production, the active areas should produce on the average 10^{18} gas molecule cm^{-2} s^{-1} at 1 AU post-perihelion. If we adopt 0.3 for the dust/gas mass ratio and $0.5 \times$ g cm^{-3} for the density, the average surface loss is about $\frac{1}{3}$ cm per hour which is quite considerable. The temperatures of the active zones were not measured directly. In the abovementioned model (Horanyi *et al.* 1984) the mantle/core temperature should be higher than 210 K to produce enough gas, and a $\frac{1}{3}$ cm thick mantle can diminish the gas output by an order of magnitude. Comparing this with the average surface loss value, we tend to believe that around preperihelion and shortly after, the active areas are almost nude (which does not exclude very thin dust partly falling back from jets or simply being left behind).

The long linear dust source observed by Vega-2 is around the 'waist' of the nucleus as seen in Fig. 6. If the two-type surface pattern were true, the dust sources would populate the surface only partly, and the nucleus would be eroded more heavily there. The visual observations of the images do not contradict that; i.e. that the irregular shape is partly due to the inhomogeneous surface activity. However, this conclusion depends very much on the number of

apparitions, whether the jet sources change from apparition to apparition, on the strength of the sources, etc. A more detailed analysis of the situation has been published by Szegö (1988).

During and after the 20th ESLAB Symposium, opinions differed as to what is the rotation period of the nucleus. Many papers were submitted which published periods close to the 2.2 d period first reported by Larson & Sekanina (1984). M.J.S. Belton *et al.* (1986), by analyzing the brightness variation of the nucleus probably during a period when it was inactive (when it was further than 5 AU from Sun) proposed a period of 54 h 07 min or 53 h 58 min $\pm$ 2 min. Celnik (1986) analyzing tail events from 17 Feb. to 17 April 1986, found variations which were consistent with an 53.5 ± 1.7 h rotation of the nucleus. In the near-nucleus zone photometric measurements were made by Lebowitz & Brosch using filters centred around the wavelengths 4845, 5140, 6840, and 7000×10^{-10} m during November and December 1985. Their results revealed periodic light variations with a 52 h period, interpreted as the effect of the rotation of the nucleus.

The flight missions also concluded that the nucleus rotates with that period. From the ultraviolet features as observed by the Lyman-alpha of Suisei spacecraft from 14 Nov. 1985 to 11 Jan. 1986 and later from 9 Feb to 14 April (with an interruption between 1–10 March) a strong breathing of Comet Halley was revealed (Kaneda *et al.* 1986) with a period of 2.2 d (52.9 h) identified with the rotation of the nucleus. Sagdeev *et al.*, analyzing the Vega spacecraft imaging data, identified the spatial orientation of the major axis of the nucleus during the Vega-1 and Vega-2 encounters, and deduced a 53.5 ± 1 h rotation period. This result made it possible to pinpoint the orientation of the nucleus at the Giotto CA as the $b = -32°$, $l = 287°$ direction. It is consistent with the Giotto observation. In Fig. 15 the orientations of the major axis are summarized for the different missions.

There were observations in which the marked variability of Halley also showed up. Cochran & Barker (1986) reported spectrophotometric observation of Halley during the period December 1984 through June 1986. In that period the brightness was quite variable on timescales of a day, with variations of a magnitude or more. However, they did not find

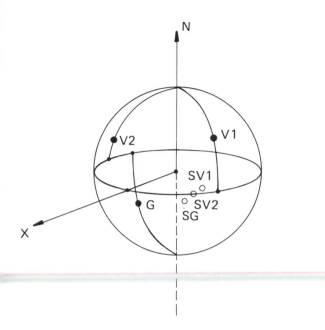

Fig. 15 — The orientation of the long axis of the nucleus and the Sun position during the encounters. V stands for VEGA, G for Giotto, S for Sun. See the text for the exact values.

any correlation in their data with a 52–53 h rotation period.

Papers have also been published in which an explicit 7.4 d rotation period was deduced. Millis & Schleicher (1986) observed the comet during two runs, March 4–18 and March 29–April 19, with filters for OH, NH, CN, C_3, CO+ and C_2 along with continuum regions. The data were reduced by the Haser model. The authors applied the method of phase dispersion minimization to find the periodicity.

Samarasinha & Klinglesmith (1986) analyzed the phase of CN gas jets during 16–29 April as they appeared on the enhanced images. The published 27 images exhibit one, two, or three jets. To phase them, the authors selected one, with definite arbitrariness. However, it is strange that one jet was always active during that period, whereas the other two do not show a clear switching on/off pattern.

To connect the jet phases with the nucleus rotation, one also has to assume that the jets originate during them. But if they were evaporated from dust jets as is generally assumed, and taking into account the difference in ion and dust motion, it is not trivial to

consider how these data can be related to the nucleus motion. We have to admit, however, that the 'rephasing' of the images for the 2.2 d period does not make them easier to understand.

We have to note however, that it is possible to accommodate the two rotation periods. A symmetric top, rotating along its long axis with 7.4 d period, and the whole system precessing around an axis inclining by 80 degrees to the first, makes possible a 2.2 d precession period. This was first proposed by Sekanina (1987) and later elaborated in detail by Julian (1987). However, based on the imaging data, this option was refuted by Smith *et al.* (1987).

Very interesting data were published by Festou *et al.* (1986), since they used the longest time base, 13 months. Applying their own corrections to the apparent brightness, they found a 7.4 d period in their data. So the case for two periods seems well documented in the groundbased observations.

Using both the groundbased and spacecraft observations we reanalyzed the rotation of the nucleus, Sagdeev *et al.* (1988). Since it is an irregular body, its rotation is the rotation of an asymmetric top. Based on the stability of the 7.4 d period we accepted that the motion is free, and we assumed that the nucleus is rigid. We also assumed that the orientation of the angular momentum vector is not far from one of the inertial axes of the nucleus, since in that case the rotation of an asymmetric top can be solved in closed form.

We accepted from the observations the existence of a short, 2.2 d and a long, 7.4 d period. The major dimensions, 16 km for the longest and 7 km for the shortest axis, were taken from the Vega observations. We identified the position of the long axis with the position of the shortest inertial momentum axis, which is reasonable from the existing data. Then, using the orientation of the long axis during the different encounters (as given earlier) we concluded that the nucleus of P/Halley rotates as a slightly asymmetric top. The orientation of the rotation vector, i.e. the orientation of the angular momentum vector, is $b = -54°, l = 39°$ in the ecliptic system. The error zone is a cone about this axis with a half opening angle of $15°$.

In the rotation of an asymmetric top the rotation axis is not fixed rigidly to the body, so while the nucleus rotates about the axis with a short period of 2.2 $\pm$ 0.05 d, its long axis 'nods' periodically with the long period of 7.4 $\pm$ 0.05 d. The amplitude of the 'nodding' is about 14° $\pm$ 3° in both directions relative to a plane perpendicular to the rotation axis. The rotation is a so-called short axis mode.

REFERENCES

Abergel, A., et al. (1987) *Applied optics,* **26** 4457, 1987

Baumbaugh, A.E., et al. (1986) preprint, *Fermi Nat. Acc. Lab.* FN-441 2100.001

Belton, M.J.S., et al. (1986) *20th ESLAB Symposium on the Exploration of Halley's Comet,* ESA SP-250, **1** 599

Catalogue of cometary orbits (1982) Marsden, B.G. ed. Cambridge

Celnik, W.E. (1986) *20th ESLAB Symposium on the Exploration of Halley's Comet,* ESA SP-250, **1** 53

Cochran, A.L., & Barker, E.S. (1986) *20th ESLAB Symposium on the Exploration of Halley's Comet* ESA SP-250 **1** 439

Emerich, C., et al. (1986) *20th Symposium on the Exploration of Halley's Comet,* ESA SP-250 **II** 381

Festou, M., et al. (1986) Proc. of the Symposium on the Diversity and Similarity of Comets (Brussels) (1987) *Astron. and Astrophys.* **169** 336

Greenberg, J.M., & Grim, R. (1986) *20th ESLAB Symposium on the Exploration of Halley's Comet,* ESA SP-250, **II** 255

Hartmann, W.K., et al. (1987) *Icarus,* **69** 33

Horanyi, M., et al. (1984) *The Astrophys. J.* **278** 449

Johnson, R.E., et al. (1986) *20th ESLAB Symposium on the Exploration of Halley's Comet,* ESA SP-250, **II** 269

Julian, W.H. (1987) *Nature,* **326** 57

Kaneda, E. et al. (1986) *20th ESLAB Symposium on the Exploration of Halley's Comet* ESA SP-250, **I** 397

Keller, H.U., et al. (1986a) *20th ESLAB Symposium on the Exploration of Halley's Comet,* ESA SP-250, **II** 347

Keller, H.U., et al. (1986b) *Nature,* **321** 320

Kissel, J. (1986) talk presented at the *12th Symposium of the 26th COSPAR Conference (Toulouse)*

Kissel, J., et al. (1986) *Nature,* **321** 280

Larson, S. (private communication)

Larson S., & Sekanina, Z. (1984) *Astron. J.* **89** 571

Larson, H.P., et al. (1986) *20th ESLAB Symposium on the Exploration of Halley's Comet,* ESA SP-250 **I** 445

Lebowitz, E.M., & Brosch, N. (1986) *20th ESLAB Symposium on the Exploration of Halley's Comet,* ESA SP-250 **1** 605

Merenyi, E. et al., (1989) KFKI 1989 35/C

Millis, R.L., & Schleicher, D.G. (1986) *Nature,* **324** 646

Mohlmann, D., et al. (1986) *20th ESLAB Symposium on the Exploration of Halley's Comet,* ESA SP-250, **II** 339

Moore, M.H., et al. (1983) *Icarus,* **54** 388

Reitsema, H.J., et al. (1986) *20th ESLAB Symposium on the Exploration of Halley's Comet,* ESA SP-250 **II** 351

Rickman, H. (1986) *Proc. of the Nucleus Sample Return Mission Workshop (Canterbury)* ESA SP-249, 149

Sagdeev, R.Z., et al. (1985) Comet Halley: nucleus and jets (results of the VEGA mission), *Adv. Space Res.* **5, 12** 95–104

Sagdeev, R.Z., et al. (1986) *20th ESLAB Symposium on the Exploration of Halley's Comet,* ESA SP-250, **2** 335

Sagdeev, R.Z., et al. (1986a) *20th ESLAB Symposium on the Exploration of Halley's Comet;* ESA SP-250, **II** 289

Sagdeev, R.Z., et al. (1986b) *20th ESLAB Symposium on the Exploration of Halley's Comet;* ESA SP-250, **II** 295

Sagdeev, R.Z., et al. (1986c) preprint, *Space Research Institute, Acad. of Sci. USSR,* Pr-1084

Sagdeev, R.Z., et al. (1986d) *20th ESLAB Symposium on the Exploration of Halley's Comet,* ESA SP-250, **II** 307

Sagdeev, R.Z., et al. (1986e) *20th ESLAB Symposium on the Exploration of Halley's Comet,* ESA SP-250, **II** 317

Sagdeev, R.Z., et al. (1987a) paper given at *Conference on the Diversity and Similarity of Comets,* Brussels (1987)

Sagdeev, R.Z., et al. (1988) The Astron. J., **97,** 546

Samarasinha, N.H., & Klinglesmith, D. (1986) *20th ESLAB Symposium on the Exploration of Halley's Comet,* ESA SP-250 **1** 487

Sekanina, Z., & Larson, S.M. (1986) *Astron. J.* **92** 462

Sekanina, Z. (1987) *Nature,* **326**

Smith, B., et al. (1986) *20th ESLAB Symposium on the Exploration of Halley's Comet,* ESA SP-250, **II** 327

Smith, B.A., et al. (1987) *Nature,* **326** 573

Sterken, C., et al. (1986) *20th ESLAB Symposium on the Exploration of Halley's Comet,* ESA SP-250 **I** 445

Szegö, K., et al. (1987) *Proceeding of the Symposium on the Diversity and Similarity of Comets, Belgium,* 6–9 April ESA SP-278, 463

Szegö, K. et al., (1989a) *Adv. Space Res.,* **9,** 89

Szegö, K. et al., (1989b) *Adv. Space Res.,* **9,** 85

Szegö, K. (1988) *Astrophys. and Space Sci.,* **144,** 439

Toth, I., Szegö, K., & Kondor, A. (1987) *Proceedings of the Symposium on the Diversity and Similarity of Comets, Brussels, Belgium,* 6–9 April ESA SP-278, 343

Vaisberg, O.L., et al. (1986) *Nature,* **321** 274

Whipple, F.L. (1986) *20th Symposium on the Exploration of Halley's Comet,* ESA SP-250, **II** 281

Whipple, F.L. (1987) *Sky and Telescope* 242

13

The mass of the nucleus of Comet Halley as estimated from nongravitational effects on the orbital motion

Hans Rickman

1 INTRODUCTION

Before the Comet Halley exploration no details were known about the structure of cometary nuclei. Even such bulk properties as the size and albedo of the Halley nucleus, as evidenced by spacecraft imaging (Keller *et al.* 1986, Sagdeev *et al.* 1986), came largely as a surprise. The widely held picture of cometary material as a two-component mixture of frozen gases and refractory grains of chondritic composition is also challenged by a diverse set of recent discoveries such as the CHON particles (Kissel *et al.* 1986a,b) and the cyano-jets (A'Hearn *et al.* 1986).

So far as the question of cometary origin is concerned, a highly diagnostic observation might be offered by an estimate of the bulk density of the Halley nucleus. Generally speaking, porous structures of low density are expected from remote sites of formation in transplanetary zones of the solar nebula, while higher accretional velocities in regions closer to the Sun might imply the formation of compact, high-density objects. Hughes (1985) estimated that a density of 0.5 g cm^{-3} together with a volume roughly estimated from the photometric cross-section, would yield a reasonable value for the mass of the nucleus, as compared with the mass loss deduced from the 1910 observations and indications about the secular behaviour of the comet from the brightness evolution and the mass of the associated meteor stream.

However, the situation regarding observational data has of course been revolutionized, and in particular a fairly reliable determination of the volume is now available. The interest of reassessing the mass determination for the cometary nucleus is thus obvious. Such an attempt was indeed made by Rickman (1986) and recently also by Sagdeev *et al.* (1988), in both cases based on the nongravitational effects in the orbital motion. Let us now review the theoretical basis and observational material used for these mass determinations in order to assess their uncertainties and arrive at an authoritative judgement on the most likely range for the density of the nucleus of Comet Halley. Such a review was recently published (Rickman 1989), and we refer to this for details.

2 NONGRAVITATIONAL MASS DETERMINATION

Following Whipple (1950), we identify the nongravitational force as a jet force acting on the nucleus as a consequence of the anisotropic outgassing. The absolute value of this force may be written:

$$F = \zeta v m Q \qquad (1)$$

where v is the typical speed of a molecule leaving the surface of the nucleus, and ζ is a dimensionless coefficient defined so that ζv is the average momentum per unit mass transferred to the nucleus. Q is the total production rate of molecules of average mass m. In the framework of our present understanding of cometary outgassing we may imagine two basic mechanisms of escape of gas molecules: direct evaporation or diffusion from subsurface layers. The relative importance of these mechanisms will be discussed in section 4, but v may generally be taken as the 'thermal speed' characterizing the point of escape from the nucleus at local temperature T:

$$v = \sqrt{\frac{8kT}{\pi m}} \qquad (2)$$

where k is Boltzmann's constant.

The momentum transfer coefficient ζ has two obvious contributions: $\zeta = \zeta_l \zeta_g$, where ζ_l characterizes the local scatter of outflow directions of single molecules from the same spot, and ζ_g describes the global distribution of outgassing flux around the nucleus. This is a generalized formalism which coincides with that of our earlier work (Rickman 1986), but let us note that Sagdeev et al. (1988) used instead of ζv the notation $K\bar{u}$ where K corresponds to our ζ_g and $\bar{u} = \zeta_l v$ is the local outflow velocity of the gas. This outflow velocity is perhaps the most meaningful physical quantity to discuss, and is highly significant for the assessment of our modelling uncertainty. We will thus pay special attention to the choice of ζ_l in section 4. As far as ζ_g is concerned, it is important to note that on many occasions the gas production from Comet Halley should be dominated by one active spot, and thus the instantaneous value of ζ_g is often near unity. However, as the nucleus rotates (precesses, nutates . . .) the location of the active spot changes with respect to the subsolar point, and the effective nongravitational force can be considered to arise from a time average during this motion. Each spot is thus replaced by a distribution of activity along its trajectory, and from a set of such distributions corresponding to the various active spots we can derive both the effective value of ζ_g and the net outgassing direction.

From numerical models of the thermal processes in the surface layers of the nucleus (heat flow, sublimation, absorption, and emission of radiation) one can obtain the flux distribution of gaseous outflow along

with the temperature distribution characterizing the outflow velocity at each point. Thus one can actually compute the force vector $\mathbf{e}$ in the framework of a certain set of modelling assumptions. This is of course equivalent to a computation of ζ_l, ζ_g, v, m, and Q, where v is to be interpreted as the average speed of all the sublimating molecules corresponding to an average sublimation temperature according to equation (2). This technique was introduced by Rickman & Froeschlé (1983a,b), and the models so far employed are essentially characterized by:

(1) chemical homogeneity, i.e., sublimation of ice occurring only at the surface;

(2) smooth surface structure and simple geometrical shape—in particular spheres or oblate spheroids in pure rotation, where the gaseous outflow is everywhere perpendicular to the surface;

(3) free molecular outflow, i.e., no further interactions between the gas in the cometary atmosphere and the nucleus—in combination with the smooth surface structure this means that we may deal with a local outflow pattern which is isotropic over the available hemisphere (hence, $\zeta_l = \frac{1}{2}$);

(4) sublimation fed only by directly absorbed sunlight, i.e., no account of additional heat sources inside the nucleus or multiple scattering and thermal emission from the dust jets;

(5) no lateral transport of heat, i.e. for the case of heat conduction, neglect of the horizontal temperature gradients in comparison with the vertical ones.

In the framework of these assumptions the models can be considered to represent spotwise activity even though the nucleus is only subdivided latitudinally into a number of strips parallel to the equator. Each latitude strip can actually be given an arbitrary percentage of activity, representing any situation from complete dust coverage (zero activity) to free sublimation at all points (full activity). It is important to note that even though Q can be computed as a function of orbital position from such models, an acceptable fit to the observed gas production curve of Comet Halley appears impossible. Thus for the mass determination we use the observed curve $Q(t)$.

Let us express the nongravitational force vector thus computed in terms of its components in the radial and transverse directions in the orbital plane. Using the polar angle η with respect to the radial axis and the

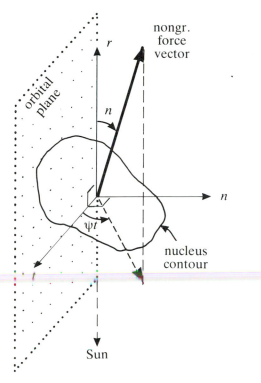

Fig. 1. — A view of the cometary nucleus with the direction to the Sun (down) and a section of the orbital plane indicated. With the aid of this a coordinate system is shown with axes marked 'r' (radial) and 't' (transverse) in the orbital plane and 'n' (normal) perpendicular to this plane. The orientation of the nongravitational force vector is thus described by means of a polar angle η counted from the radial axis and an azimuthal angle ψ_t counted from the transverse meridian.

azimuthal angle ψ_t measured from the transverse direction in the orbital plane (Fig. 1), we can write the nongravitational acceleration components:

$$j_r = FM^{-1} \cos \eta \atop j_t = FM^{-1} \sin \eta \cos \psi_t \Big\} \qquad (3)$$

where M is the mass of the cometary nucleus. j_r and j_t can be inserted into the Gauss equations (see e.g. McCuskey 1963) in order to compute the corresponding rate of change of any function of the orbital semimajor axis (a) and eccentricity (e). In particular, the major nongravitational effect observed in the orbital motions of comets—the delay or advance at perihelion passage—is obtained from the perturbation of the orbital period P by integrating over one orbital revolution. Thus we obtain:

$$\Delta P = \frac{6\pi\sqrt{1-e^2}\,m\langle \zeta v \rangle}{n^2 M} \left\{ \frac{e}{p} \int_0^p Q \cos \eta \sin f \, dt \right.$$

$$\left. + \int_0^p \frac{Q}{r} \sin \eta \cos \psi_t \, dt \right\} \qquad (4)$$

where n is the mean motion, $p = a(1-e^2)$ is the semilatus rectum of the orbit, r is heliocentric distance, f is true anomaly and t is time. N.B.: we have dropped ζv from both integrals and replaced it by a common orbital average $\langle \zeta v \rangle$. This would of course be incorrect if ζv would vary drastically with orbital position. According to the modelling results referred to above, however, that is not the case: equation (4) should indeed be a very good approximation. It will be used to compute M, taking all the other parameters from a combination of observational and modelling data, as described in the following section.

Let us emphasize an important aspect of equation (4) that was not often noticed in earlier work i.e., before (Rickman 1986). Owing to the factor $\sin f$ in the first (radial) term within the brackets there is a cancellation effect reducing the influence of this term, but the cancellation is not complete unless $Q\cos\eta$ is perfectly symmetric around perihelion. Since η is usually envisaged as a small angle, even a moderate perihelion asymmetry of the gas production rate may have a strong influence on the nongravitational effect, which thus cannot be solely ascribed to a rotational thermal lag.

3 OBSERVATIONAL AND MODELLING DATA

In order to evaluate the integrals in equation (4) one should know the variations of Q, η, and ψ_t with orbital position. Unfortunately only $Q(t)$ can be determined directly from observations—for the other quantities we have to rely on circumstantial evidence and theoretical arguments.

From the recent apparition of Comet Halley we have a wealth of data on the function $Q(t)$, and in particular there is very good evidence that the gas production from the nucleus is strongly dominated by H_2O. Krankowsky et al. (1986) found from Giotto-NMS data that H_2O contributes at least 80% of the parent molecules. The remainder appeared to be mainly CO with a production rate of 10–20% that of H_2O (Festou *et al.* 1986, Woods *et al.* 1986). However, further analysis of the NMS results (Eberhardt *et al.* 1986) showed that most of this CO is produced from a distributed source in the coma, so that only 5% or less of the molecules sublimating from the nucleus might be CO. For the earlier mass determination, Rickman (1986) added 15% of CO to the observed H_2O production rates in view of the preliminary results, but let us now first take only the water molecules into account. We thus have: $m = 18$ amu, and $Q \equiv Q_{H_2O}$.

Fig. 2 shows a plot of $Q_{H_2O}(t)$ with data points and ranges from various investigations using ultraviolet emissions of H and OH, the forbidden line of $O[^1D]$ at 630 nm, and near-infrared bands of H_2O. A local scatter due to activity variations connected with the rotation of the nucleus is obvious mainly in the IUE and Suisei data sets. This scatter is not of major concern for the definition of a mean production curve. The most serious problem is rather the discrepancy between the different data sets, as already recognized by Sagdeev *et al.* (1988). A reasonable though uncertain fit by a continuous mean production curve is possible by restricting attention to the OH(309 nm) data, as shown in Fig. 2 by the full-drawn curve fitted to the open circles. The Ly-α data are neither continuous, nor extensive enough to define an independent production curve, but it is seen to deviate systematically from the OH curve in February–March 1986 shortly after perihelion passage. On an earlier occasion hydrogen production rates from the Ly-α line observed by the IUE in Comet P/Crommelin gave

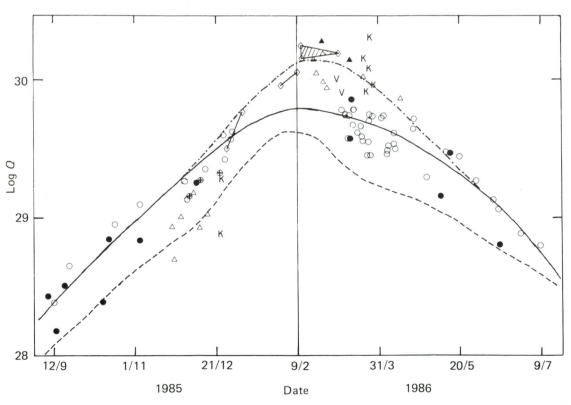

Fig. 2. — A summary of observations of H_2O production rate (*Q*) for Comet Halley during September 1985–July 1986. Open circles (○) denote IUE observations (Feldman *et al.* 1986); filled circles (●) denote observations of the $O[^1D]$ emission at 630 nm (Spinrad *et al.* 1986); crossed circles denote OH observations by the Soviet UV orbital telescope 'Astron' (Boyarchuk *et al.* 1986). Open triangles (△) denote Suisei observations (Kaneda *et al.* 1986), and filled triangles (▲) denote sounding rocket Ly-α observations (McCoy *et al.* 1986). Open diamonds (◇) delimit domains covered by Pioneer Venus observations (Stewart 1987). 'K' indicates Kuiper Airborne Observatory observations (Weaver *et al.* 1986); 'V' indicates the results from the two VeGa probes (Combes *et al.* 1986, Krasnopolsky *et al.* 1986); 'G' denotes the Giotto observation (Krankowsky *et al.* 1986). The fitted curves are explained in the text.

water production rates considerably exceeding $Q(OH)$ for some time after perihelion in 1984, and these rates were interpreted as arising from the dissociation of other molecules than H_2O (Wallis & Carey 1985). However, one should be aware of specific problems relating to the interpretation of cometary Ly-α brightnesses (see e.g. Festou *et al.* 1985), and even though the geocoronal problem does not affect the present Pioneer Venus or Suisei data, the g-factor is uncertain since it depends on the variable flux at the centre of the solar Ly-α line. This may explain why the trend is not unique in the present material: on the preperihelion branch the Suisei data points fall below the OH curve.

Near-IR measurements by the VeGa spacecraft and the Kuiper Airborne Observatory are in agreement with the Ly-α data after perihelion, but again the KAO observations before perihelion give remarkably low production rates. The Giotto-NMS measurement of actual H_2O number density in the coma is in very good agreement with the OH curve. Furthermore, the $O[^1D]$ data follow closely the variation of the OH production rate. Observations of the OH 18-cm lines were performed continually and led to production rates over the whole period covered by Fig. 2 (Gérard *et al.* 1986, Schloerb *et al.* 1986, Mirabel *et al.* 1986). A mean production curve resulting from these observations is indicated by the dashed curve, but this has to be regarded as highly uncertain owing to a large scatter of individual measurements and possible inconsistencies between different observers. The shape of the curve agrees fairly well with the corresponding IUE data wherever available, and in addition a near-perihelion peak is indicated where IUE data are absent owing to the solar conjunction. As to the different levels of the two OH curves, the IUE data set is considered as the most reliable one. Recently, however, Bockelée-Morvan *et al.* (1989) have analyzed the radio data and found evidence for a variation of the OH parent expansion velocity with heliocentric distance and gas production rate. Taking this into account in the IUE reductions means a change of slope of the production curve, and the true curve may thus have a somewhat larger post-perihelion excess than the full-drawn or dashed curves in Fig. 2.

The visual light curve (Green & Morris 1986) provides another nearly continuous data set, but it cannot be used to resolve the question of the shape of the gas production curve. Whether interpreted by means of Festou's (1986) correlation or Newburn's (1981) photometric theory, it does not correlate well with the data in Fig. 2. This may be due to shortcomings of the modelling of the influence on the visual brightness by dust coma scattering (Fischer 1986), or destruction of visible grains by sublimation of organic constituents such as POM (Huebner *et al.* 1987) as the comet comes close enough to the Sun.

When modelling the direction of the nongravitational force in terms of $\eta(t)$ and $\psi_t(t)$, the basic observational data are the spin properties of the nucleus. Even though the question of the actual state of rotation of the nucleus of Comet Halley is not yet settled (e.g. Sekanina 1986, Julian 1987, Smith *et al.* 1987, Wilhelm 1987, Festou *et al.* 1987, Kamél 1988), there seems to be little doubt that the angular momentum axis makes an angle of typically 20–30° with the orbital plane and corresponds to a spin period of slightly more than two days. This axis is close to that of maximum moment of inertia.

As shown by Sekanina (1986), the coexistence of two photometric periods of 2.2 d and 7.4 d can be explained if this spin is actually a precession of the long body axis at an angle of 77° with 2.2 d period while the nucleus rotates about its long axis with a 7.4 d period. Such wobbling is likely to be forced by the nongravitational torque on the strongly non-spherical nucleus. It has the interesting consequence that most of the active spots will describe complicated rotational trajectories with respect to the subsolar point covering most latitudes nearly uniformly (see Rickman 1989). Thus the thermal lag of Comet Halley, on the average, is likely to occur in directions near the orbital plane. In these circumstances it is a fair approximation to identify η with the thermal lag angle and to consider small absolute values (< 20–$30°$) for ψ_1. Let us thus treat $\cos\psi_t$ as a constant with an expected value of approximately 0.9.

The thermal lag angle varies with orbital position in a way which is fairly easy to model. It increases with heliocentric distance, qualitatively because heat flow has an increasing influence on the thermal regime as sublimation buffering of the surface temperature is reduced (Froeschlé & Rickman 1986), and quantitatively according to the approximate relation:

$$n \simeq 45° \frac{x^2 r^2}{1.25 + x^2 r^2} \qquad (5)$$

where r is to be expressed in AU and x is a constant dimensionless parameter that depends on spin period P_s and thermal inertia I_t (Rickman et al. 1987). The actual value of x for any particular comet is unknown since we have no accurate estimates of the thermal inertia. But an order-of-magnitude estimate of I_t can be made since the thermal conductivity and heat capacity data for compact H_2O ice given by Klinger (1981) leads to: $I_t \simeq 2000$ Ws m^{-2}K^{-1}. The more or less porous ice–dust mixture envisaged as a typical cometary material should have a value of I_t in the range of 100–1000 Ws m^{-2}K^{-1} (Weissman & Kieffer 1981, Rickman & Froeschlé 1983a). In combination with $P_s \simeq 2$ d the results of Rickman et al. (1987) then show that x should be in the range of 0.17–0.47.

We can obviously compute the mass M as a function of x from equation (4), given a certain production curve $Q(t)$ and the above estimate of $<\cos\psi_t> = 0.9$, if we also have a good estimate of $<\zeta v>$. From the above-described thermal model calculations of the nongravitational force using $\zeta_1 = \frac{1}{2}$ we find: $<\zeta v> = 0.16 + 0.1q$ km s^{-1} as a function of the perihelion distance q (AU) (Rickman et al. 1987), which gives $<\zeta v> \simeq 0.17$ km s^{-1} for Halley's Comet.

As a general comment on the function $M(x)$, let us note that $x = 0$ would correspond to a limiting case with no thermal lag (i.e., outgassing always in the solar direction), hence the nongravitational effect would arise solely from the radial term in equation (4). This means that ΔP would be a measure of the perihelion asymmetry of the gas production curve, and the observed positive value for Halley's Comet: $\Delta P = 4.1$ d (Kiang 1972, Yeomans & Kiang 1981) would require an excess of gas production after perihelion passage. The smaller this excess, the lower the mass of the nucleus would have to be. As expected from Fig. 2, the uncertainty of the actual shape of the gas production curve renders the limiting mass $M(0)$ relatively uncertain, but with increasing values of x an ever more important contribution to ΔP comes from the transverse term in equation (4). Thus M increases with x, and at the same time it becomes less sensitive to the perihelion asymmetry of the gas production curve.

4 RESULT AND DISCUSSION

Fig. 3 shows the result of a computation of $M(x)$ using the three different gas production curves indicated in Fig. 2. The full-drawn and dashed curves were described above, and the dot-dashed curve has been fitted to the IUE data both before perihelion and on the outer part of the post-perihelion branch, but was forced to agree with the Ly-α data in between. It does not give a particularly good fit to either of the data sets on the whole—it is aimed only at estimating the maximum mass that can reasonably be expected on the basis of these observations. From the slope of each of the curves in Fig. 3 we can conclude that the uncertainty of x would prevent an accurate mass determination even if the gas production curve were known in detail. However, the major uncertainty of our mass determination arises from the choice of gas production curve.

If we trust our maximum-mass curve, then the result comes out to be $(2.5 \pm 0.4) \times 10^{14}$ kg; if we take the IUE production curve, we find $(1.0 \pm 0.3) \times 10^{14}$ kg; and if we base our estimate on the radio-OH data, we get $(0.4 \pm 0.1) \times 10^{14}$ kg. The latter curve, however, surely gives an underestimate being based on reductions which underestimated the quenching of the OH maser in the inner coma (see Schloerb 1989), so we will pay no further attention to the corresponding mass estimate. These results are insensitive to the average value chosen for $<\cos\psi_t>$ and remain virtually identical for $<\cos\psi_t> = 0.8$ or 1.0. Dividing by an estimated volume of 550 km^3 for the nucleus (Keller 1987), we might thus conservatively state that the density of the nucleus is around 0.6 g cm^{-3} at a maximum. This is already a much sharper limit than that found by Sagdeev et al. (1988), evidently because these authors gave their maximum curve a more distinct perihelion asymmetry since they did not consider the near-perihelion Pioneer Venus data. Furthermore, in view of the above mentioned uncertainties of the Ly-α hydrogen production rates, the maximum-mass curve indeed appears highly unreliable.

The above error bars on the mass estimate reflect

Fig. 17* (Lamy, Chapter 1, p. 13) — Comet Halley photographed on 12–13 March 1986 by Dr. William Liller, stationed on Easter Island as part of the International Halley Watch Large-Scale Phenomena Network's (L-SPN) 'Island Network'. The exposure time was 7 minutes commencing at 1150:00 UT, on Fujichrome 400 ISO film using an 8-inch f/1.5 Celestron Schmidt. The bluish plasma tail and yellow, curving dust tail are well shown.

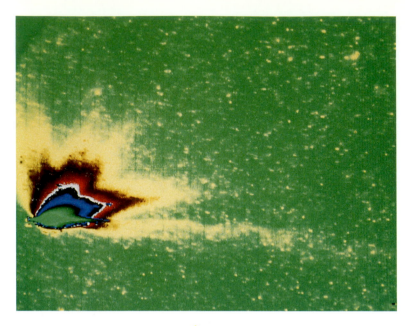

Fig 7* (Lamy, Chapter 1, p. 6) — After perihelion passage, Halley's comet developed a broad, fan-shaped dust tail, highly structured by bright streamers 1° to 2° long and spread over a sector angle of more than 120°, as shown in the false-colour isophote picture here, showing the comet as it appeared on 1986 February 22.4. (Photograph courtesy of the European Southern Observatory.)

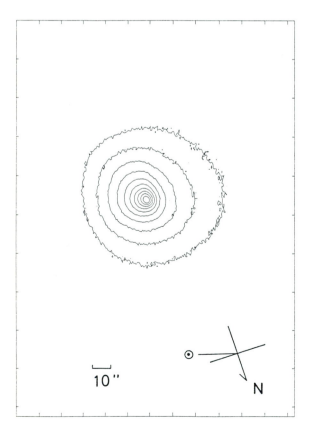

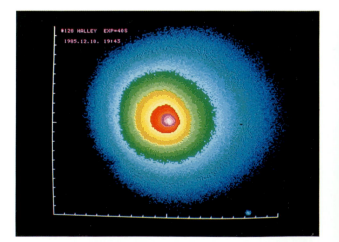

Fig. 2a, 2b* (Watanabe, Chapter 7, p. 83) — A CCD image of comet Halley taken at 1043 UT on 10 December 1985, obtained with the 1.8-metre telescope at Okayama Astrophysical Observatory. The colour photograph (Fig. 2a* at right) shows the central part of the image on the computer display. Alongside, to the same scale, is a contour map (Fig. 2b*) of the near-nucleus image shown in the photograph. The contour level step is 0.25 magnitude. The field of view is 2.'5 × 3.'5; the apparent motion of the comet is leftward. ⊙ and N on the contour map show the sunward and north pole directions. The inner contour has a fairly round shape.

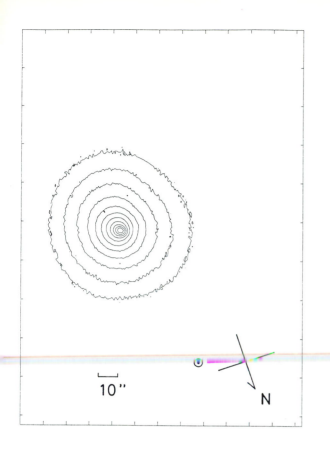

10 "

N

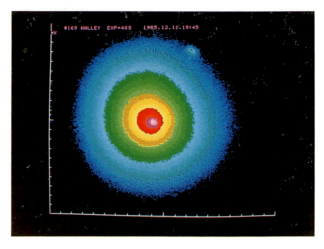

Fig. 3a, 3b* (Watanabe, Chapter 7, p. 83) — A CCD image of comet Halley taken at 1045 UT on 11 December 1985, obtained with the 1.8-metre telescope at Okayama Astrophysical Observatory. A star image which appears at upper right on the colour photograph (Fig. 3a*) has been removed from the corresponding contour map (Fig. 3b*). Other details are as given in caption for Fig. 2*. Note the inner contour of Fig. 3b* still has a fairly round shape.

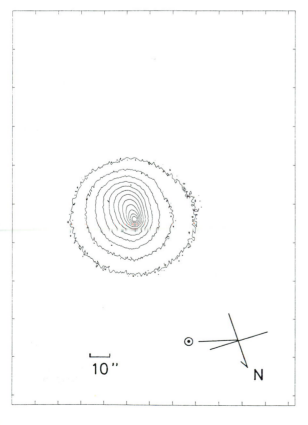

10 "

N

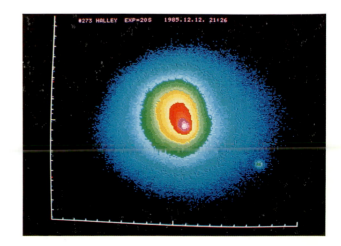

Fig. 4a, 4b* (Watanabe, Chapter 7, p. 83) — A CCD image of comet Halley taken at 1226 UT on 12 December 1985, obtained with the 1.8-metre telescope at Okayama Astrophysical Observatory. A star image which appears at lower right on the colour photograph (Fig. 4a*) has been removed from the corresponding contour map (Fig. 4b*). Other details are as given in caption for Fig. 2*. Note the inner contour of Fig. 4b* shows an elongated jet feature.

Fig. 3* (Greenberg, Chapter 9, p. 106) — A piece of a fluffy comet: Model of an aggregate of 100 average interstellar dust particles each of which consists of a silicate core, an organic refractory inner mantle, and an outer mantle of predominantly water ice in which are embedded the numerous very small (< 0.01 μm) particles responsible for the interstellar 216 nm absorption and the far ultraviolet extinction. Each particle as represented corresponds to an interstellar grain 0.5 μm thick and about 1.5 μm long. The mean mantle thickness corresponds in reality to a size distribution of thicknesses starting from zero. The packing factor of the particles is about 0.2 (80% empty space) and leads to a mean mass density of 0.3 g cm^{-3} and an aggregate diameter of 5 μm.

HMC COMPOSITE IMAGE

IMAGES 3457,3475,3480,3491,3496,3500

Front cover illustration and colour reproduction of Fig. 3 (Keller, Chapter 11, p. 136) — This image of the nucleus of comet Halley is composed of six images taken by the Halley Multicolour Camera during the encounter of the Giotto spacecraft on 13 March 1986. Images #3475, #3480, #3491, #3496 and #3500 replace parts of the outermost image #3457. The images were taken at 14 420, 9510, 8150, 5150, 3800 and 2730 km from the nucleus. The resolution improves from 325 to 60 m per pixel while the brightest part of the image is approached. The Sun is 30° above the horizontal on the left and between 15° and 12° below the image plane. Images are radiometrically and geometrically corrected. Noise is removed and the PSF is deconvolved. The image is displayed in pseudo colours. (Images courtesy Max-Planck-Institut für Aeronomie.)

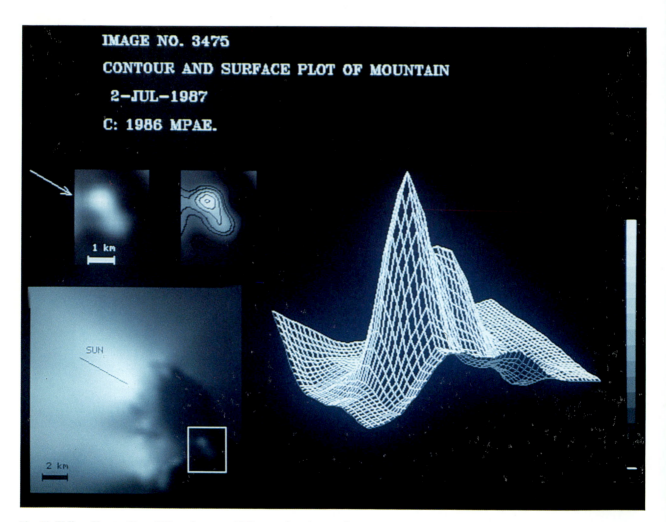

Fig. 5* (Keller, Chapter 11, p. 136) — Image #3475 was taken from a distance of 9510 km, yielding a resolution of 210 m per pixel. The Sun is 14° behind the image plane. The cut-out indicated by the white box has been magnified by two and contrast stretched to produce the display on the upper left. To the right of it is the same image with isophotes. On the right is an intensity surface plot of the cut-out, the viewing angle being approximately from the sunward direction as shown by the arrow. The cut-out is of the area around the mountain, and it includes the upper part of the central depression. Note the very steep gradient on the northern tip. (Image courtesy Max-Planck-Institut für Aeronomie.)

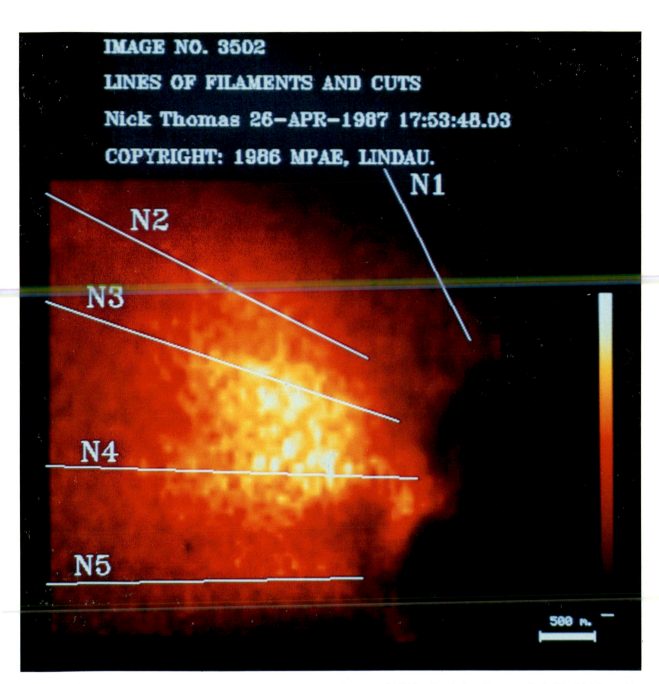

Fig. 13* (Keller, Chapter 11, p. 141) — Image #3502 was taken from a distance of 2195 km (resolution 50 m per pixel). The high intensities are stretched to enhance the contrast. The five dust filaments N1–N5 are numbered as in Fig. 12 (see page 141). The filament N4 emanates from the source on the surface. Note, however, that the bright spot to the right and below this does not appear to be related to a specific dust activity (from Thomas & Keller 1987b). (Image courtesy Max-Planck-Institute für Aeronomie.)

Fig. 14* (Keller, Chapter 11, p. 142) — Image #3461 (for further details, see caption of Fig. 2 on page 135). The isophotes are labelled in units of mW m^{-2} sr^{-1}. The increase in intensity on the right side of the nucleus (green colour) towards the northern tip is obvious. The strongest dust emission points south of the solar direction. The intensity values of this figure are based on the preliminary calibration of HMC (Wilhelm, 1986) and have to be multiplied by a factor of 1.24 to reflect the present calibration standard (see Thomas & Keller 1987b). (Image courtesy Max-Planck-Institut für Aeronomie.)

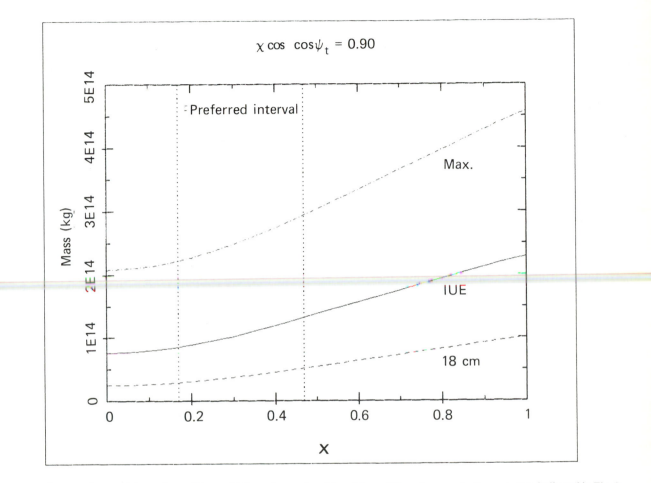

$$\chi \cos \cos\psi_t = 0.90$$

Fig. 3. — Computed mass of the nucleus of Comet Halley using $\cos\psi_t \equiv 0.9$ and three different gas production curves as indicated in Fig. 2. The preferred interval of 0.17–0.47 for x is shown by vertical dotted lines.

only the uncertainty of the x parameter, and including also a reasonable estimate of the coma modelling uncertainty affecting the IUE production curve of OH, we may state the mass of the nucleus as $(1.0 \pm 0.5) \times 10^{14}$ kg. Inclusion of other volatiles may increase this by up to 20% (Rickman 1989), so a better estimate is $(1.2 \pm 0.5) \times 10^{14}$ kg. The maximum estimate likewise becomes $(2.8 \pm 0.6) \times 10^{14}$ kg. The most likely density is 0.22 g cm^{-3}, and its probable error slightly exceeds 50% or 0.1 g cm^{-3} since the uncertainty of the volume estimate is also involved. Let us now pay attention to the modelling uncertainty associated with the momentum transfer factor $<\zeta v>$.

First, concerning the mechanism of gas production the model applied in this analysis is based on free

sublimation from the surface of the nucleus. The alternative picture of subsurface sublimation has not been considered. There are two basic variants of the latter picture, both of which may be discarded for different reasons. A very thin and porous dust layer would present no obstacle to the gas flow, and this would influence the production rate only by its thermal insulation (Mendis & Brin 1977). This would have implications still to be explored for the thermal lag angle, but the dust layer in question appears unstable as Comet Halley passes inside the Earth's orbit, and so should not be important during the most active period of the comet. An alternative is a thin but relatively compact dust layer where gas molecules penetrate by diffusion, colliding many times with the constituent grains (Fanale & Salvail 1984, Rickman & Fernández 1986). Such a layer may be stable during the perihelion passage of Halley's Comet (Rickman & Fernández 1986, Ip & Rickman 1986), and it would modify the present picture significantly by increasing the surface temperature T and thus the speed of escape of the gas molecules according to equation (2). However, the gas flux is drastically reduced by such a layer, and there is no question that a predominant part of Comet Halley's gas production must come from a non-diffusive regime prevailing in the active spots.

Second, let us recognize that the assumption of free molecular outflow is not justified for an active comet close to the Sun (Gombosi et al. 1986, and references therein), and that in future modelling this short-coming has to be remedied. However, based on hydro-dynamic calculations Crifo (1987) states that the local jet force contribution is indeed given by:

$$(\mathrm{d}F/\mathrm{d}A)_l = \zeta_l \, v \, m \, Z_f \qquad (6)$$

where $\mathrm{d}F/\mathrm{d}A$ is the force per unit surface area and Z_f is the sublimation flux of free molecular outflow. Furthermore, the Mach number of the initial gas flow should be in the range 0.75–1, implying that ζ_l is in the range 0.53–0.67, just above the value used here. A qualitative explanation of this difference is that the hydrodynamic approach allows for some return flow and recondensation on the surface of the nucleus. Let us also note that such low values of ζ_l as here discussed imply that there is no strong collimation of the gaseous outflow, and that this expectation seems verified by the coma analyses performed during the spacecraft encounters with Halley's Comet: the dust jets emanat-ing from the nucleus are not manifestations of well-confined gas jets. We thus conclude that the best estimate of $<\zeta v>$ is 1.2 times the value here employed, and by direct proportionality the resulting mass and density should be likewise corrected. The most likely value of the density is hence 0.26 ± 0.15 g cm^{-3}, and the maximum value is 0.6 ± 0.3 g cm^{-3}.

One further point concerning ζ_l has to be comment-ed upon. Sagdeev et al. (1988), following Strelnitsky et al. (1987), use a value of $\zeta_l = \pi/4$ for the case of free molecular outflow. This arises from an incorrect averaging to compute the momentum flux of the gas leaving the nucleus. Their averaging is based on a

Maxwellian velocity distribution of the gas molecules at some distance from the nucleus surface, but under the assumption of free molecular outflow the true distribution is heavily biased toward small vertical velocities. They hence overestimate the efficiency of the momentum transfer and the density of the nucleus by a factor $\pi/2$. A recent observational verification came from the work of Colom et al. (1989), who analyzed the 18 cm line profiles and deduced a velocity offset of ~ 0.17 km s^{-1} for the parent H$_2$O molecules with respect to the nucleus, in perfect agreement with the value here derived.

5 IMPLICATIONS FOR COMETARY ORIGIN

The density of the nucleus of Halley's Comet appears to be between 0.1 and 0.4 g cm^{-3}, with a most likely value of 0.26 g cm^{-3}. It is possible to extend these limits by invoking extreme combinations of correc-tions to the nominal (IUE) gas production curve and the volume, but 1 g cm^{-3} appears as a tentative lower limit. Unfortunately it is not yet possible to deduce accurate information about the formation circum-stances of the nucleus from this result. However, remote sites of formation are clearly indicated, and the transplanetary region appears much more likely than any planetary accretion zone, since very low velocities are required.

The mechanism of formation of cometary nuclei is not yet known. Random accretion leading to a fractal structure has been computer-simulated by Daniels & Hughes (1981) and further discussed in the context of cometary formation by Donn & Hughes (1986). This mechanism leads naturally to a high porosity (see Lang 1987), but if it was operative all the way to final accretion of cometary nuclei from cometesimals of $> \sim 100$ m size, then according to Donn (1987) the gravitational acceleration should have produced com-paction to densities of 0.4–0.5 g cm^{-3}. It thus seems that a density below 0.4 g cm^{-3} for Halley's Comet is not consistent with this picture.

More theoretical and laboratory work is undoubted-ly needed to give a definitive answer to the question whether km-sized bodies could be expected to have formed with densities of only 0.2 g cm^{-3}, and also, possibly, to set limits on the physical conditions under which this might have occurred. It now appears likely that comets are made up of interstellar dust that did not suffer chemical processing in the solar nebula (Greenberg & Grim 1986), and on the basis of such a model Greenberg (1986a,b) used observed meteor densities to show that a very fluffy structure should be expected for cometary nuclei. This expectation is put on a more firm basis by the above estimate of the density of Comet Halley. To explain the origin of such objects remains a challenge for future investigations.

REFERENCES

A'Hearn, M.F., Hoban, S., Birch, P.V., Bowers C., Martin, R., & Klinglesmith, D.A. III. (1986) Gaseous jets in Comet P/Halley, In: *Exploration of Halley's Comet* I Battrick, B., Rolfe, E.J. & Reinhard, R. eds, ESA SP-250, 483

Bockelée-Morvan, D., Crovisier, J., & Gérard, E. (1989) Retrieving the coma gas expansion velocity from the ultraviolet OH line shapes, In: *Asteroids, comets, meteors* III Lagerkvist, C.-I., Rickman, H., Lindblad, B.A., & Lindgren, M. eds., Uppsala University, in press

Boyarchuk, A.A., Grinin, V.P., Zvereva, A.M., & Sheihet, A.I. (1986) Estimations of water production rate in the Halley Comet using the ultraviolet data obtained on the space station 'Astron', In: *Exploration of Halley's Comet* III Battrick, B., Rolfe, E.J., & Reinhard, R. eds, ESA SP-250, 193

Colom, P., Gérard, E., & Crovisier, J. (1989) A study of the non-gravitational forces on Comet Halley nucleus from the observations of the OH radio lines, In: *Asteroids, comets, meteors* III Lagerkvist, C.-I., Rickman, H., Lindblad, B.A., & Lindgren, M. eds., Uppsala University, in press

Combes, M., *et al.* (1986) Detection of parent molecules in Comet Halley from the IKS-VeGa experiment. In: *Exploration of Halley's Comet.* I Battrick, B., Rolfe, E..J., & Reinhard, R. eds, ESA SP-250, 353

Crifo, J.F. (1987) Improved gaskinetic treatment of cometary water sublimation and recondensation, with application to Comet P/Halley. *Astron. Astrophys.* **187** 438

Daniels, P.A., & Hughes, D.W. (1981) The accretion of cosmic dust—a computer simulation, *Mon. Not. R. Astr. Soc.* **195** 1001

Donn, B. (1987) The formation of fractal-like cometary nuclei by random accretion, preprint

Donn, B., & Hughes, D. (1986) A fractal model of a cometary nucleus formed by random accretion, In: *Exploration of Halley's Comet* III Battrick, B., Rolfe, E.J., & Reinhard, R. eds, ESA SP-250, 523

Eberhardt, P., *et al.* (1986) On the CO and N_2 abundance in Comet Halley, In: *Exploration of Halley's Comet* I (eds B. Battrick, E.J. Rolfe & R. Reinhard), ESA SP-250, 383

Fanale, F.P., & Salvail, J.R. (1984) An idealized short-period comet model: surface insolation, H_2O flux, and mantle evolution, *Icarus* **60** 476

Feldman, P.D., *et al.* (1986) IUE observations of Comet Halley: evolution of the UV spectrum between September 1985 and July 1986, In: *Exploration of Halley's Comet* I Battrick, B., Rolfe E.J., & Reinhard, R. eds, ESA SP-250, 325

Festou, M.C. (1986) The derivation of OH gas production rates from visual magnitudes of comets, In: *Asteroids, comets, meteors* II Lagerkvist C.-I., *et al.* eds, Uppsala University, 299

Festou, M.C., Carey, W.C., Evans, A., Wallis, M.K., & Keller, H.U. (1985) IUE observations of Comet P/Crommelin (1983n), *Astron. Astrophys.* **152** 170

Festou, M.C., Feldman, P.D., A'Hearn, M.F., Arpigny, C., Cosmovici, C.B., Danks, A.C., McFadden, L.A., Gilmozzi, R., Patriarchi, P., Tozzi, G.P., Wallis, M.K., & Weaver, H.A. (1986) IUE observations of Comet Halley during the VeGa and Giotto encounters, *Nature* **321** 361

Festou, M.C., Drossart, P., Lecacheux, J., Encrenaz, T., Puel, F., & Kohl-Moreira, J.L. (1987) Periodicities in the light curve of P/Halley and the rotation of its nucleus, *Astron. Astrophys.* **187** 575

Fischer, D. (1986) Comet Halley's confusing brightness—early results from worldwide visual observations. In: *Exploration of Halley's Comet* III (eds B. Battrick, E.J. Rolfe, & R. Reinhard), ESA SP-250, 303

Froeschlé, Cl., & Rickman, H. (1986) Model calculations of nongravitional forces on short-period comets. I. Low-obliquity case, *Astron. Astrophys.* **170** 145

Gérard, E., Bockelée-Morvan, D., Bourgois, G., Colom, P., & Crovisier, J. (1986) 18 cm wavelength radio monitoring of the OH radical in Comet P/Halley 1982i, In: *Exploration of Halley's Comet* I (eds B. Battrick, E.J. Rolfe & R. Reinhard), ESA SP-250, 589

Gombosi, T.I., Nagy, A.F., & Cravens, T.E. (1986) Dust and neutral gas modelling of the inner atmospheres of comets, *Rev. Geophys.* **24** 667

Green, D.W.E., & Morris, C.S. (1986) The visual brightness behaviour of P/Halley during 1981–1986, In: *Exploration of Halley's Comet* I Battrick, B., Rolfe E.J., & Reinhard R. eds, ESA SP-250, 613

Greenberg, J.M. (1986a) Fluffy comets, In: *Asteroids, comets, meteors* II Lagerkvist, C.-I., *et al.* eds, Uppsala University, 221

Greenberg, J.M. (1986b) Evidence for the pristine nature of Comet Halley, In: *The comet nucleus sample return mission* (ed. O. Melita), ESA SP-249, 47

Greenberg, J.M., & Grim, R. (1986) The origin and evolution of comet nuclei and Comet Halley results, In: *Exploration of Halley's Comet* II Battrick, B., Rolfe E.J., & Reinhard R. eds, ESA SP-250, 255

Huebner, W.F., Boice, D.C., & Sharp, C.M. (1987) Polyoxymethylene in Comet Halley, *Astrophys. J.* **320** L149

Hughes, D.W. (1985) The size, mass, mass loss and age of Halley's Comet, *Mon. Not. R. Astr. Soc.* **213** 103

Ip, W.-H., & Rickman, H. (1986) A comparison of nucleus surface models to space observations of Comet Halley, In: *The comet nucleus sample return mission* Melita, O., ed. ESA SP-249, 181

Julian, W.H. (1987) Free precession of the Comet Halley nucleus, *Nature* **326** 57

Kamél, L. (1988) The rotation, precession and nutation of P/Halley's nucleus, *Astron. Astrophys.* **193** L4

Kaneda, E., Hirao, K., Shimizu, M., & Ashihara, O. (1986) Activity of Comet Halley observed in the ultraviolet, *Geophys. Res. Letters* **13** 833

Keller, H.U. (1987) The nucleus of Comet Halley, In *Diversity and similarity of comets* Battrick, B., & Rolfe E.J. eds, ESA SP-278, 447

Keller, H.U., Arpigny, C., Barbieri, C., Bonnet, R.M., Cazes, S., Coradini, M., Cosmovici, C.V., Delamere, W.A., Huebner, W.F., Hughes, D.W., Jamar, C., Malaise, D., Reitsema, H.J., Schmidt, H.V., Schmidt, W.K.H., Seige, P., Whipple, F.L., & Wilhelm, K. (1986) First Halley multicolour camera imaging results from Giotto, *Nature* **321** 320

Kiang, T. (1972) The past orbit of Halley's Comet, *Mem. Roy. Astron. Soc.* **76** 27

Kissel, J., Sagdeev, R.Z., Bertaux, J.-L., Angarov, V.N., Audouze, J., Blamont, J.E., Büchler, K., Evlanov, E.N., Fechtig, H., Fomenkova, M.N., von Hoerner, H., Inogamov, N.A., Khromov, V.N., Knabe, W., Krueger, F.R., Langevia, Y., Leonas, V.B., Levasseur-Regourd, A.C., Managadze, G.G., Podkolzin, S.N., Shapiro, V.D., Tabaldyev, S.R., & Zubkov, B.V. (1986a) Composition of Comet Halley dust particles from VeGa observations, *Nature* **321** 280

Kissel, J., Brownlee, D.E., Büchler, K., Clarke, B.C., Fechtig, H., Grün, E., Hornung, K., Igenbergs, E.B., Jessberger, E.K., Krueger, F.R., Kuczera, H., McDonnell, J.A.M., Morfill, G.M., Rahe, J., Schwehm, G.H., Sekanina, Z., Utterback, N.G., Völk, H.J., & Zook, H.A. (1986b) Composition of Comet Halley dust particles from Giotto observations, *Nature* **321** 336

Klinger, J. (1981) Some consequences of a phase transition of water ice on the heat balance of comet nuclei, *Icarus* **47** 320.

Krankowsky, D., Lämmerzahl, P., Herrwerth, I., Woweries, J., Eberhardt, P., Dolder, U., Herrmann, U., Schulte, W., Berthelier, J.J., Illiano, J.M., Hodges, R., & Hoffman, J.H. (1986) *In situ* gas and ion measurements at Comet Halley *Nature* **321** 326

Krasnopolsky, V.A., *et al.*, (1986) Near infrared spectroscopy of Comet Halley by the VeGa-2 three-channel spectrometer, In: *Exploration of Halley's Comet* I (eds B. Battrick, E.J. Rolfe, & R. Reinhard), ESA SP-250, 459

Lang, B. (1987) Porosity and fractality of a cometary nucleus, In: *Diversity and similarity of comets* (eds B. Battrick & E.J. Rolfe), ESA SP-278, 483

McCoy, R.P., Opal, C.B., & Carruthers, G.R. (1986) H Lyman-α imagery of Comet P/Halley from sounding rockets, In: *Exploration of Halley's Comet* I (eds B. Battrick, E.J. Rolfe, & R. Reinhard), ESA SP-250, 403

McCuskey, S.W. (1963) *Introduction to celestial mechanics* Addison-Wesley Publ. Comp.

Mendis, D.A., & Brin, G.D. (1977) Monochromatic brightness variations of comets, II, core-mantle model, *The Moon* **17** 359

Mirabel, I.F., *et al.* (1986) Post-perihelion radio monitoring of the OH in Comet Halley In: *Exploration of Halley's Comet* I (eds B. Battrick, E.J. Rolfe & R. Reinhard). ESA SP-250, 595

Newburn, R.L. (1981) A semi-empirical photometric theory of cometary gas and dust production: application to P/Halley's gas production rates, In: *The Comet Halley dust and gas environment* (eds B.Battrick & E. Swallow), ESA SP-174, 3

Rickman, H. (1986) Masses and densities of Comets Halley and Kopff, In: *The comet nucleus sample return mission* Melita, O., ed. ESA SP-249, 195

Rickman, H. (1989) The nucleus of Comet Halley: surface structure, mean density, gas and dust production, *Adv. Space Res.* **9** (3) 59

Rickman, H., & Fernández, J.A. (1986) Formation and blowoff of a cometary dust mantle, In: *The comet nucleus sample return mission* Melita, O. ed. ESA SP-249, 185

Rickman, H., & Froeschlé, Cl. (1983a) Thermal models for the nucleus of Comet P/Halley, In *Cometary exploration* I (ed. T.I. Gombosi), Hungarian Acad. Sci., 75

Rickman, H., & Froeschlé, Cl. (1983b) Model calculations of nongravitational effects on Comet P/Halley, In: *Cometary exploration* III (ed. T.I. Gombosi), Hungarian Acad. Sci., 109

Rickman, H., Kamél, L., Festou, M.C., & Froeschlé, Cl. (1987) Estimates of masses, volumes and densities of short-period comet nuclei, In: *Diversity and similarity of comets* (eds B. Battrick & E.J. Rolfe), ESA SP-278, 471

Sagdeev, R.Z., Szabó, F., Avanesov, G.A., Cruvellier, P., Szabó, L., Szegö, K., Abergel, A., Balazs, A., Barinov, I.V., Bertaux, J.-L., Blamont, J., Detaille, M., Demarelis, E., Dul-Nev, G.N., Endröczy, G., Gardes, M., Kanyo, M., Kostenko, V.I., Krasikov, V.A., Nguyen-Trong, T., Nyitrai, Z., Remy, I., Rusznyak, P., Shamis, V.A., Smith, B., Sukhanov, K.G., Szabó, F., Szalai, S., Tarnopolsky, V.I., Toth, I., Tsukanova, G., Valnícek, B.I., Varhalmi, L., Zaiko, Yu.K., Zatsepín, S.I., Ziman, Ya.L., Zsenei, M., & Zhukov, B.S. (1986) Television observations of Comet Halley from VeGa spacecraft, *Nature* **321** 262

Sagdeev, R.Z., Elyasberg, P.E., & Moroz, V.I. (1988), Is the nucleus of Comet Halley a low density body? *Nature* **331** 240

Schloerb, F.P. (1989) UV and radio observations of OH in comets, In: *Asteroids, comets, meteors* III Lagerkvist, C.-I., Rickman, H., Lindblad, B.A., & Lindgren, M. eds., Uppsala University, in press

Schloerb, F.P., Claussen, M.J., & Tacconi-Garman, L. (1986) OH radio observations of Comet Halley In: *Exploration of Halley's Comet*. I (eds B. Battrick, E.J. Rolfe & R. Reinhard), ESA SP-250, 583

Sekanina, Z. (1986) Nucleus of Comet Halley as a torque-free rigid rotator, *Nature* **325** 326

Smith, B.A., Larson, S.M., Szegö, K., & Sagdeev, R.Z. (1987) Rejection of a proposed 7.4-day rotation period of the Comet Halley nucleus, *Nature* **326** 573

Spinrad, H., McCarthy, P.J., & Strauss, M.A. (1986) Oxygen production rates for P/Halley over much of the 1985–86 apparition, In: *Exploration of Halley's Comet* I (eds B. Battrick, E.J. Rolfe, & R. Reinhard), ESA SP-250, 437

Stewart, A.I.F. (1987) Pioneer Venus measurements of H, O, and C production in Comet P/Halley near perihelion, *Astron. Astrophys.* **187** 369

Strelnitsky, V.S., Bisikalo, D.V., & Shematovich, V.I. (1987) *Astron. Tsirk.* No. 1488, 3 (in Russian)

Wallis, M.K., & Carey, W.C. (1985) Observations of Comet Crommelin–V. Anomalous hydrogen source, *Mon. Not. R. Astr. Soc.* **217** 673

Weaver, H.A., Mumma, M.J., Larson, H.P., & Davis, D.S. (1986) Airborne infrared investigation of water in the coma of Halley's Comet, In: *Exploration of Halley's Comet,* I Battrick, B., Rolfe, E.J., & Reinhard, R. eds, ESA SP-250, 329

Weissman, P.R., & Kieffer, H.H. (1981) Thermal modelling of cometary nuclei, *Icarus* **47** 302

Whipple, F.L. (1950) A comet model. I. The acceleration of Comet Encke, *Astrophys. J.* **111** 375

Wilhelm, K. (1987) Rotation and precession of Comet Halley, *Nature* **327** 27

Woods, T.N., Feldman, P.D., Dymond, K.F., & Sahnow, D.J. (1986) Rocket ultraviolet spectroscopy of Comet Halley and abundance of carbon monoxide and carbon, *Nature* **324** 436

Yeomans, D.K., & Kiang, T. (1981) The long term motion of Comet Halley, *Mon. Not. R. Astr. Soc.* **197** 633

14

Models of cometary nuclei

Harry L. F. Houpis

SYMBOLS

A_B	surface bolometric albedo
A_{IR}	surface infrared albedo
A_V	surface visual albedo
AU	astronomical unit
$C(T)$	specific heat of nucleus material at temperature T
$C_i(T)$	specific heat of solid ice or dust species i at temperature T
r_H	heliocentric distance in AU
g_i	volume fraction of solid ice or dust species i
E_s	heat content emission rate
i	angle between comet's rotation axis and Sun-comet line
J_{bb}	radiation at comet surface due to atmospheric dust thermal reradiation
J_d	solar constant
J_{ms}	radiation at comet surface due to multiple scattering of atmospheric dust
k	Boltzmann constant
$K(T)$	thermal conductivity of nucleus material at temperature T
$K_b(T)$	bulk conductivity of solid ice and/or dust at temperature T
$K_c(T)$	conductivity of core
$K_m(T)$	conductivity of mantle
$L_i(T)$	heat of sublimation for volatile species i at temperature T
n_s	surface gas particle density
Q	heat generation rate per unit volume
R	radial distance from centre of nucleus
t	time

T	temperature, generally a function of R and t
T_c	temperature at mantle-core interface
T_s	temperature at the surface of the nucleus
v_s	surface gas velocity
y	distance from nucleus surface
$Z_i(T)$	gas production rate of volatile species i at temperature T
Z^+	solar zenith angle Z for $Z \leq \pi/2$; otherwise, $\pi/2$
α,β,γ	coefficients in conductivity relation
Δ	thickness of mantle
ε_s	infrared emissivity of surface
η_i	surface fraction of volatile species i
$\rho(T)$	mass density of nucleus material at temperature T
$\rho_i(T)$	mass density of solid ice or dust species i at temperature T
σ	Stephan-Boltzmann constant
τ	optical depth at comet surface

1 INTRODUCTION

Models of the comet nucleus come in two forms. The first is the descriptive construct based primarily on observations, but guided by theoretical intuition. For about twenty-five years now, this type of modelling has followed the icy-conglomerate formalism of Whipple (1950). As observations have accumulated and improved, the icy-conglomerate has undergone a number of changes with regard to the distribution of ice and dust, the chemical composition, the creation and destruction of mantles, the global structure of the entire nucleus, and the existence of such surface structures as 'mountains', 'valleys' and line and jet sources. However, as vividly displayed in the images of Comet Halley sent back to Earth by the Giotto and Vega spacecraft, the basic idea of a comet as a single body of ice and dust still remains valid.

The second form of comet nucleus models is the theoretical construct. In a sense, this is the opposite of the descriptive construct in that the emphasis is on theoretical calculations from equations tailored by physical intuition gained from observations. Theoretical modelling has been rather slow to progress owing to the lack of known physical parameters associated with a comet, but the recent inner solar system visit of Comet Halley and the large entourage of spacecraft to meet it have accelerated this kind of model development.

The purpose of this review is twofold. First, we review the theoretical models that have been developed to date. We will concern ourselves here with only the main points of these models, since more detailed reviews can be found elsewhere (see e.g., Mendis et al. 1985, Gombosi et al. 1986). We then continue with our second purpose; namely, pointing out the deficiencies in the theoretical models in the light of the recent Comet Halley observations. Finally, we discuss various proposals (both empirical and theoretical) that have been put forward to guide future modelling efforts.

2 THEORETICAL MODELS

The primary purpose of theoretical modelling has been the calculation of thermal and production rate profiles over the entire nucleus surface and/or interior. To this end, two equations have served as the main vehicle. The first is the energy balance equation at the surface of the comet:

$$\left(\frac{1-A_B}{r_H^2}e^{-\tau}\cos Z^+\right)J_d + (1-A_B)J_{ms}$$

$$+ (1-A_{IR})J_{bb} = \varepsilon_s\sigma T_s^4 + \sum_i \eta_i Z_i(T_s)L_i(T_s)$$

$$+ K(T_s)\nabla T|_s + E_s \qquad (1)$$

where the equation is satisfied at any given point on the surface and all the terms are defined in the Table of Symbols. The terms on the LHS represent the energy influx from direct and multiply-scattered solar radiation and thermal reradiation from dust in the cometary atmosphere. The energy terms on the RHS are from surface thermal reradiation, gas sublimation, conduction into the interior, and heat content lost owing to surface erosion.

We note that the direct solar radiation term is modified by the surface albedo, which accounts for the optical effects of the ice, dust and/or pores at the surface, the solar zenith angle, and the optical depth, which is a function of the comet atmosphere gas and dust distribution along the line-of-sight to the Sun. The multiple-scatter and thermal reradiation terms are important only for comets with a significant dust production, in which case a complicated angular dependence can be expected.

With regard to the RHS of equation (1), we make a special note of the sublimation term. For the case of a pure dust surface, naturally all the η_i terms are zero. In general, however, the determination of the production rate Z_i, which is the product of the gas surface density of that species and its velocity, requires a few important assumptions. For the surface density, the use of the Clausius–Clapeyron equation together with the ideal gas law is widely acknowledged, while the determination of the velocity is more complicated. For a comet with very low gas production, the kinetic approach of Delsemme & Swings (1952) is satisfactory. However, for a rather gaseous comet a fluid approximation is needed. With little dust production, the sonic speed may be used, since the gas expands virtually unhindered into a vacuum, but once the dust production becomes significant, a numerical estimate is the best that is available (see Gombosi et al. 1986). This is because the atmosphere and the gas flow within and from the nucleus are dynamically linked, and technically cannot be divorced from one another. Thus, in the case of a rather dusty and gaseous comet, where the gas mean-free-path is much smaller than the nucleus radius, the dust-fluid equations of the atmosphere must also be considered when determining the flow from the surface.

The term E_s is usually not mentioned in the literature. It comes from considering the loss of energy due to entrainment of surface dust and ice by the gas, and hence from a change in the radius of the nucleus. This process was suggested by Whipple to Smoluchowski (1981b), who showed it may account for about 5% of the heat flux, $K(T_s)\nabla T|_s$, not reaching the deeper layers of the nucleus. In practice, however, the term is not directly included in equation (1), but rather in some numerical codes after a number of time steps, a layer of surface is removed from the nucleus. The lost energy content and the newly exposed underlying layer with a lower temperature simulate the E_s term.

The conductivity $K(T)$ is usually written in the form (Mendis & Brin 1977):

$$K(T) = \alpha K_b(T) + \beta T^3 + \gamma T^{1/2} \qquad (2)$$

where the first term is the bulk conductivity of the dust and ice modified by the Hertz factor, α, the second term is radiative conduction by photon transmission along the pores, and the last term is energy transport via the gas escaping along the pores. The Hertz factor accounts for the possibility that a real connection between adjacent material may be limited to small contact points. The bulk conductivity is usually the dominant term to within 0.5–1.0 AU from the Sun, with the radiative term becoming more important with increasing temperature and pore size. We point out, however, that equation (2) is rather simplistic; multi-component systems (see Houpis et al. 1986) and/or the existence of voids (see Smoluchowski 1985) complicates the picture. Also, γ may be temperature dependent (see Fanale & Salvail 1984).

The second important equation for the modelling of the thermal and production rate profiles is the heat conduction equation:

$$\rho(T)C(T)\frac{\partial T}{\partial t} = \frac{1}{R^2}\frac{\partial}{\partial R}\left(R^2 K(T)\frac{\partial T}{\partial R}\right) - Q \qquad (3)$$

where again the terms are defined in the Table of Symbols. Equation (2) implicitly assumes that there is little or no heat flow transverse to the radial direction, R; hence, the temperature is a function of only R and t. Except perhaps near the surface at the terminators, this is probably a reasonable assumption.

The mass density, $\rho(T)$, and the specific heat, $C(T)$, are shown as explicit functions of temperature. In practice, recent research efforts have assumed a constant mass density and a specific heat given by the 'dominant' component of ice or dust. In fact, as shown by Houpis et al. (1986), the product of the mass density and specific heat is given by:

$$\rho(T)C(T) = \sum_i g_i \rho_i(T)C_i(T) \qquad (4)$$

where the sum is over all the ice and dust components, and the g_i factors are the volume fractions. The g_i factors implicitly allow for porosity.

The point we wish to make here is that one must be careful when dealing with multi-component systems. Equation (4) shows that an 'average' mass density times an 'average' specific heat is not a correct product for $\rho(T)C(T)$. This idea should also be kept in mind for the surface albedo, the emissivity, etc. An excellent case is discussed by Mendis et al. (1985), who point out that the use by some researchers of an 'average' latent heat in equation (1) to replace the summation has no physical justification. The species are not mixed on a molecular level, in which case one would have a new chemical compound with a latent heat of its own, and if the mixture is more granular, then the sublimation term in equation (1) is the proper choice.

Finally, the energy generation term, Q, in equation (3) is the sum of any combination of the following: condensation and sublimation within the pores (Houpis et al. 1986), energy exchange from gas flow accommodation (Horanyi et al. 1984), energy release from solid phase transitions (Smoluchowski 1981a, Herman & Podolak 1985) and chemical reactions involving unstable ices, and energy deposition from radioactive decay and cosmic ray bombardment (Whipple & Stefanik 1966, Draganic et al. 1984). All these processes have been dealt with in one way or another, but usually only one at a time.

In general, the solution of equation (3) is obtained by assuming some initial temperature distribution throughout the nucleus. Usually this temperature is that characteristic of a comet in the Oort cloud, or it is a value close to the orbital mean temperature (see Klinger 1983). The choice depends on the problem being investigated. Equation (1) is then used as one of the spatial boundary conditions, while the other boundary condition is taken as zero heat flux at some predetermined interior point. Finally, the complete nucleus evolution can be followed by incorporating the orbital elements of some particular comet.

The complete forms of equations (1) and (3) are rarely used. In fact, over the years, three distinct methods have evolved for calculating the thermal and production rate profiles. We call these calculation styles: the homogeneous surface method, the mantle–core method, and the complete nucleus method. We discuss each one of these approaches in turn.

2.1 The homogeneous surface method

The homogeneous surface method, as the name implies, strictly concentrates its calculations to the comet surface or a surface layer, while holding constant over the entire comet the albedo, the emissivity and the dust-to-ice mass ratio. It is implicitly assumed that any dust freed at the surface is carried away by the escaping gases. The earliest theoretical modelling of the thermal and production rate profiles are along these lines (e.g., see Squires & Beard 1961, and Watson *et al.* 1963). In these calculations, the dust, conduction, and heat content erosion terms are completely ignored, and only one species of ice is used. The neglect of energy exchange with the interior negates the use of equation (3). As a result, an easily solved transcendental equation is derived from equation (1). A number of researchers have used this technique since then (see Gombosi *et al.* 1986 for a list of references), with the most recent examples being that of Cowan & A'Hearn (1979) and Houpis & Mendis (1981a). Figs (1) and (2) are typical results. The important point to notice here is that a maximum surface temperature of about 200K can be expected at 1 AU for H_2O ice. Also, it has been a matter of choice as to whether the solar zenith angle is accounted for, as in the case of Cowan & A'Hearn (1979), or is completely averaged out, as in the case of Houpis & Mendis (1981a).

The homogeneous surface method has become more sophisticated in the last few years through the efforts of Weissman & Kieffer (1981, 1984) and Rickman & Froeschlé (1982); see also Froeschlé & Rickman (1983) and Squyres *et al.* (1985). These models have in effect defined the 'surface' to be a few thermal skin depths in order to make use of the

thermal lag properties of the heat conduction equation (3). At the lower interface of this surface, zero net heat flux into the interior is assumed. Homogeneity is maintained throughout this surface layer with the conductivity and specific heat either constant or explicitly functions of temperature only. The mass density is assumed constant, while the energy source-/sink term Q is ignored. Fig. (3) gives an example from this type of modelling (see the paper by P.R. Weissman, later in this volume, for more details). Again we note that the typical surface temperatures at around 1 AU is about 200K.

2.2 The mantle–core method

The idea of an evolving layer of dust on a comet surface was first proposed by Whipple (1950) in connection with the nongravitational effects of short-period comets. However, a quantitative analysis had to wait until the research effort of Mendis & Brin (1977); see also Brin & Mendis (1979). The primary effects of the dust layer are to provide insulation for the volatile core, therefore modulating its sublimation and to inhibit the flow of gas as it diffuses to the surface. While the sublimating surface of the core promotes the growth of the mantle thickness, the entrainment of dust particles from the mantle by the gas decreases the mantle thickness. Fig. (4) illustrates this schematically.

The basic procedure used to solve the mantle–core problem is to use two surface energy equations similar to equation (1) together with a steady-state form of equation (3) with no source/sink terms. One surface equation is at the atmosphere–mantle boundary, and ignores atmospheric dust, sublimation, and heat content loss. The second equation applies to the mantle–core surface and equates sublimation to the energy conducted down from the upper surface; it is assumed that conduction into the core is zero (see Fig. 4). Furthermore, a sonic gas flow speed is used throughout the dust mantle. Finally, the mantle evolution is followed by not only assuming some comet orbital parameters, but also assuming that all dust smaller than a given critical size escapes from the nucleus. The critical size is determined as that radius below which a particle can be lifted by escaping gases against the force of gravity. This critical size is

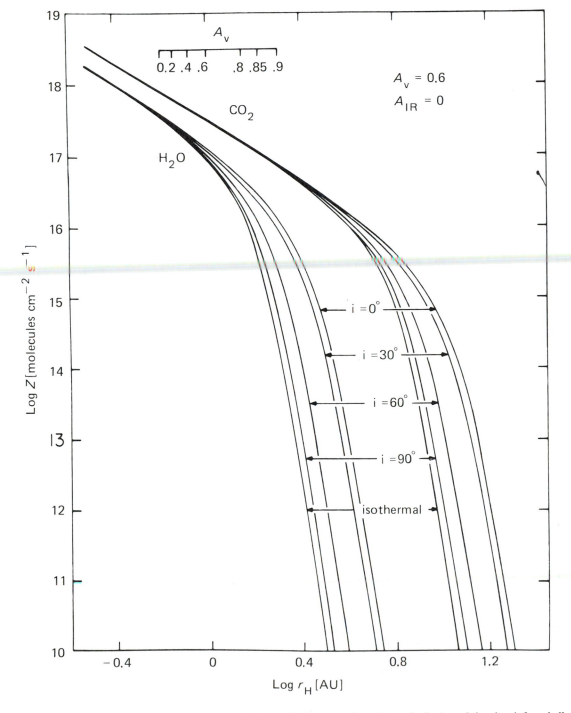

Fig. 1 — H_2O and CO_2 vaporization rates as a function of heliocentric distance r_H for a visual albedo $A_V = 0.6$ and an infrared albedo A_{IR} = 0.0. i is the angle between the comet's rotation axis and the Sun–comet line. An isothermal case is also given for comparison. The albedo scale at the top gives the horizontal translation of the production rate at 1 AU for different visual albedos. (From Cowan & A'Hearn 1979).

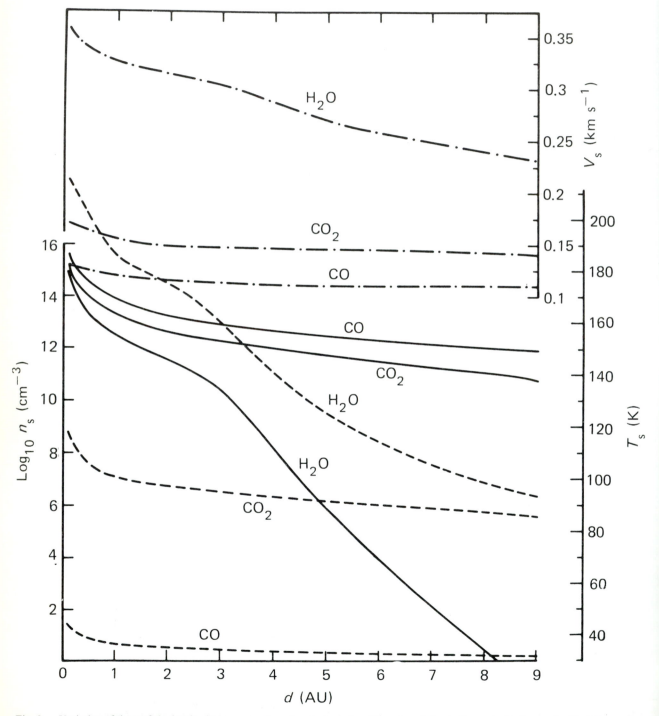

Fig. 2 — Variation of the surface density ($\log_{10} n_s$ ——), surface temperature (T_s, ---), and the surface velocity (v_s, -·-·-·) with heliocentric distance for comets dominated by H_2O, CO, and CO_2. (From Houpis & Mendis 1981a).

obviously dependent on heliocentric distance.

A number of improvements in the mantle–core method have been developed recently. Two notable examples are the friable sponge mantle of Horanyi *et al.* (1984) and the chemical differentiation scheme of Houpis *et al.* (1985). The former improvement suggests that dust emission is only from the comet surface as opposed to the entire mantle, that the dust mass production rate is proportional to the momentum flux of the gas, and that the gas flow through the pores exchanges heat with the solid mantle. Also, all dust at the surface is assumed to be smaller than the critical size. The latter improvement (first hinted at by Whitney 1955, with regard to comet outbursts) makes allowances for ices other than clathrate hydrates, and therefore permits several mantles, each with the degree of volatility increasing with distance from the surface.

We also note the mantle–core model of Fanale & Salvail (1984). This model follows the Mendis–Brin idea for escaping dust particles, but vastly improves the details. This includes the variation of solar insolation with respect to cometo-longitude and -latitude and the effects of a non-zero obliquity. More important, though, is that they include a gas diffusion process through the mantle as opposed to the previous assumptions of free-flow, and they adopt the gas flow accommodation to the pore wall temperature as proposed by Horanyi *et al.* (1984). Figs (5) and (6) are typical results from these improvements. Recently, Fanale and Salvail (1987) have also modified their model to include the effects of chemical differentiation as discussed by Houpis *et al.* (1985).

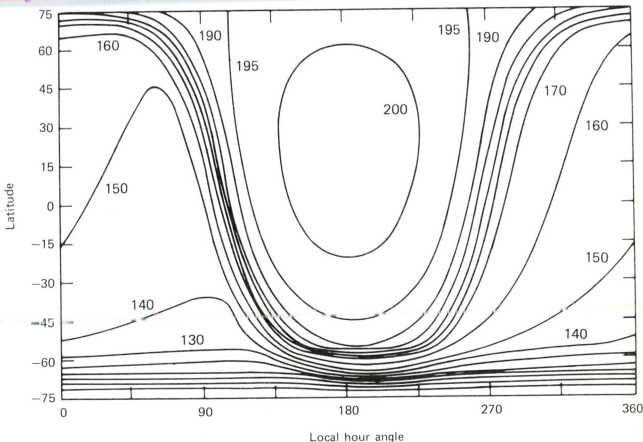

Fig. 3 — Temperature distribution on the surface of Comet Halley at 1 AU with no dust coma. Temperature contours are every 10K below 170K and every 5K above. (From Weissman & Kieffer 1984).

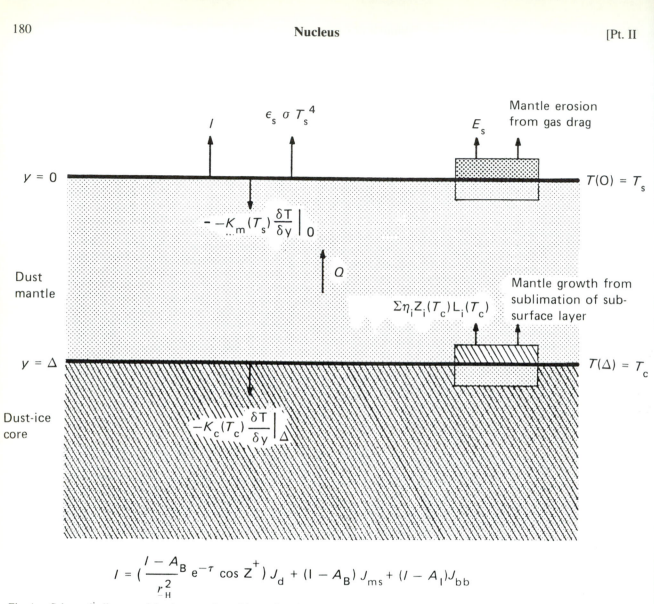

Fig. 4 — Schematic diagram of the dust mantle and inner dust–ice core indicating the mantle growth and erosion processes together with the various energy flow contributions.

2.3 The complete nucleus method

The final method we wish to point out is the complete nucleus approach, which attempts to look at some aspect of the nucleus as a whole. Although equations (1) and (3) are used in this method, there is not any typical manner in which the equations are used; the equations are specifically tailored to a particular problem of interest. The earliest use of this method appears to be Whipple & Stefanik (1966), who were primarily interested in the heating effects of radioactivity in the Oort cloud comets and subsequent splitting. More recently, Klinger (1981) and Herman & Podolak

(1985) have studied the effects of water-ice phase changes, Kührt (1984) has looked into thermal stresses and cracking of the surface, and Smoluchowski (1981a,b) and his colleagues (1984a,b) have investigated various aspects of pore migration and conductivity. In addition, Dobrovolskii & Markovich (1972), Horanyi et al. (1984), and Podolak & Herman (1985) have included a dust mantle in their analysis of a nonvarying mantle thickness, diurnal variations, and phase changes, respectively.

The point to be made here is that no concerted effort has been made to take into account more than

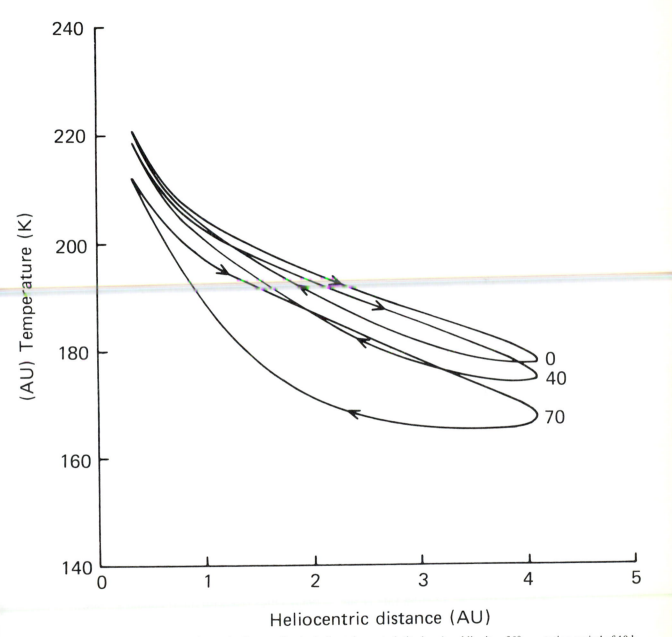

Fig. 5 — Ice surface temperature vs heliocentric distance for the indicated cometo-latitudes. An obliquity of 0°, a rotation period of 10 h, and a dust-to-ice mass ratio of 0.5 are assumed. (From Fanale & Salvail 1984).

one or two important physical processes at once. For instance, many of these models do not include an evolving dust mantle, and those that do, neglect the variability of the pores and hence the conductivity. This criticism can also be levelled against the two previously mentioned modelling approaches.

Finally, the name 'complete nucleus' may be a misnomer in some cases, since computer limitations usually require a variable grid size which, while only a few centimetres or less at the surface, may be hundreds of metres toward the centre. This is acceptable so long as there are no physical processes

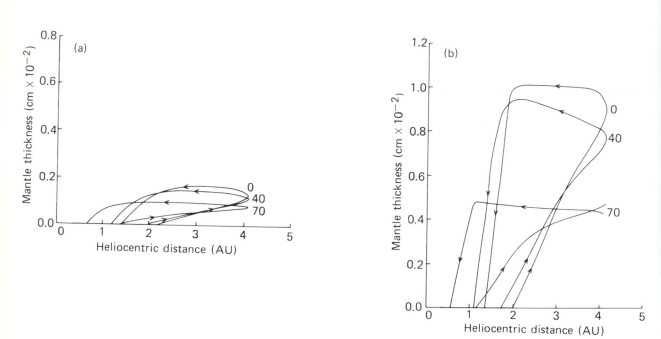

Fig. 6 — Mantle thickness Δ vs heliocentric distance for the indicated cometo-latitudes. Parameters are the same as Fig. 5 with the grain size distribution exponents given as (a) 4.2 and (b) 3.5. (From Fanale & Salvail 1984).

with a scale length that requires a better resolution; e.g. radioactive heating in the central core or gas flow from very large temperature gradients at the surface.

3 MODEL DEFICIENCIES AND THEIR SOLUTION

To further discuss the deficiencies in the theoretical constructs of the past and even the present, it is necessary to briefly review the more important observations, especially those recently made of Comet Halley. Up until the encounter with Halley, the major observational characteristics that comet modelling attempted to explain were the production rate of the various chemical species and its asymmetric behaviour with respect to perihelion, the 'turn-on' of most short and intermediate period comets ($P < 200$ years) inside about 3 AU, and the onset of gaseous activity of some 'new' comets at very large heliocentric distances (5–10 AU). Except perhaps for the last characteristic, the theoretical reasons for these observations appear to be reasonably understood. For example, the asymmetry can be explained as a combination of mantle development and the thermal lag of the

nucleus; whereas, the turning-on of comets inside 3 AU can be attributed to the preponderance of water ice (which may or may not be in clathrate form) in the upper volatile layer of the comet.

Comet modelling becomes more tenuous when 'transient-type' behaviour is considered. Splitting of comets is a well known example. Disruption appears to be completely random with respect to dynamical age, orbital parameters, and heliocentric distance, although most comets appear to split within about 2.5 AU (Pittich 1971, Kresak 1981). Comets also show outbursts even when they do not split. Suggested reasons for this type of activity are many: tidal disruptions and rotational break-up (see Sekanina 1982), collisions with interplanetary boulders (Harwit 1968), radioactive heating and cracking (Whipple & Stefanik 1966), ice phase transitions (Smoluchowski 1981a, Herman & Podolak 1985), exothermic chemical reactions (Donn & Urey 1956), sudden release of large sub-surface gas pressure (Whitney 1955, Barnun et al. 1985), and electrostatic blow-off (Opik 1956, Mendis et al. 1981). Although a few of these processes have been incorporated into the theoretical models discussed above, most of them are not easily included in this type of analysis. Hence it is still too soon to be able to theoretically discriminate between the various observed transient events.

This situation is complicated further when we review the major results from the Comet Halley encounter. First of all, the shape of the comet is anything but spherical, more potato-like with altitude variations of the order of 100 metres and features labelled 'mountains' and 'valleys' (Reitsema *et al.* 1986). The comet is also much larger than previously expected, being of the order of $16 \times 10 \times 9$ km^3 (Wilhelm *et al.* 1986). Activity is confined to about 10% of the surface at any given time, and this activity is primarily in the form of jet bundles and fans, and is confined for the most part to the day side (Keller *et al.* 1986). Additionally, each individual jet and fan source appears to have a rather intermittent life expectancy, being active on some rotations and inactive on others (Sekanina 1986). Finally, the surface albedo is a few percent and rather uniform (Sagdeev *et al.* 1986) with the surface temperature far in excess of 360K (Emerich *et al.* 1986).

Consequently, there are two basic problems with the three standard forms of theoretical modelling. The first is one of philosophy. The observations quite clearly show a comet that is non-uniform and non-steady in activity both local to the comet and in its orbital path about the Sun. The underlying theoretical belief, or hope, is that if one starts with a spherical nucleus with some initial uniform composition, then after many revolutions about the Sun, the comet will deform, through the creation of mantles, phase changes, and/or other processes, into a surface configuration similar to what is observed. Although none of the more sophisticated numerical approaches have really run their full course or explored a large portion of their parameter space, it is probably a safe bet that the observed features mentioned above will not occur in the theoretical construct as it is employed today. One could argue that non-linear equations are rich in the unexpected, and that if we allow our programs enough time to run with the appropriate physical parameters and assumptions, we might see the desired results. However, there appears to be another trend in the theoretical arena, which we will review shortly, that lends itself to more credence.

The second problem with our theoretical models at present is that each of them is inherently restricted in

some fashion. For example, Comet Halley at the time of encounter definitely had a very low albedo porous mantle of very hot (>360K) material covering most if not all of the surface. This fact essentially shows that the homogeneous surface modelling approach with its predicted ~200K surface is inadequate, if not inapplicable, when applied to the entire comet surface. Although the mantle–core approach solves this problem, the neglect of heat flow to and from the core ignores the important thermal history of the interior, a point clearly made by the complete nucleus approach. This is especially important when one considers that the mantle may be only a few centimetres thick at most. However, even when this last type of calculation includes a dust mantle, the methods usually are quite specialized, and tend to exclude a number of important processes; for instance, proper equations for gas diffusion and gas/solid energy exchange, as well as pore evolution and irradiation chemistry. Finally, none of the three methods attempts to self-consistently couple the nucleus surface conditions with the expanding atmosphere; usually a sonic speed or kinetic approximation is invoked.

The need to change our theoretical approach was pointed out by Houpis & Gombosi (1986), who suggest that a new methodology of comet nucleus modelling must be invoked. The new methodology must be able to address two major observations. One is the division of the comet into active and inactive zones. The second is the linear structure of the active regions. With regard to the division of the activity into two zones, Houpis & Gombosi (1986) suggest a separate theoretical analysis for each zone. The zones themselves can either be located in some random fashion, or perhaps they can be linked with the source mapping technique of Sekanina & Larson (1986).

Actually, the division of the surface into active and inactive zones has been around for some time under the descriptive construct (e.g., see Shul'man 1972 and Whipple 1978). It is only now that it is becoming obvious that theoretical modelling must account for it, much like modelling oceans and land separately, rather than that of some 'average' terrain. In fact, if the various types of dust (CHON, silicate, mineral, mixed, and exotic; see Sekanina 1986) are any indication, there might be a variety of land forms that need to be addressed separately.

Two recent suggestions that the comet surface is made up of two types of surface are those of Sekanina (1986) and Wallis (1986). Sekanina separates the surface into irradiation-driven mantles and sublimation-driven mantles. The former, as the result of cosmic ray proton bombardment, is dominated by polymerization, which seals off any volatile activity; whereas, the latter occurs every place else where little or no polymerization exists and the structure is fragile. Wallis (1986) follows a similar line, but instead uses surface forces and chemical transformations of less volatile material to create the non-active zones. Both of these researchers then rely to some extent on surface cracking to help maintain the vitality of the active zones, although in Wallis's snowball version, gas pressure blow-off is sufficient.

We must add, however, that this type of speculation, especially when it involves irradiation chemistry, still requires more experimental study (see Strazzulla *et al.* 1983, Draganic *et al.* 1984, Pirronello & D'Arrigo 1985, Johnson *et al.* 1986). For example, Shul'man (1972) and Whipple (1977) suggest a highly reactive surface formed by cosmic ray bombardment as the reason for pre-perihelion brightening of 'new' comets, whereas, Donn (1976), like Sekanina (1986), suggest that the cosmic rays seal off potential activity and hence require an alternative mechanism for pre-perihelion brightening (e.g., the existence near the remaining active surfaces of highly volatile ices like CO and CH_4, see Houpis *et al.* 1985).

Another approach to the two-phase nature of the comet surface has been proposed by Gombosi & Houpis (1986). In their icy-glue model, the zones of activity are a natural consequence of the global structure of the nucleus. This structure is constructed of large (hundreds of metres) porous refractory boulders 'cemented' together by an ice–dust mix. The boulders near the surface provide the base for the inactive regions, while the ice-dust mix between these boulders are the potential linear active zones. Although irradiation chemistry and similar processes may still account for some variations on the surface, the icy-glue model builds the general two-phase nature into the nucleus from the start as opposed to requiring its evolution. A consequence of this global structure is the long lifetime of the linear features, which appears

to be the case (see Sekanina 1986). We note that Mohlmann *et al.* (1986) and Reitsema *et al.* (1986) come to essentially the same conclusion as Gombosi & Houpis (1986).

Finally, with regard to the linear nature of the active regions, we would like to address the issue of thermal cracking more directly. Kührt (1984) and Kührt *et al.* (1986) have looked at the theoretical implications of a compact nucleus, a nucleus composed of cometesimals and a nucleus with inclusions. For the physical parameters that these authors assumed for a comet entering the inner solar system, it is found that cracks due to thermal stress are easily formed to a depth of 10 metres in a compact object, in cometesimals larger than about 5 metres, and from CO_2 inclusions of the order of 0.5 metre that can remove an appreciable amount of overlying material upon outgassing. However, it is important to notice that this analysis requires the object that is cracking to be monolithic and of a size much larger than that characterized by grains (sub-centimetre).

We wish to emphasize this last point since it is very important to future modelling efforts. From past observations of comets and the recent results from Comet Halley data, we know that comets have a large reservoir of dust that manifests itself by its intimate relation with the outgassing volatiles. It appears likely, both from observations and theory, that the *active* surface is of the mantle–core model type. Also, the recent calculations of Rickman (1986) for Comet Halley, as well as the analyses of Donn and his colleagues (see Donn 1963, Donn & Rahe 1982, Donn & Hughes 1986) and Greenberg (1985) suggest that the comet has a rather low density implying a porous (fractal, fluffy) global structure. This all naturally leads to the following questions: What part of the nucleus is cracking; i.e. certainly a fine loose dust mantle will not crack, but what can be said of a very porous substructure? How often does cracking occur; i.e. can cracking and the intermittent activity of the line sources be linked by fracture mechanics (a field of study which determines how fast a crack propagates; e.g., see Kanninen & Popelar 1985)? And are cracks needed at all; i.e. does an alternative model such as the global structure of Gombosi & Houpis (1986) and/or the 'loose assemblage' cracking caused from the small relative motion between cometesimals

(Bertaux & Abergel 1986) provide a better explanation for the line sources?

4 CONCLUSION

Since the introduction of the icy-conglomerate model by Whipple (1950), cometary nuclear science has progressed, albeit slowly at times, from advancements made in the theoretical and descriptive constructs. In particular, the theoretical construct has evolved into three distinct modelling approaches: the homogeneous surface, the mantle–core, and the complete nucleus. The descriptive construct has also changed, particularly in its handling of transient behaviour and the two-phase nature of the nucleus. Up until the recent apparition of Comet Halley, the two constructs could remain rather detached, owing to the lack of known physical parameters and structures. The *in situ* measurements of the VeGa and Giotto spacecraft have changed this relationship, and it is apparent that future rendezvous missions will require better understanding and modelling of the physics and chemistry at work on a comet.

The primary theoretical efforts should now be aimed at modelling two distinct land forms: active and inactive. Whether a global structure or a surface evolutionary process is responsible for this dichotomy

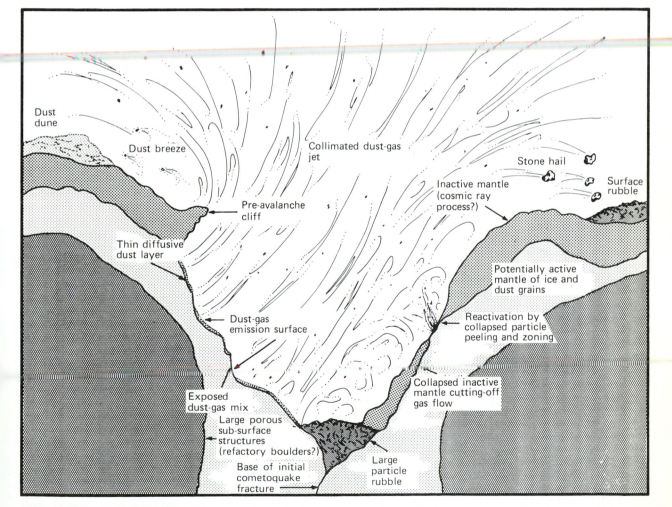

Fig. 7 — Schematic representation of various processes which may be responsible for the inactive and intermittently active regions of a comet surface.

should be investigated. Furthermore, the linear nature of the active regions must be studied. As with the land forms, are they a consequence of a global structure or an evolutionary process, or does thermal cracking play the dominant role?

In our endeavour to model these problems, we must keep the observations in mind. In particular, we know that the dust activity is confined primarily to the day side and that the surface is primarily very dark and hot. Activity also progresses in a 'chain' activation sequence toward the setting Sun (Sekanina 1986). This is all indicative of a low-conductivity porous dust mantle, but the observation by Smith *et al.* (1986) of a dust jet emanating from the night side surface is a puzzle. Is there nocturnal heating from atmospheric dust as discussed by Hellmich (1981), or is the heat source internal?

The dust jets are also collimated (Sekanina 1986), suggesting that they emanate from depressed regions. From the Giotto images, we know that the surface has features on the order of 100 metres or less. Can the theoretical spherical nucleus still give adequate results, or must we incorporate altitude variations in our models?

The intermittent life of the active zones, with a frequency of the order of a rotation period, and the tendency for activity to be more pronounced where the linear sources approach or meet (Sekanina 1986), also calls into question the dynamics of the dust mantle and the diffusive gas. Is thermal and/or dynamic cracking responsible for this, or can a modified form of our dust mantle models be employed? For example, our two primary dust mantle constructs are the implied fluidized bed of Mendis & Brin (1977) and the surface-only friable erosion model of Horanyi *et al.* (1984). Both of these models represent extremes in the manner in which they handle the dust entrainment. Also, neither of these schemes includes 'external' sources of dust transported from or to other regions by either gravity (avalanches and dust slides caused by sublimation undermining of inactive zones, Wallis 1986), zoning (dust motion from centrifugal and coriolis forces, Houpis & Mendis 1981b), or dust breezes and stone hail (from gas pressure differentials, Huebner *et al.* 1986). Kührt *et al.* (1986) suggest that a fissure may maintain its activity along the insides of the crack by

allowing all loose dust to fall to the bottom of the fissure and collect there as rubble. However, it appears that some type of transport, such as those just mentioned, may be needed to acquire the intermittent nature of the activity zones. Furthermore, no theoretical work has gone into the adhesion of the dust particles comprising the mantle. Fig. 7 gives an idea of how a number of the present descriptive constructs may interplay. The problem now is whether a theoretical model can be adequately constructed.

Finally, the link between observation and cosmic irradiation chemistry has not yet progressed far enough to make a clear statement for future modelling efforts. The rich variety of dust particles determined from the Halley encounters, together with a number of laboratory experiments, suggest more study is needed, particularly with regard to carbon chemistry and its effect on polymerization vs the creation of highly reactive radicals. It may be that different processes dominate in different areas of the comet depending on the initial heterogeneity of the surface and the path of the comet.

ACKNOWLEDGEMENTS

The author wishes to acknowledge NASA grant NAGW-15.

REFERENCES

Bar-nun, A., Herman, G., Laufer, D., & Rappaport, M.L. (1985) Trapping and release of gases by water ice and implications for icy bodies, *Icarus*, **63** 317

Bertaux J.-L., & Abergel, A. (1986) Some physical characteristics of Halley's nucleus as inferred from VeGa and Giotto pictures, in the proceedings of *20th ESLAB Symposium on the Exploration of Halley's Comet*, Heidelberg, Germany, October 27–31, 1986, **II** p. 341

Brin, G.D., & Mendis, D.A. (1979) Dust release and mantle development in comets, *Astrophys. J.*, **229** 402

Cowan, J.J., & A'Hearn, M.F. (1979) Vaporization of comet nuclei: light curves and life times, *Moon and Planets*, **21** 155

Delsemme, A.H., & Swings, P. (1952) Hydrates de gaz dans les noyaux cométaires et les grains interstellaires, *Ann. Astrophys.*, **15** 1

Dobrovolskii, O.V., & Markovich, M.Z. (1972) On non-gravitational effects in two classes of models of cometary nuclei, In: *The motion, evolution of orbits and origin of comets*, Chebotarev, G.A., Kazimirchak-Polonskaya E.I., & Marsden, B.G. eds, D. Reidel Pub. Co., Dordrecht, Holland, 287

Donn, B. (1963) The origin and structure of icy cometary nuclei, *Icarus*, **2** 396

Donn, B. (1976) The nucleus: panel discussion, In: *Study of comets*, Donn, B., Mumma, M., Jackson, W., A'Hearn, M., & Harrington, R. eds, Part 2, 611

Donn, B., & Hughes, D. (1986) A fractal model of a cometary nucleus formed by random accretion, in the proceedings of *20th ESLAB Symposium on the Exploration of Halley's Comet*, Heidelberg, Germany, October 27–31, 1986, **III** 523

Donn, B., & Rahe, J. (1982) Structure and origin of cometary nuclei, in *Comets*, L.L. Wilkening (ed.), University of Arizona Press, Tucson, Arizona, 203

Donn, B., & Urey, H.C. (1956) On the mechanism of comet outbursts and the chemical composition of comets, *Astrophys. J.*, **123** 339

Draganic, I.G., Draganic, Z.D., & Vujosevic, S. (1984) Some radiation chemical aspects of chemistry in cometary nuclei, *Icarus*, **60** 464

Emerich, C., Lamarre, J.M., Moroz, V.I., Combes, M., Sanko, N.F., Nikolsky, Yu V, Rocard, F., Gispert, R., Coron, N., Bibring, J.P., Encrenaz, T., & Crovisier, J. (1986) Temperature and size of the nucleus of Halley's Comet deduced from I.K.S. infrared VeGa-1 measurements, in the proceedings of *20th ESLAB Symposium on the Exploration of Halley's Comet*, Heidelberg, Germany, October 27–31, 1986, **II** 381

Fanale, F.P., & Salvail, J.R. (1984) An idealized short-period comet model: surface insolation, H_2O flux, dust flux, and mantle evolution, *Icarus*, **60** 476

Fanale, F.P., & Salvail, J.R. (1987) The loss and depth of CO_2 ice in comet nuclei, *Icarus*, **72** 535

Froeschlé, C., & Rickman, H. (1983) Surface temperature distributions of the nucleus of Comet P/Halley, In: *Asteroids, comets and meteors*, Lagerkvist, C.-I., & Rickman, H. eds, 225

Greenberg M. (1985) Fluffy comets, In: *Asteroids, comets, meteors II*, Lagerkvist, C.-I., Lindblad, B.A., Lundstedt H., & Rickman, H. eds, 221

Gombosi, T.I., & Houpis, H.L.F. (1986) An icy-glue model of cometary nuclei, *Nature*, **324** 43

Gombosi, T.I., Nagy, A.F., & Cravens, T.E. (1986) Dust and neutral gas modeling of the inner atmospheres of comets, *Rev. Geophys*, **24** 667

Harwit, M. (1968) Spontaneously split comets, *Astrophys. J.*, **151** 789

Hellmich, R. (1981) The influence of the radiation transfer in cometary dust halos on the production rates of gas and dust, *Astron. Astrophys.*, **93** 341

Herman, G., & Podolak, M. (1985) Numerical simulation of comet nuclei. I. Water-ice comets, *Icarus*, **61** 252

Horanyi, M., Gombosi, T.I., Cravens, T.E., Korosmezey, A., Kecskemety, K., Nagy, A.F., & Szegö, K. (1984) The friable sponge model of a cometary nucleus, *Astrophys. J.* **278** 449

Houpis, H.L.F., & Gombosi, T.I. (1986) An icy-glue nucleus model of Comet Halley, in the proceedings of *20th ESLAB Symposium on the Exploration of Halley's Comet*, Heidelberg, Germany, October 27–31, 1986, **II** 397

Houpis, H.L.F., & Mendis, D.A. (1981a) The nature of the solar wind interaction with CO_2/CO-dominated comets, *Moon and Planets*, **25** 95

Houpis, H.L.F., & Mendis, D.A. (1981b) On the dust zoning of rapidly rotating cometary nuclei, *Astrophys. J.*, **251** 409

Houpis, H.L.F., Ip, W.-H., & Mendis, D.A. (1985) The chemical differentiation of the cometary nucleus: the process and its consequences, *Astrophys. J.*, **295** 654

Houpis, H.L.F., Gombosi, T.I., & Korosmezey, A. (1986) A self-consistent continuous approach to thermal history calculations of cometary nuclei, preprint from paper read at the *20th ESLAB Symposium on the Exploration of Halley's Comet*, Heidelberg, Germany, October 27–31,

Huebner, W.F., Keller, H.U., Wilhelm, K., Whipple, F.L., Delamere, W.A., Reitsema, H.J., & Schmidt, H.U. (1986) Dust-gas interaction deduced from Halley multicolour camera observations, in the proceedings of *20th ESLAB Symposium on the Exploration of Halley's Comet*, Heidelberg, Germany, October 27–31, 1986, **II** 363

Johnson, R.E., Cooper, J.F., & Lanzerotti, L.J. (1986) Radiation formation of a non-volatile crust, in the proceedings of *20th ESLAB Symposium on the Exploration of Halley's Comet*, Heidelberg, Germany, October 27–31, 1986 **II** 269

Kanninen, M.F., & Popelar, C.H. (1985) *Advanced fracture mechanics*, Oxford engineering series

Keller, H.U., Arpigny, C., Barbieri, C., Bonnet, R.M., Cazes, S., Coradini, M., Cosmovici, C.B., Delamere, W.A., Huebner, W.F., Hughes, D.W., Jamar, C., Malaise, D., Reitsema, H.J., Schmidt, H.U., Seige, P., Whipple, F.L., & Wilhelm, K. (1986) First Halley multicolour camera imaging results from Giotto, *Nature*, **321** 320

Klinger, J. (1981) Some consequences of a phase transition of water ice on the heat balance of comet nuclei, *Icarus*, **47** 320

Klinger, J. (1983) Classifications of cometary orbits based on the concept of the orbital mean temperature, *Icarus*, **55** 169

Kresak, L. (1981) Evolutionary aspects of the splits of cometary nuclei, *Bull. Astron Inst. Czechoslovakia*, **32** 19

Kührt, E. (1984) Temperature profiles and thermal stresses in cometary nuclei, *Icarus*, **60** 512

Kührt, E., Mohlmann, D., Giese B., & Tauber, F. (1986) Thermal stresses and dust dynamics on comets, in the proceedings of *20th ESLAB Symposium on the Exploration of Halley's Comet*, Heidelberg, Germany, October 27–31, 1986, **II** 385

Mendis, D.A., & Brin, G.D. (1977) The monochromatic brightness variations of comets, II, The core–mantle model, *Moon and Planets*, **17** 359

Mendis, D.A., Houpis, H.L.F., & Marconi, M.L. (1985) The physics of comets, *Fundamentals of Cosmic Physics*, **10** 1

Mendis, D.A., Hill, J.R., Houpis, H.L.F., & Whipple, Jr., E.C. (1981) On the electrostatic charging of the cometary nucleus, *Astrophys. J.*, **249** 787

Mohlmann, D., Borner, H., Danz, M., Elter, G., Mangoldt, T., Rubbert, B., & Weidlich, U. (1986) Physical properties of P/Halley—derived from VeGa images, in the proceedings of *20th ESLAB Symposium on the Exploration of Halley's Comet*, Heidelberg, Germany, October 27–31, 1986, **II** 339

Opik, E.J. (1956) Interplanetary dust and terrestrial accretion of meteoric matter, *Irish Astron. J.*, **4** 84

Pirronello, V., & D'Arrigo, C. (1985) Particle bombardment of minor bodies during eruptive phases of the early solar system, in *Asteroids, comets, meteors*, II, Lagerkvist, C.-I., Lindblad, B.A., Lundstedt, H., & Rickman, H. eds, 235

Pittich, E.M. (1971) Space distribution of the splitting and outbursts of comets, *Bull. Astron Inst. Czechoslavakia*, **22** 143

Podolak, M., & Herman, G. (1985) Numerical simulations of comet nuclei, II, The effect of the dust mantle, *Icarus* **61** 267

Reitsema, H.J., Delamere, W.A., Huebner, W.F., Keller, H.U., Schmidt, W.K.H., Wilhelm, K., Schmidt, H.U., & Whipple, F.L. (1986) Nucleus morphology of Comet Halley, in the proceedings of *20th ESLAB Symposium on the Exploration of Halley's Comet*, Heidelberg, Germany, October 27–31, 1986, **II** 351

Rickman, H. (1986) paper read at the *ESA Workshop—Comet nucleus sample return*, Canterbury, UK, July 15–17

Rickman, H., & Froeschlé, C. (1982) Thermal models for the nucleus of Comet P/Halley, in *Cometary Exploration*, Gombosi T.I., (ed.), **1** 75

Sagdeev, R.Z., Blamont, J., Galeev, A.A., Moroz, V.I., Shapiro, V.D., Shevchenko, V.I., & Szegö, K. (1986) VeGa spacecraft encounters with Comet Halley, *Nature*, **321** 259

Sekanina, Z. (1982) The problem of split comets in review, In: *Comets*, Wilkening L.L., (ed.), University of Arizona Press, Tucson, Arizona, 251

Sekanina, Z. (1986) Dust environment of Comet Halley, in the proceedings of *20th ESLAB Symposium on the Exploration of Halley's Comet*, Heidelberg, Germany, October 27–31, 1986, **II** 131

Sekanina, Z., & Larson, S.M. (1986) Coma morphology and dust-emission pattern of periodic Comet Halley. IV. Spin vector refinement and map of discrete dust sources for 1910, *Astron. J.* **92** 462

Shul'man, E.M. (1972) The chemical composition of cometary nuclei, in: *The motion, evolution of orbits and origin of comets*, Chebotarev, G.A., Kazimirchak-Polonskaya, E.I., & Marsden, B.G. eds, D. Reidel Pub. Co., Dordrecht, Holland, 265

Smoluchowski, R. (1981a) Amorphous ice and the behavior of cometary nuclei, *Astrophys. J.*, **244** L31

Smoluchowski, R. (1981b) Heat content and evolution of cometary nuclei, *Icarus*, **47** 312

Smoluchowski, R. (1985) Brightness curve and porosity of cometary nuclei, In: *Asteroids, comets, meteors II*, Lagerkvist, C.-I., Lindblad, B.A., Lundstedt, H., & Rickman, H., eds, 305

Smoluchowski, R., & McWilliam, A. (1984a) Structure of ices on satellites, *Icarus*, **58** 282

Smoluchowski, R., Marie, M., & McWilliam, A. (1984b) Evolution of density in solar system ices, *Earth, moon and planets*, **30** 281

Squires, R.E., & Beard, D.B. (1961) Physical and orbital behavior of comets, *Astrophys. J.*, **133** 657

Squyres, S.W., McKay, C.P., & Reynolds, R.T. (1985) Temperatures within cometary nuclei, *J. Geophys. Res.*, **90** 12381

Strazzulla, G., Pirronello, V., & Foti, G. (1983) Physical and chemical effects induced by energetic ions on comets, *Astron. Astrophys.*, **123** 93

Wallis, M.K. (1986) Composition and evolution of cometary nuclei, In: *Comet nucleus sample return*, proceedings of ESA Workshop, Canterbury, UK, July 15–17

Watson, K., Murray, B.C., & Brown, H. (1963) The stability of volatiles in the solar system, *Icarus*, **1** 317

Weissman, P.R., & Kieffer, H.H. (1981) Thermal modelling of cometary nuclei, *Icarus*, **47** 302

Weissman, P.R., & Kieffer, H.H. (1984) An improved thermal model for cometary nuclei, *J. Geophys. Res.*, **89** C358

Whipple, F.L. (1950) A comet model, I, The acceleration of Comet Encke, *Astrophys. J.*, **III** 375

Whipple, F.L. (1977) The constitution of cometary nuclei, in *Comets, asteroids, meteorites*, Delsemme, A.H., ed. University of Toledo Press, Toledo, Ohio, 25

Whipple, F.L. (1978) Cometary brightness variation and nucleus structure, *Moon and Planets*, **18** 343

Whipple, F.L., & Stefanik, R.P. (1966) On the physics and splitting of cometary nuclei, *Mem. Soc. Roy. Liége* (ser. 5), **12** 33

Whitney, C.A. (1955) Comet outbursts, *Astrophys. J.*, **122** 190

Wilhelm, K., Cosmovici, C.B., Delamere, W.A., Huebner, W.F., Keller, H.U., Reitsema, H., Schmidt, H.U., & Whipple, F.L. (1986) A three-dimensional model of the nucleus of Comet Halley, in the proceedings of *20th ESLAB Symposium on the Exploration of Halley's Comet*, Heidelberg, Germany, October 27–31, 1986, **II** 367

15

Rotation vector of Halley's Comet

Z. Sekanina

1 INTRODUCTION

Rotation is one of the fundamental properties of solar-system objects. Knowledge of it yields information that is useful for constructing models of the solar nebula, and provides clues to understanding the subsequent evolution of the individual objects. Having spent most of the 4.6 thousand million years since their origin near the solar system's periphery, comets can be expected to have rotational characteristics potentially very different from those typical for objects confined to the inner region, such as the planets or asteroids.

Rotation of a solid body in a heliocentric orbit produces two types of change on its surface. Its spinning brings about *diurnal* variations, whereas the tilt of the spin axis to the orbital plane causes *seasonal* variations. Mankind's experience with diurnal and seasonal variations on Earth, however profound, is very limited in comparison with the possible range of extremes that one would encounter elsewhere, on comets in particular, for several reasons: (i) Earth orbits the Sun at a nearly constant distance, so that the seasonal variations are not amplified by the insolation's inverse-square power law of heliocentric distance; (ii) Earth's spin axis makes only a relatively small angle with the normal to the orbital plane (low-obliquity configuration), and the overwhelming majority of the world's population lives either in one of the temperate zones or in the equatorial belt, thus being protected from harsh effects of the seasonal variations; and (iii) Earth's oceans and its permanent, dense atmosphere and relatively rapid rotation moderate diurnal variations and keep their effects within tolerable limits.

By contrast, most comets orbit the Sun in extremely elongated ellipses (often barely distinguishable from a parabola), appear to have a more or less random distribution of rotation-pole orientations (which includes one with the most unequal surface-insolation distribution possible, with one pole pointing at the Sun at perihelion), have a tendency to spin slowly, and have no permanent atmosphere of their own to shield the surface of the nucleus. All these circumstances add up to yield a large-amplitude, highly variable surface-temperature profile, with major implications for the surface morphology, activity, and, indeed, survival of comets.

The reader should be aware of the fact that this review was written in October 1987 and that it reflects the state of our knowledge at that time. Only at several places, where deemed most urgent, were a few lines of updated infomration added in October 1989, to alert the reader of recent progress. The development of Halley's rotation models since 1987 has been so explosive that updating this chapter fully would be equivalent to rewriting it almost completely, a solution that for practical reasons could not be implemented. It is hoped that the following review can be of interest as an introduction to a specific topic of cometary science that has probably been studied more intensively than any other issue of comparably limited scope. In early October 1989, a workshop on the rotation of Halley's Comet was organized by S.M. Larson and M.J.S. Belton in Tucson, sponsored jointly by the University of Arizona's Lunar & Planetary Laboratory and the National Optical Astronomy Observatories. Two dozen invited participants from all over the world, each of them heavily involved in Halley's rotation studies, were still unable to reach a consensus on the matter, although tighter constraints on the range of feasible models were agreed upon. At the same time, announcements were made that new major efforts are in progress or planned for the near future in an attempt to resolve the problem. In the meantime, the reader is referred to two review papers presented in April 1989 at the Bamberg, West Germany, colloquium 'Comets in the Post-Halley Era', which are scheduled to appear in the written form in the proceedings from this conference, to be published in 1990 under the same title by Kluwer, with R.L. Newburn, Jr., J. Rahe, and M. Neugebauer as editors. The two papers are titled 'Characterization of the Rotation of Cometary Nuclei' by M.J.S. Belton and 'Cometary Activity, Discrete Outgassing Areas, and Dust-Jet Formation' by Z. Sekanina and cover, from different angles, the various issues closely related to the problem of Comet Halley's rotation. Either of them refers to the results of about 40 diverse studies which were published in 1988 or 1989, following the completion of this review, or which, by the end of 1989, were still in press.

2 INTERACTION BETWEEN ACTIVITY OF A COMET AND ITS ROTATION

Activity and rotation of comets affect each other in a most unusual and complicated way. The story begins with Whipple's (1950) classical study, in which the fundamental concept of a rotating ice–dust conglomerate comet nucleus was conceived. One of the strongest points of Whipple's model is its ability to explain the well-known nongravitational perturbation of a comet's motion as a result of backthrust on the nucleus from its anisotropic outgassing. Whipple has

shown that rotation must play a major role, because heat-transfer lags make the sublimation rate peak not at local noon, but later in the comet's day. The resulting thrust on the nucleus is either in the general direction of the comet's heliocentric motion or in the opposite direction (thus bringing about an orbital deceleration or acceleration, that is, lengthening or shortening the orbital period), depending upon whether the sense of rotation is prograde or retrograde. The approximately equal numbers of comets known to be affected by nongravitational deceleration and acceleration are believed to imply an essentially random distribution of cometary spin-vector orientations. One may add that outgassing asymmetry relative to perihelion must also result in a nongravitational change in the orbital period, even if sublimation should peak at noon. Perturbations of the orbit are not the only dynamical effect imparted by outgassing on a nonspherical nucleus. An off-centre component of the momentum that is transferred to the nucleus exerts a torque on the spin axis, forcing its precession. And if the momentum vector does not pass through the spin axis, the rotation rate is affected (spin-up or spin-down). The magnitudes of these effects depend strongly upon the mass, shape, rotation period, and degree of surface roughness of the nucleus, as well as upon the mass distribution in its interior (via the moments of inertia). Sophisticated *in situ* experiments on rendezvous missions to comets are required to determine these effects.

Additional complications arise from significant morphological heterogeneity of the nucleus surface, now universally recognized as the primary cause for the highly variable activity of Periodic Comet Halley and most other comets. As a rule, major outgassing proceeds only from discrete emission centres (vents) and only when they face the Sun. As a result, activity is generally confined to a rather small fraction of the comet's surface and is spin modulated, as the individual sources rotate into and out of sunlight. Depending upon the obliquity of the spin axis, some sources may be exposed to the Sun continuously over a number of rotations (polar-day regime), while others may experience long periods with the Sun below the horizon (polar-night regime). Obviously, comet activity *is* substantially affected by rotation. But as outgassing patterns change, so do their effects on the spin vector: Rotation is thus, in turn, affected by the activity. Very irregular, low-mass, and/or slowly spinning nuclei are especially susceptible to rapid precession, which must entail major changes in sublimation on relatively short time scales (months or even weeks). To make things worse, comet nuclei may also precess freely as a result of having been spun (perhaps during catastrophic events, such as a splitting; or by other, more benign processes) about an axis that does not coincide with one of the principal axes of inertia. Free precession is gradually damped out because of internal energy dissipation, as the rotation vector has a tendency to align with the angular momentum vector. However, the alignment time required is believed to be extremely long for comets, in spite of their poorly cemented, nonrigid structure (section 5.4).

3 TECHNIQUES FOR DETERMINING THE SPIN VECTOR OF COMETS

The complexities of the relationship between rotation and activity, outlined qualitatively in section 2, are certain to require an enormous amount of information on the nucleus and sophisticated methods of analysis in order to be described quantitatively. The major role of rotation contrasts sharply with the fact that until recently it was ignored in virtually all theoretical models of comets, because no information was available on either the spin-axis orientation or the rotation rate. In fact, the issue had been addressed on only a few occasions before 1950, based on evidence that could not measure up to present-day criteria. Because of the size of comet nuclei, indirect approaches must be developed that identify and exploit the expected diagnostic effects of rotation on observable properties of comets.

3.1 Rotation period

Beginning in the 1970s, efforts aimed at determining the period of rotation have largely concentrated on finding an appropriate attribute of a comet, or, preferably, of its nucleus condensation, which recurs at demonstrably regular intervals. Two types of attributes have been explored: (i) distinct dust features in the coma, and (ii) the brightness. The first

feature-based techniques exploited either the curvature of dust jets or the separation pattern of dust halos. Larson & Minton (1972) pioneered the 'lawn-sprinkler' model, applying it to Comet Bennett 1970 II (cf. also Larson 1978). Whipple (1978) developed a halo approach to derive the spin period of Comet Donati 1858 VI. By contrast, the light curve analysis represents an extension of a routine technique used to determine the rotation periods of asteroids; it was first applied to a comet by Fay & Wisniewski (1978).

Comet activity makes the light-curve approach vulnerable to criticism, in part because the contributions to the observed brightness from the nucleus and from emitted dust cannot usually be separated from each other, especially not when the comet is at large heliocentric distance. On the other hand, one can argue that a *regular* activity from a discrete emission source should be spin modulated and is therefore also diagnostic of rotation. However, if the brightness variations are due to sunlight scattered from the nucleus whose shape deviates from sphericity, the light curve should be dominated by a second harmonic. If the variations are due to the surface albedo distribution or are caused by activity of a solitary emission centre, the fundamental should prevail. If there is a multitude of active areas on the surface, a number of peaks per rotation would be expected. This approach may break down completely, if the comet's spin axis points Earthward, depending upon the phase angle and other circumstances. An additional disadvantage of this approach, if based on photoelectric measurements (rather than on integration of light from CCD images), is that the contributions from the nucleus condensation and from the various parts of the coma remain indeterminate; only their total is known.

Because of all these problems, a rotation-period determination based on positional observations of distinct features in the head is generally preferred. If the data points are dense enough, one can identify the individual features and follow their motions. The projected velocities can then be determined, rather than assumed. The major weakness of the halo approach, the most prolific among the feature-based techniques (Whipple 1982), remains its inability to ascertain the number of active areas involved. Since this weakness is shared by the light-curve method, they both often yield submultiples of the true rotation period.

3.2 Rotation pole and complete spin vector

The first method for determining rotation-pole orientations of comets was developed and applied to four objects by Sekanina (1979). It was pertinent only to comets that display an emission fan, which was perceived as a projection onto the plane of the sky of expanding ejecta, emerging from the point of maximum activity. This point was assumed to be located on the parallel of subsolar latitude and lagging behind the subsolar point because of thermal inertia of the sublimation process, the fan's axis measuring the thermal lag angle of maximum outgassing. This concept, subsequently incorporated into a forced-precession model for Periodic Comet Encke (Whipple & Sekanina 1979), regarded the nucleus as an essentially homogeneous snowball, on which the surface distribution of sublimation is controlled by the insolation. An alternative model of fan formation, emphasizing the existence of discrete centres of activity, was formulated two years later (Sekanina 1981a). More recently, a conceptually different paradigm has been conceived, which attributes fan-shaped features to a collimated continuous flow of ejecta from high-latitude sources on the nucleus rotating about an axis with a relatively high obliquity of equator to orbital plane (Sekanina 1987a). This paradigm is part of a general framework which offers an interpretation for a broad variety of dust features observed in comet heads, including spiral jets, halos, plumes, spikes, and envelopes (in addition to fans).

Since the late 1970s, work on rotational properties of comets has kept accelerating. Cowan & A'Hearn (1979) concluded that the light curve of Periodic Comet d'Arrest, whose nearly-constant plateau extends for almost three months after perihelion, may be explained if the spin axis has a particular orientation in the comet's orbital plane (obliquity 90°). A complete spin vector for Periodic Comet Schwassmann–Wachmann 1 was determined by Whipple (1980a) from a large volume of data on its asymmetric coma's expansion and orientation. Sekanina (1981b) calculated a complete spin vector for Periodic Comet Swift–Tuttle and constructed the first map ever of

active regions on the surface of a cometary nucleus. One year later, following its recovery in October 1982, Halley's Comet entered the scene.

4 ROTATION OF HALLEY'S COMET

By 1982 it had been clearly demonstrated that the structure of the dust coma contains information on the comet's spin vector, but it was recognized that different comets display vastly different structural characteristics whose analysis may require custom-made methods.

4.1 Spin vector from observations made at earlier apparitions

From the drawings made in 1835/36 (Bessel 1836, Schwabe 1836, Maclear 1838, Struve 1839, Herschel 1847) and from Bobrovnikoff's (1931) detailed descriptions of photographic and visual observations made in 1910, it was long known that the dust-coma morphology of Halley's Comet is extremely complex, and that it changes from day to day. From variations in the apparent size of the coma in 1835/36 and 1910, Whipple (1980b, 1982) derived a rotation period of 0.43 day. The efforts to determine the spin-axis position also began with an estimate by Whipple (1983). However, his analysis was not concerned with the various structural changes in the coma, and he did not discriminate between dust and gas. There was also a possibility that the 0.43-day cycle represented a submultiple of the true rotation period.

A major issue that needed to be addressed was that of how to 'read' the comet's dust-morphology patterns. The clue to their understanding was eventually offered by evidence on the evolution of spiral jets, 'unwinding' from the nucleus condensation on the Sunward side of the coma, into expanding quasi-parabolic envelopes. This development was shown to be diagnostic of dust ejected continuously from discrete emission regions on the sunlit side of Halley's rotating nucleus (Sekanina & Larson 1983). The first identification of structural changes in the dust coma was made possible by computer processing (Larson & Sekanina 1984) of the comet's images on the high-resolution photographs exposed by G.W. Ritchey with the Mount Wilson Observatory's 152 cm reflector in May–June 1910.

The initial set of rotation parameters, derived as a byproduct of analysis of dust emission from three active regions, placed the northern pole of rotation (defined as the pole from which the nucleus is seen to rotate counterclockwise) at R.A. 19° and Decl. $-27°$ (equinox 1950.0) and the spin period at 1.73 ± 0.4 days (Sekanina & Larson 1984). The large uncertainty in the rotation rate was a result of undersampling and a low sensitivity of the observed curvature of jets. It was concluded that the period could be as short as three and as long as five Whipple periods. The subsequent work (Sekanina 1985a, Sekanina & Larson 1986a), which eventually extended the analysis to 55 features on 34 computer processed images, yielded a refined position for the pole, implying an obliquity of nucleus equator to orbital plane of 30°. The rotation period came out to be very close to five Whipple periods. These results were soon to be compared with the first determinations based on observations from the comet's current apparition.

4.2 The 2.2-day period from observations made in 1982–86

The detection of a 2.2-day periodicity in late 1985 was first reported by Itoh (1985) and by Kaneda et al. (1986a) from a pulsation (or 'breathing') of Halley's hydrogen halo recorded by the Lyman Alpha imager on the Suisei spacecraft. Besides demonstrating for the first time that the comet's volatile fraction displays the same periodicity that had previously been shown to apply to dust, this experiment also detected the presence of two major and several minor events during the recurrence period (Kaneda et al. 1986b,c) that appear to correlate with some of the dust sources (cf. Sekanina 1986a for details). The 2.2-day periodicity was also supported by the Vega and Giotto spacecraft observations (Wilhelm et al. 1986, Sagdeev et al. 1986a, Bertaux & Abergel 1986); by observations of recurring expanding CN shells (Schlosser et al. 1986); by separations of discrete streamers (synchrones) in the dust tail (Sekanina 1986b); and by motions of condensations in the plasma tail (Celnik 1986).

As expected (section 3), photometric observations yielded very discordant results. A 2.2-day periodicity was found by Leibowitz & Brosch (1986a,b) in the intensity ratio of C_2 to H_2O^+, but not in the

continuum light; and by Belton et al. (1986), who, however, restricted their search to periods between 50 and 60 h, and used selected data sets. On the other hand, a number of authors concluded that no preferred solution can be found from Halley's brightness data beyond 4.5 AU (Jewitt & Danielson 1984, West & Pedersen 1984, Le Fèvre et al. 1984, Morbey 1985, Sekanina 1985b, Festou et al. 1986; Meech et al. 1986), even though some of the solutions listed by Le Fèvre et al., by Morbey, and by Sekanina were near 2.2 days. By contrast, from their postperihelion photoelectric observations, in March–April 1986, Millis & Schleicher (1986) detected a recurrence period of 7.37 days, or about 3.3 times longer, in the brightness pertaining to the passbands of several minor species (including C_2 and CN) as well as to the continuum. Their light curves exhibited three peaks per period in March, but only two in April.

4.3. The 7.4-day period

Millis & Schleicher's (1986) discovery played havoc with the rotation models, which were already running into difficulties with regard to the determination of the spin-axis orientation (section 4.4). Support for the 7.4-day periodicity was subsequently expressed by Samarasinha et al. (1986), who studied the distribution of CN jets in the coma in April 1986; by Schleicher et al. (1986), who constructed a composite light curve from observations at three observatories and analyzed data from the International Ultraviolet Explorer (IUE) in the time interval covering the spacecraft encounters with Halley's Comet; by Williams et al. (1987), who conducted ground-based narrow-band photometry of the comet in March and May 1986; by Stewart (1987), who analyzed hydrogen, oxygen, and carbon atomic abundances measured by the ultraviolet spectrometer aboard the Pioneer Venus Orbiter; and by Festou et al. (1987), who tested Millis & Schleicher's period on their IUE observations and who also reinvestigated the comet's light curve at large heliocentric distances along the inbound branch of the orbit. Festou et al.'s conclusions were that the IUE observations could be fitted with a period of about 7.4 days, and that the light curve indicated a period of 7.307 days, displaying three peaks per period. The curve, however, looks unusual in the sense that it shows steep slopes between the individual extrema, while the latter are very flat. Belton *et al.*'s (1986) and Festou *et al.*'s (1987) results appear to be mutually unreconcilable.

4.4 Spin axis in 1985–86

The first estimate of the position of the spin axis from observations of the current apparition was published by Grün *et al.* (1986), who interpreted stationary jets visible on several exposures in November 1985 as ejecta from the area of the sunlit rotation pole on the nucleus. Their constraints placed the pole some 30° to 50° away from Sekanina & Larson's (1986a) solution based on the comet's images from 1910. Sekanina & Larson (1986b) commented that Grün *et al.*'s pole coordinates would not be compatible with the motions of dust features in 1910, and that the discrepancy could be explained by nucleus precession.

Independent determinations of the spin-axis orientation were offered by Samarasinha *et al.* (1986) from their analysis of the cyano jets in April 1986 and by Sekanina *et al.* (1986) from their study of the Sunward spike in April–June 1986. These agree with each other within 14°, but differ from Grün *et al.*'s results by anywhere from 22° to over 50°. However, in view of an alternative interpretation of the spike (Fulle 1987), that solution must be regarded with caution.

Finally, the spin vector was also determined from images of the nucleus obtained with the cameras aboard Vega-1, Vega-2, and Giotto. Assuming that the polar axis was perpendicular to the long body axis of the nucleus, Sagdeev *et al.* (1986a) found an obliquity of nucleus equator to orbit of about 20° from the Vega 1 and 2 images. Wilhelm *et al.* (1986) derived the spin vector in the process of their three-dimensional modelling of the nucleus, based on available images from both Giotto and the two Vega spacecraft. They constructed cross-sectional contours of the nucleus at different points along an axis parallel to a spacecraft's relative-velocity vector as functions of the spin period and the rotation-pole position, and searched for the best correspondence with information on the existing images. Wilhelm *et al.* pointed out that the nucleus could be approximated by a triaxial ellipsoid, with 'some protrusions and indentations.' Their spin-vector solution differed by about 10° from

that by Sagdeev *et al.* and by 5° to 30° from that by Grün *et al.* Wilhelm *et al.* remarked that the images from Vega-1 would be fitted much better with 'slightly different' sets of parameters than the images from Vega-2. An iterative procedure, involving a variation in the orientation of the vector of maximum moment of inertia, led them to conclude that the rotation axis of the nucleus nutates with a half-cone angle of 5° to 10° about the angular momentum vector, and that the principal body axes of the ellipsoid were 16, 10, and 9 km.

4.5 Summary

For clarity, Tables 1–3 list the available determinations of, respectively, the shorter period, the longer period, and the rotation-pole orientation. In cases of the same approach yielding gradually refined solutions, only the most updated result is presented. It should be noted that all relevant investigations led to the conclusion that Halley rotates in a prograde sense with respect to its orbital motion, that is, in a retrograde sense relative to Earth's rotation and orbital motion. This is reflected in Table 3, where the obliquity of equator to orbit is always smaller than 90° and the northern pole's declination is always negative.

5 MORE COMPLICATED ROTATION MODELS FOR HALLEY'S COMET

Before the nucleus shape and the rotation-period dichotomy became known, there had been no compelling reason to question as reasonable an assumption that Halley's nucleus is in pure spin. It is true that forced-precession models were developed for some comets (Whipple & Sekanina 1979, Sekanina 1984, 1985c,d, 1986c; Sekanina & Yeomans 1985), but all these objects exhibit significant variations, over periods of several revolutions about the Sun, in the transverse component of the nongravitational acceleration that affects their orbital motion. These variations were believed to be diagnostic of nucleus precession. For Halley's Comet this parameter is known to have stayed essentially constant over a span of some 2000 years (Yeomans & Kiang 1981), which was interpreted to indicate the lack of measurable precessional motion of the comet's spin axis (Yeomans & Kiang 1981, Sekanina 1983).

Table 1. List of the 2.2-day recurrence-period determinations for P/Halley

Apparition	Time span used	Recurrence period (h)[a]	Approach/Event(s)	Reference(s)
1910	May–June 1910	52.1	Recurring dust jets	Sekanina & Larson (1986a), Sekanina (1985a)
1986	November 1985–April 1986	52.9 ± 1.0	Ly-α intensity variation (Suise)	Kaneda et al. (1986c)
1986	November–December 1985	52^{+3}_{-4}	Near-nucleus variation in C_2/H_2O^+	Leibowitz & Brosch (1986b)
1986	January–February 1986	52.5 ± 1.6	Dust outbursts (multiple tail)	Sekanina (1986b)
1986	February–April 1986	53.5 ± 1.7	Condensations in plasma tail	Celnik (1986)
1986	March–April 1986	53.4 ± 0.9	Expanding cyano shells	Schlosser et al. (1986)
1986	March 1986	53.5 ± 1.0	Nucleus images (Vega 1 and 2)	Sagdeev et al. (1986a)
1986	March 1986	54 ± 1	Nucleus images (Giotto and Vega)	Willhelm et al. (1986)
1986	March 1986	(54)	Nucleus images (Vega and Giotto)	Bertaux & Abergel (1986)
1986	January 1984–January 1985	53.96 ± 0.03 54.12 ± 0.03	Total-brightness variation beyond 5.09 AU from Sun preperihelion	Belton et al. (1986)[b]

[a] Parenthesized value indicates that the period was confirmed, but not independently determined.
[b] The two solutions are considered essentially equivalent by the authors.

Table 2. List of the 7.4-day recurrence-period determinations for P/Halley

Apparition	Time span used	Recurrence period (days)[a]	Approach/Event(s)	Reference
1986	March–April 1986	7.37 ± 0.07	Narrow-band photometry (C_2, CN, etc.)	Mills and Schleicher (1986)
1986	March 1986	(7.4)	Composite light curve; abundances (IUE)	Schleicher et al. (1986)
1986	March and May 1986	(7.4)	Narrow-band photometry (C_2, CN, etc.)	Williams et al. (1987)
1986	April 1986	(7.4)	Cyano jets	Samarasinha et al. (1986)
1986	January 1984–February 1985	7.307 ± 0.005	Light curve at large solar distances	Festou et al. (1987)
1986	February–March 1986	7.7 to 8.1,[b] (7.4)	Molecular abundances (IUE) Atomic abundances (Pioneer Venus)	Stewart (1987)

[a] Parenthesized value indicates that the period was confirmed, but not independently determined.
[b] These are referred to as synodic periods (Sun oriented); for a prograde rotation they represent upper limits to a sidereal period.

Table 3. List of the rotation-pole determinations for P/Halley

Apparition(s)	Time span used	North pole of rotation (deg) for eg. 1950.0		Obliquity of equator to orbit (deg)	Approach/Event(s)	Reference(s)
		R.A.	Decl.			
1835/36, 1910	May–June 1910	225 ± 25	−70 ± 10	40 ± 10		Whipple (1983)
1910		357	−49	30	Recurring dust jets	Sekanina & Larson (1986a), Sekanina (1985a)
1986	November 1985	34 to 43	−15 to −40	30 to 55	Stationary dust jets	Grün et al. (1986)
1986	March 1986	60	−49	21	Nucleus images (Vega 1 and 2)	Sagdeev et al. (1986a)
1986	March 1986	50 ± 10	−40 ± 5	30^{+5}_{-10}	Nucleus images (Giotto and Vega)	Willhelm et al. (1986)
1986	April 1986	8 ± 20	−53 ± 20	23 ± 20	Cyano jets	Samarasinha et al. (1986)
1986	April–June 1986	20	−66	9	Sunward spike	Sekanina et al. (1986)

The response to the existence of two recurrence periods has ranged from rejection of one or the other to attempts to reconcile them by assuming that the nucleus not only spins but also wobbles. The proposed interpretations aimed at accommodating the two periods are briefly described below, in the order of their publication.

5.1. Investigation by Sekanina (1986d, 1987b)

Sekanina's model was based on the assumption of a torque-free, homogeneous, rigid nucleus and on the fact, known from closeup imaging (e.g., Wilhelm *et al.* 1986), that the nucleus shape can be approximated by a triaxial ellipsoid, with one long and two short body axes.

The dynamically significant property of Halley's nucleus is the fact that the two short axes are of equal lengths, within the limits of uncertainty. In this case, Euler's equations indicate that the nucleus could rotate freely about the long axis, which itself would precess freely about the inertially fixed angular-momentum axis. The solution requires that the shorter period refer to the precessional motion and the longer period to the rotation about the long axis. Assuming that the periods are 2.2 and 7.4 days, the long axis subtends an angle of 77° with the angular-momentum vector, if the nucleus approximated by a homogeneous prolate spheroid has an axial ratio of 1.9. The instantaneous spin axis is always located in the plane defined by the long body axis and the angular-momentum vector, and in the described scenario it wobbles about the angular-momentum axis with a

ROTATION MODEL FOR HALLEY'S COMET

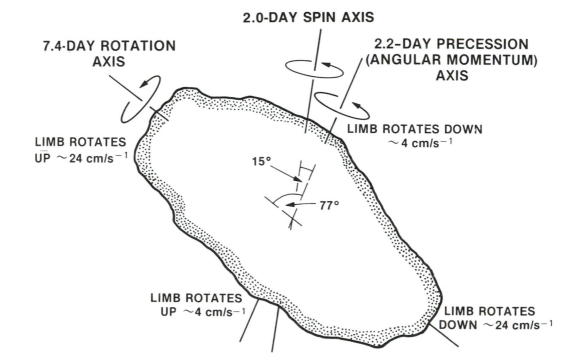

Fig. 1 — Schematic representation of the rotation model of Halley's Comet proposed by Sekanina (1986d, 1987b). To an outside observer, the nucleus rolls on its long axis with a period of 7.4 days, the axis making an angle of 77° with the angular-momentum vector and gyrating about it with a period of 2.2 days. The intrinsic spin period is 2 days about an axis that wobbles about the angular-momentum axis with a semiamplitude of 15°. Conceptually similar models were considered by Julian (1987), by Belton (1987), and by Festou *et al.* (1987). The models preferred in 1989 are those for which the long-axis roll is confined to oscillations with a semi-amplitude of several tens of degrees. (Sketch adapted from Sekanina 1987b.)

semiamplitude of approximately 15°. The model is schematically depicted in Fig. 1.

It should be pointed out that the terminology for complicated rotational motions is not fully standardized, and that the discussed model can equivalently be described by saying that the nucleus spins with a period of 2.2 days about an axis which is oblique to the principal axes of inertia, and that it rolls on its long axis with a period of 7.4 days. The interplay between the two periods may reflect the surface distribution of vents, major elevation variations, and other regolith irregularities. If the 7.4-day period should, instead, be a product of an (unknown) intrinsic rolling period about the long axis and the 2.2-day cycle, the model's numerical parameters would change, but not its basic properties. It would especially be desirable to have the two periodicities locked in a resonance.

Sekanina remarked that the 2.2-day period, being by far the shorter of the two, usually determines the initiation and termination of jet activity on the sunlit side of the nucleus, while the rotation about the long axis can at least qualitatively account for the temporary 'dormancy' and subsequent reactivation of discrete emission sources on the surface on a time scale of weeks, as noticed by Sekanina & Larson (1986a) from their analysis of the comet's jet patterns during the 1910 apparition. However, the concept of rotation about the long axis was criticized by Smith *et al.* (1987) for being incompatible with the spacecraft results of nucleus imaging.

5.2. Investigations by Julian (1987) and by Smith *et al.* (1987)

Julian extended the modelling of Halley's nucleus shape to include an ellipsoid with principal axes of unequal lengths. He remarked that a triaxial ellipsoid can precess freely in either of two qualitatively different modes: (i) a long-axis mode, which is equivalent to the prolate spheroid's precession of the long axis about the angular momentum vector; or (ii) a short-axis mode, which has no equivalent solution when the two short axes are equal. According to Julian, a triaxial-ellipsoid approximation for Halley's nucleus predicts that in the short-axis mode the angular-velocity vector should wobble about the shortest body axis with a semiamplitude of 11°.

The short-axis free-precession mode was proposed

in direct response to Smith *et al.*'s criticism of the long-axis mode. Smith *et al.* rejected the notion, based on Millis & Schleicher's (1986) work, that the nucleus spins with a period of 7.4 days. Using the nucleus images from Vega-1, Vega-2, and Giotto, they argued convincingly that the asymmetrical outlines of the body rule out the 7.4 days as the rotation period about an axis approximately perpendicular to the long axis.

Unfortunately, Smith *et al.*'s arguments are much less compelling when it comes to their rejection of 7.4 days as the rotation period about the long axis. The basis for their conclusion was a comparison of two nucleus profiles, one from a composite image by Giotto (Reitsema *et al.* 1986), the other from an image taken 1.5 s before closest approach by Vega-2. Smith *et al.* remarked that the relative positions of the spacecraft and the long axis of the nucleus on the two occasions were such that the respective nucleus profiles should be mirror images of each other, if the nucleus did not rotate about the long axis, or if it did so with a period of 4.7 days, the time difference between the encounters of the two spacecraft with Halley, or a submultiple of this interval. Their conclusion on the unacceptability of the 7.4-day rolling period was not based on a similarity of outlines of the whole nucleus, but only of its large end. The contour of the southern limb on the Giotto image is interrupted on Smith *et al.*'s chart (their Fig. 3), ostensibly because strong jet activity made it indefinite. However, the reconstruction of nucleus shape from the Giotto images (Reitsema *et al.* 1986) shows the nucleus bulging out in the region where the Vega-2 image reveals a deep depression. There are other problems with Smith *et al.*'s reasoning, including unequal aspect angles which imply that the viewing on the two occasions under consideration was not symmetrical relative to the orientation of the long axis. Constraints should be based on positional identification of surface features, an approach which cannot be pursued because the Vega images are not of sufficient quality to show any clear-cut features.

Festou *et al.* (1987) expressed doubts whether the implied semiamplitude of 11° of Julian's proposed short-axis mode of nucleus precession would be sufficient to produce light-curve variations as dramatic as those observed by Millis & Schleicher (1986) and by others.

5.3 Investigation by Belton (1987)

In a short semipopular review of facts and speculations on Halley's rotation, Belton brought forward some ideas that deserve to be reported. In several respects—such as the basic nucleus-shape approximation, the relationship between the 2.2-day and 7.4-day periods, and the angle between the long body axis and the angular-momentum axis—his model strongly resembles Sekanina's (1986d, 1987b; cf. section 5.1). Belton distinguished between the apparent rotational motion, as perceived by an outside observer; and a 'true' motion, which presumably would be experienced by the comet's imaginary 'resident.' Belton found, for example, that a 'true' spin period is 49 h, and that the component of spin about the long axis has a period of approximately 4.4 days, almost coinciding with the interval between the Vega-2 and Giotto encounters with the comet (cf. section 5.2). Slight deviations in the nucleus figure from a symmetric top should, according to Belton, bring about an additional motion, a 'nodding' of the spin axis with a period of 3.7 days, or exactly half the apparent rotation period about the long axis. He also remarked that the other periods should in this case experience slight modulations about their mean values, with amplitudes diagnostic of the distribution of mass in the nucleus interior.

5.4 Investigation by Wilhelm (1987)

In his model, Wilhelm considered both free and forced precession, solving Euler's equations with terms that included the components of the torque exerted on Halley's nucleus by jets of escaping ejecta. The model-dependent results showed that the forced precession entailed a spin-axis displacement equivalent to about 30° or more per perihelion passage. This result may explain the differences between the rotation-pole determinations derived for the apparitions of 1910 and 1986 (Table 3). Free precession was generally found by Wilhelm to be unimportant except in the cases of a random jet distribution. In an effort to explain the 7.4-day periodicity, he adopted the concept of a 'moderately deformable nucleus,' advocated by Bertaux & Abergel (1986) and by Julian (1987) to find that a nonrigid, fragile nucleus would be subjected to stress variations that could control the gas

production. According to Wilhelm, the highest production rates would be expected twice per precession period, when the terminus of the angular velocity vector shifts most rapidly in the body-fixed coordinate system. This scenario implied a free-precession period of 14.8 days and a general increase of activity after perihelion.

A worrisome aspect of Wilhelm's model is the unrealistically short damping time that he assumed for the free-precession motion, of the order of a few tens of years at most. This time scale contrasts sharply with Burns & Safronov's (1973) theory, according to which Halley should not have in fact aligned the rotation axis with its body axis on time scales longer by several orders of magnitude. According to Burns & Safronov, the alignment time due to internal energy dissipation varies as the cube of the rotation period and inversely as a product of the bulk density and the square of the nucleus size. It also depends on the object's shape and rigidity modulus and on the material's quality factor. The two most important effects that make the alignment process so slow for comets is their slow rotation and very low density. These two alone account for a factor of 10^4(!) difference in the damping rate between small asteroids and Halley, if one adopts Rickman's (1986) density estimate for the latter object. The quality factor and the rigidity modulus are of course unknown for comets, but even if their product is much smaller than Burns & Safronov's value for asteroids (based presumably on terrestrial rock analogs), Wilhelm would still under-estimate the damping time considerably. This line of reasoning leads inevitably to a somewhat disturbing conclusion that either the magnitude of random jet action produces no measurable wobbling of the spin axis, in which case Wilhelm's model provides no information on the source of the 7.4-day period, or the wobbling has been a runaway process for a long time, in which case the remarkable regularity of modulation of the comet's activity, whether persistent or only temporary, remains equally puzzling.

5.5. Investigation by Festou et al. (1987)

Festou et al.'s strong emphasis on the 7.3-day periodicity (section 4.3 and Table 2) was the starting point of their model. Its dynamical concept was derived from that of Sekanina's (1986d, 1987b) model (cf. section 5.1), with a few modifications. In the first place, Festou *et al.* concluded that the rotation period about the long axis was not 7.3 days, but twice this value. They found that a 14.6-day rotation period would produce a 7.3-day modulation of the light-curve variations, with the correct number of peaks and troughs, if superimposed on a precession period of 50 h, or precisely 1/7 of the rotation period. To accommodate the longer period, the angle between the angular-momentum vector and the long body axis must be increased from 77° in Sekanina's model to 84°. Assuming that there is only one active area on the nucleus surface and that it is located on the equator, Festou *et al.* calculated that the light curve should have three maxima per 7.3 days, separated from each other by either 2.05 days or 3.2 days.

An obvious problem with this model is the 50 h period. The best among the determinations listed in Table 1 suggest strongly that a periodicity of less than 52 h is unlikely, and that a periodicity of less than 51 h is virtually unacceptable.

The author of this review performed calculations of the expected insolation variations at various points on Halley's nucleus approximated by a rotating and precessing prolate spheroid, and found that the number and separations of maxima on the light curve depend dramatically upon the location of the active region on the surface and upon the positions of the angular-momentum and long body axes relative to the Sun. In addition, several sources can be—and in all probability often are—active at the same time, thus further complicating the light curve. It appears that more work is needed to clarify this issue.

6 CONCLUDING REMARKS

The brief summary of recent efforts to solve the dilemma of Halley's rotation shows that the situation is less than satisfactory. No specific model is widely accepted, and there is no consensus as to the true character of the rotational motion.

Most efforts aimed at developing a realistic rotation model have been frustrated by the lack of a resonance between the 2.2-day and 7.4-day periods on the one hand and by the apparent persistence of the 7.4-day periodicity, at least throughout March and much of April 1986, on the other hand. Nominally, the two periods are at a ratio of 10:3. A model in which the longer period would be three times the 7.4 days, or about 22 days, would formally imply a resonance of 10:1, but preliminary calculations suggest that the resulting light curve would be inconsistent with the observations.

The rotation models that are preferred for Halley's Comet in 1989 are those based on the short-axis mode (primarily because they imply an energetically less excited state), with a fairly fast oblique spinning of the long axis, a small nodding oscillation of the long axis relative to the angular-momentum vector, and a large oscillation (with an amplitude of many tens of degrees) in the long-axis roll.

Our continuing failure to understand fully the rotation of Halley's Comet in spite of the fact that *three* spacecraft intercepted the nucleus within 10 000 km of the surface illustrates that flyby-type missions are incapable of providing adequate information on even very fundamental properties of a cometary nucleus. The problem of Halley's rotation will either be solved primarily on the basis of Earth-based observations, or remain a puzzle until at least 2061, the comet's next return to the Sun. Even though significant discoveries were made thanks to several experiments carried out by the Halley armada, the future of space exploration of comets belongs to rendezvous-type missions, which allow extensive manoeuvering in the proximity of the target, guarantee comprehensive nucleus-surface reconnaissance, and offer adequate time for implementation of a sophisticated strategy of data acquisition. A complete description of a comet's state of rotation should certainly pose no problem for this type of mission.

This research was carried out by the Jet Propulsion Laboratory, California Institute of Technology, under contract with the National Aeronautics and Space Administration.

REFERENCES

Belton, M.J.S. (1987) The wobbling nucleus of Halley's Comet. *Planet. Rep.* **7** No. 2, 8–10

Belton, M.J.S., Wehinger, P., Wyckoff, S., & Spinrad, H. (1986) A precise spin period for P/Halley. In: *20th ESLAB Symposium on the Exploration of Halley's Comet*, ESA SP-250, Battrick, B., Rolfe, E.J., & Reinhard, R. eds, (ESTEC, Noordwijk, the Netherlands), vol. I, 599–603

Bertaux, J.L. & Abergel, A. (1986) Some physical characteristics of Halley's nucleus as inferred from Vega and Giotto pictures. In: *20th ESLAB Symposium on the Exploration of Halley's Comet*, ESA SP-250, Battrick, B., Rolfe, E.J., & Reinhard, R. eds, (ESTEC, Noordwijk, the Netherlands), vol. II, 341–345

Bessel, F.W. (1836) Beobachtungen über die physische Beschaffenheit des Halleyschen Kometen und dadurch veranlasste Bemerkungen. *Astron. Nachr.* **13** 185–232

Bobrovnikoff, N.T. (1931) Halley's Comet in its apparition of 1909–1911. *Publ. Lick Obs.* **17** 305–482

Burns, J.A., & Safronov, V.S. (1973) Asteroid nutation angles. *Mon. Not. Roy. Astron. Soc.* **165**, 403–411

Celnik, W.E. (1986) The acceleration within the plasma tail, the rotational period of the nucleus, and the aberration of the plasma tail of Comet P/Halley 1986. In: *20th ESLAB Symposium on the Exploration of Halley's Comet*, ESA SP-250, Battrick, B., Rolfe, E.J., & Reinhard, R. eds, (ESTEC, Noordwijk, the Netherlands), vol. I, 53–58

Cowan, J.J., & A'Hearn, M.F. (1979) Vaporization of comet nuclei: light curves and lifetimes. *Moon Planets* **21** 155–171

Fay, T.D., Jr., & Wisniewski, W. (1978) The light curve of the nucleus of Comet d'Arrest. *Icarus* **34** 1–9

Festou, M.C., Lecacheux, J., Kohl, J.L., Encrenaz, T., Baudrand, J., Combes, M., Despiau, R., Laques, P., Le Fèvre, O., Lemonnier, J.P., Lelièvre, G., Mathez, G., Pierre, M., & Vidal, J.L. (1986) Photometry and activity of the nucleus of P/Halley at heliocentric distances larger than 4.6 AU, pre-perihelion. *Astron. Astrophys.* **169** 336–344

Festou, M.C., Drossart, P., Lecacheux, J., Encrenaz, T., Puel, F., & Kohl-Moreira, J.L. (1987) Periodicities in the light curve of P/Halley and the rotation of its nucleus. *Astron. Astrophys.* **187** 575–580

Fulle, M. (1987) A possible neck-line structure in the dust tail of Comet Halley. *Astron. Astrophys.* **181** L13–L14

Grün, E., Graser, U., Kohoutek, L., Thiele, U., Massonne, L., & Schwehm, G. (1986) Structures in the coma of Comet Halley. *Nature* **321** 144–147

Herschel, J.F.W. (1847) *Results of Astronomical Observations at the Cape of Good Hope*. Smith & Elder, London. Chapter 5, 393–413

Itoh, T. (1985) Periodic Comet Halley (1982i) *IAU Circ.* No. 4155

Jewitt, D., & Danielson, G.E. (1984) Charge coupled device photometry of Comet P/Halley. *Icarus* **60** 435–444

Julian, W.H. (1987) Free precession of the Comet Halley nucleus. *Nature* **326** 57–58

Kaneda, E., Ashihara, O., Shimizu, M., Takagi, M., & Hirao, K. (1986b) Observation of Comet Halley by the ultraviolet imager of Suisei. *Nature* **321** 297–299

Kaneda, E., Hirao, K., Takagi, M., Ashihara, O., Itoh, T., & Shimizu, M. (1986a) Strong breathing of the hydrogen coma of Comet Halley. *Nature* **320** 140–141

Kaneda, E., Takagi, M., Hirao, K., Shimizu, M., & Ashihara, O. (1986c) Ultraviolet features of Comet Halley observed by Suisei. In: *20th ESLAB Symposium on the Exploration of Halley's Comet*, ESA SP-250, Battrick, B., Rolfe, E.J., & Reinhard, R. eds, (ESTEC, Noordwijk, the Netherlands), vol. I, 397–402

Larson, S. (1978) A rotation model for the spiral structure in the coma of Comet Bennett (1970 II). *Bull. Am. Astron. Soc.* **10** 589

Larson, S.M., & Minton, R.B. (1972) Photographic observations of Comet Bennett, 1970 II. In: *Comets: Scientific Data and Missions*, Kuiper, G.P., & Roemer, E. eds, (University of Arizona, Tucson) 183–208

Larson, S.M., & Sekanina, Z. (1984) Coma morphology and dust emission pattern of Periodic Comet Halley. I. High-resolution images taken at Mount Wilson in 1910. *Astron. J.* **89** 571–578, 600–606

Le Fèvre, O., Lecacheux, J., Mathez, G., Lelièvre, G., Baudrand, J., & Lemonnier, J.P. (1984) Rotation of Comet P/Halley: recurrent brightening observed at the heliocentric distance of 8 AU. *Astron. Astrophys.* **138** L1–L4

Leibowitz, E.M., & Brosch, N. (1986a) Periodic Čomet Halley (1982i) *IAU Circ.* No. 4162

Leibowitz, E.M., & Brosch, N. (1986b) Photoelectric discovery of a 52-hr periodicity in the nuclear activity of P/Halley. *Icarus* **68** 418–429

Maclear, T. (1838) Observations of Halley's Comet, made at the Royal Observatory, Cape of Good Hope, in the years 1835 and 1836. *Mem. Roy. Astron. Soc.* **10** 91–155

Meech, K.J., Jewitt, D., & Ricker, G.R. (1986) Early photometry of Comet P/Halley: development of the coma. *Icarus* **66** 561–574

Millis, R.L., & Schleicher, D.G. (1986) Rotational period of Comet Halley. *Nature* **324** 646–649

Morbey, C.L. (1985) Brightness variations in Comet P/Halley determined by the least scatter algorithm. *Astron. Express* **1** 133–136

Reitsema, H.J., Delamere, W.A., Huebner, W.F., Keller, H.U., Schmidt, W.K.H., Wilhelm, K., Schmidt, H.U., & Whipple, F.L. (1986) Nucleus morphology of Comet Halley. In: *20th ESLAB Symposium on the Exploration of Halley's Comet*, ESA SP-250, Battrick, B., Rolfe, E.J, & Reinhard, R. eds, (ESTEC, Noordwijk, the Netherlands), vol. II, 351–354

Rickman, H. (1986) Masses and densities of Comets Halley and Kopff. In: *The Comet Nucleus Sample Return Mission*, ESA SP-249, Melita, O. ed (ESTEC, Noordwijk, the Netherlands), 195–205

Sagdeev, R.Z., Krasikov, V.A., Shamis, V.A., Tarnopolski, V.I., Szegö, K., Tóth, I., Smith, B., Larson, S., & Merényi, E. (1986a) Rotation period and spin axis of Comet Halley. In: *20th ESLAB Symposium on the Exploration of Halley's Comet*, ESA SP-250, Battrick, B., Rolfe E.J., & Reinhard, R. eds, (ESTEC, Noordwijk, the Netherlands), vol. II, 335–338.

Sagdeev, R.Z., Avanesov, G.A., Shamis, V.A., Ziman, Ya.L., Krasikov, V.A., Tarnopolsky, V.I., Kuzmin, A.A., Szegö, K., Merényi, E., & Smith, B.A. (1986b) TV experiment in Vega mission: image processing technique and some results. In: *20th ESLAB Symposium on the Exploration of Halley's Comet*, ESA SP-250, Battrick, B., Rolfe, E.J., & Reinhard, R. eds, (ESTEC, Noordwijk, the Netherlands), vol. II, 295–305

Samarasinha, N.H., A'Hearn, M.F., Hoban, S., & Klinglesmith, D.A. (1986) CN jets of Comet P/Halley—rotational properties. In: *20th ESLAB Symposium on the Exploration of Halley's Comet*, ESA SP-250, Battrick, B., Rolfe, E.J., & Reinhard, R. eds, (ESTEC, Noordwijk, the Netherlands), vol. I, 487–491

Schleicher, D.G., Millis, R.L., Tholen, D., Lark, N., Birch, P.V.,

Martin, R., & A'Hearn, M.F. (1986) The variability of Halley's Comet during the Vega, Planet-A, and Giotto Encounters. In: *20th ESLAB Symposium on the Exploration of Halley's Comet*, ESA SP-250, Battrick, B., Rolfe, E.J., & Reinhard, R., eds, (ESTEC, Noordwijk, the Netherlands), vol. I, 565–567

Schlosser, W., Schulz, R., & Koczet, P. (1986) The cyan shells of Comet P/Halley. In: *20th ESLAB Symposium on the Exploration of Halley's Comet*, ESA SP-250, Battrick, B., Rolfe, E.J., & Reinhard, R. eds, (ESTEC, Noordwijk, the Netherlands), vol. III, 495–498

Schwabe, H. (1836) Der Halley'sche Komet. *Astron. Nachr.* **13** 145–152

Sekanina, Z. (1979) Fan-shaped coma, orientation of rotation axis, and surface structure of a cometary nucleus. *Icarus* **37** 420–442

Sekanina, Z. (1981a) Rotation and precession of cometary nuclei. *Annu. Rev. Earth Planet. Sci.* **9** 113–145

Sekanina, Z. (1981b) Distribution and activity of discrete emission areas on the nucleus of Periodic Comet Swift-Tuttle. *Astron. J.* **86** 1741–1773

Sekanina, Z. (1983) The Halley dust model. *Adv. Space Res.* **2** No. 12, 121–131.

Sekanina, Z. (1984) Precession model for the nucleus of Periodic Comet Kopff. *Astron. J.* **89** 1573–1586

Sekanina, Z. (1985a) Periodic Comet Halley (1982i). *IAU Circ.* No. 4151

Sekanina, Z. (1985b) Light variations of Periodic Comet Halley beyond 7 AU. *Astron. Astrophys.* **148** 299–308

Sekanina, Z. (1985c) Precession model for the nucleus of Periodic Comet Giacobini–Zinner. *Astron. J.* **90** 827–845

Sekanina, Z. (1985d) Nucleus precession of Periodic Comet Comas Solá. *Astron. J.* **90** 1370–1381

Sekanina, Z. (1986a) Nucleus studies of Comet Halley. *Adv. Space Res.* **5** No. 12, 307–316

Sekanina, Z. (1986b) Periodic Comet Halley (1982i). *IAU Circ.* No. 4187

Sekanina, Z. (1986c) Effects of the law for nongravitational forces on the precession model of Comet Encke. *Astron. J.* **91** 422–431

Sekanina, Z. (1986d) Periodic Comet Halley (1982i). *IAU Circ.* No. 4273

Sekanina, Z. (1987a) Anisotropic emission from comets: fans versus jets. I. Concept and modelling. In: *Diversity and Similarity of Comets*, ESA SP-278, Rolfe, E.J., & Battrick, B. eds, (ESTEC, Noordwijk, the Netherlands;) 315–322

Sekanina, Z. (1987b) Nucleus of Comet Halley as a torque-free rigid rotator. *Nature* **325** 326–328

Sekanina, Z., & Larson, S.M. (1983) Rotating comets: development of dust spirals into envelopes. *Bull. Am. Astron. Soc.* **15** 651

Sekanina, Z., & Larson, S.M. (1984) Coma morphology and dust emission pattern of Periodic Comet Halley. II. Nucleus spin vector and modelling of major dust features in 1910. *Astron. J.* **89** 1408–1425, 1446–1447

Sekanina, Z., & Larson, S.M. (1986a) Coma morphology and dust emission pattern of Periodic Comet Halley. IV. Spin vector refinement and map of discrete dust sources for 1910. *Astron. J.* **92** 462–482

Sekanina, Z., & Larson, S.M. (1986b) Dust jets in Comet Halley observed by Giotto and from the ground. *Nature* **321** 357–361

Sekanina, Z., Larson, S.M., Emerson, G., Helin, E.F., & Schmidt, R.E. (1986) Sunward spike and equatorial plane of Halley's Comet. In: *20th ESLAB Symposium on the Exploration of Halley's Comet*, ESA SP-250, Battrick, B., Rolfe, E.J., & Reinhard, R. eds, (ESTEC, Noordwijk, the Netherlands), vol. II, 177–181

Sekanina Z., & Yeomans D.K. (1985) Orbital motion, nucleus precession, and splitting of Periodic Comet Brooks 2. *Astron. J.* **90** 2335–2352

Smith, B., Szegö, K., Larson, S., Merényi, E., Tóth, I., Sagdeev, R.Z., Avanesov, G.A., Krasikov, V.A., Shamis, V.A., & Tarnopolski, V.I. (1986) The spatial distribution of dust jets seen at Vega-2 fly-by. In: *20th ESLAB Symposium on the Exploration of Halley's Comet*, ESA SP-250, Battrick, B., Rolfe, E.J., & Reinhard, R. eds, (ESTEC, Noordwijk, the Netherlands), vol. II, 327–332

Smith, B.A., Larson, S.M., Szegö, K., & Sagdeev, R.Z. (1987) Rejection of a proposed 7.4-day rotation period of the Comet Halley nucleus. *Nature* **326** 573–574

Stewart, A.I.F. (1987) Pioneer Venus measurements of H, O, and C production in Comet Halley near perihelion. *Astron. Astrophys.* **187** 369–374

Struve, F.G.W. (1839) *Beobachtungen des Halleyschen Cometen auf der Dorpater Sternwarte*. Kaiserl. Akad. Wiss., St Petersburg, 132 pp.

West, R.M., & Pedersen, H. (1984) Variability of P/Halley. *Astron. Astrophys.* **138** L9–L10

Whipple, F.L. (1950) A comet model. I. The acceleration of Comet Encke. *Astrophys. J.* **111** 375–394

Whipple, F.L. (1978) Rotation period of Comet Donati. *Nature* **273** 134–135.

Whipple, F.L. (1980a) Rotation and outbursts of Comet Schwassmann-Wachmann 1. *Astron. J.* **85** 305–313

Whipple, F.L. (1980b) Periodic Comet Halley. *IAU Circ.* No. 3459.

Whipple, F.L. (1982) The rotation of comet nuclei. In: *Comets*, ed Wilkening, L.L. (University of Arizona, Tucson), 227–250

Whipple, F.L. (1983) Cometary nucleus and active regions. In: *Cometary Exploration*, ed Gombosi, T.I (Hung. Acad. Sci., Budapest), vol. I, 95–110.

Whipple, F.L., & Sekanina, Z. (1979) Comet Encke: precession of the spin axis, nongravitational motion, and sublimation. *Astron. J.* **84** 1894–1909.

Wilhelm, K. (1987) Rotation and precession of Comet Halley. *Nature* **327** 27–30.

Wilhelm, K., Cosmovici, C.B., Delamere, W.A., Huebner, W.F., Keller, H.U., Reitsema, H., Schmidt, H.U., & Whipple, F.L. (1986) A three-dimensional model of the nucleus of Comet Halley. In: *20th ESLAB Symposium on the Exploration of Halley's Comet*, ESA SP-250, Battrick, B., Rolfe, E.J., & Reinhard, R., eds, (ESTEC, Noordwijk, the Netherlands), vol II, 367–369

Williams, I.P., Andrews, P.J., Fitzsimmons, A., & Williams, G.P. (1987) The variation in the brightness of Halley's Comet from ground-based narrow-band photometry. *Mon. Not. Roy. Astron. Soc.* **226** 1P–4P

Yeomans, D.K., & Kiang, T. (1981) The long term motion of Comet Halley. *Mon. Not. Roy. Astron. Soc.* **197** 633–646.

Part III
Evolution and orbital motion

16

The identification of early returns of Comet Halley from ancient astronomical records

F. Richard Stephenson

1 INTRODUCTION

Comet Halley is the only known comet which periodically becomes a bright naked-eye object. It is thus unique in having a long recorded history—extending over more than two millennia. The comet returns to the inner solar system at quasi-regular intervals varying from about 74 to 79 years (depending upon the extent of planetary perturbations). Around the time of perihelion passage, Halley's Comet often becomes a prominent object in the night sky, and typically remains visible to the unaided eye for about two months. Although it has probably never rivalled some of the brilliant daylight comets of very long period (of the order of a million years) which appear from time to time, it usually attracts widespread attention. Since the first telescopic observations in AD 1682, Comet Halley has been extensively observed at each return. In the pre-telescopic period, several independent reports are extant for most apparitions since AD 1000, while at least one record can be traced at *every* return since 240 BC. The number of successive returns which are chronicled (including the 1986 apparition) is thus as many as 30.

Identification of ancient returns of P/Halley (the prefix denoting the periodic nature of the comet) serves two main purposes. Foremost of these is the opportunity to test the accuracy of current theories of the long-term motion of the comet. In addition, such investigations provide valuable information on the reliability of ancient astronomical records. The most recent studies have been particularly concerned with the history of the comet in the period anterior to the Christian Era and this will be a major theme of the present paper.

2 HISTORICAL OUTLINE

It is convenient to divide the pre-telescopic history of Halley's Comet into three distinct periods, working in reverse chronological order. These are (i) AD 1607 to 1456, (ii) AD 1378 to 12 BC, and (iii) the period before 12 BC. The effects of the European Renaissance feature prominently in the recorded history of P/Halley. In AD 1378, although the comet was widely noted in Europe, the most accurate observations were made in China. However, by the following apparition in AD 1456, European positional measurements had surpassed the traditional-style observations of the imperial astronomers of the Orient. The disparity between occidental and oriental observations became even more marked at subsequent returns. Consequently, in period (i) Chinese and other Far Eastern (i.e. Japanese and Korean) reports of the comet, although often quite detailed, prove to be of little more than academic interest today. It is for this reason that most of these records have yet to be translated into a Western language.

Throughout much of interval (ii), European accounts of P/Halley are numerous, but few texts contain useful factual information. Most records are found in chronicles compiled by annalists who had no special interest or expertise in astronomy. The best source of European data—such as it is—at this period is still the catalogue of cometary sightings compiled by Pingre (1783). Undoubtedly, Arabic astronomical records have not been adequately exploited in the search for medieval references to P/Halley. Few translations of appropriate texts have ever been published. As noted briefly by Stephenson & Walker (1985), several chroniclers of Baghdad, Cairo, etc. report sightings of Halley's Comet. However, little positional information is supplied. On the evidence available, it would appear that throughout most of period (ii) observations from China and its cultural satellites Japan and Korea are consistently superior to those from any other part of the world. An extensive translation of virtually all known Far Eastern accounts of Comet Halley from ancient times to AD 1378 is given by Stephenson & Yau (1985). This is partly based on the general catalogue of oriental cometary sightings compiled by Ho Peng Yoke (1962), but contains many additional records.

The principal sources of Oriental observations of Halley's Comet are the astronomical treatises of the various dynastic histories of China. These extensive compilations cover the entire period from about 200 BC to the 17th century AD, and contain summaries of the observations made by the imperial astronomers. Apart from comets, the treatises regularly note celestial phenomena such as eclipses, lunar and planetary movements, stellar outbursts, and aurorae. Toward the end of the first millennium AD, similar observations began in both Japan and Korea. Hence from about AD 1000 it is not unusual to find two or even three independent reports of the same celestial event from the Far East. These often supplement one another to a considerable degree. Korean observations from about AD 1000 to 1400 are summarized in the astronomical treatise of the *Koryo-sa* ('History of Koryo'), a work which is modelled on a typical Chinese dynastic history. However, Japanese celestial phenomena tend to be much less systematically recorded. Some of the most detailed reports are found in privately compiled histories and diaries. The

observations contained in many of these works would have been relatively inaccessible had not Kanda (1934, 1935) made an exhaustive compilation of the astronomical records of his country.

The careful investigation of Kiang (1971), supplemented by the work of Stephenson & Yau (1985), shows that throughout period (ii) there are few difficulties in recognizing observations of Comet Halley among the 500 or so records of other bright comets. Often the description of the motion of the comet through the constellations is sufficiently detailed not only to identify P/Halley but to enable an important check to be made on the accuracy of modern numerical integrations of its past orbit. This latter aspect will be discussed further in section 2. It seems appropriate to give here an example of one of the very best Chinese descriptions of the comet. Although rather long, it illustrates just what the Chinese astronomers could achieve. The following account of the AD 837 apparition of P/Halley is taken from the astronomical treatise (chapter 36) of the *Chiu-t'ang-shu* ('Old History of the T'ang Dynasty'). All dates have been converted directly from the lunar calendar to the Julian system.

'On the night of Mar 22, a broom star appeared at the E direction. Its length was more than 7 ft and it was situated in the 1st deg of *Wei* (the 12th lunar mansion); it pointed W. On the night of Mar 24 it was SW of *Wei*. The length of the broom was more than 7 ft and the brightness of its rays was becoming fierce. It again pointed W. On the night of Mar 29 the broom was 8 deg in *Wei*. On the night of Apr 5 it was 3.5 deg in *Hsu* (11th lunar mansion). On the night of Apr 6 the length of the broom was more than 10 ft. It moved W in a straight line, pointing slightly S. It was 1.5 deg in *Hsu*. On the night of Apr 7 its length was more than 20 ft and its width was 3 ft. It was 9 deg in *Nu* (10th lunar mansion). On the night of Apr 9 its length was more than 50 ft. It branched into two tails, one pointing towards *Ti* (3rd lunar mansion) and the other concealing *Fang* (4th lunar mansion). It was 10 deg in *Nan-tou* (10th lunar mansion). On the night of Apr 11, the length of the broom was 60 ft. The tail was without branches and it pointed N. The Emperor summoned the Astronomer Royal and asked him the reason for these star changes ... That night the length of the broom was 50 ft and its width was 5 ft. It was moving towards the NW and pointing E. On the night of Apr 13 the length of the broom was more than 80 ft. It was still moving NW and pointing E; it was 14 deg in *Chang* (26th lunar mansion) . . . On the night of Apr 28 the broom was 3 ft in length. It appeared to the right of *Hsuan-yuan* (in Leo) and was pointing E. It was 7 deg in *Chang* (26th lunar mansion)'.

In the above translation (by Stephenson & Yau, 1985), the term rendered 'degree' is *tu*, which was actually equivalent to about 0.99 deg (there were 365.25 *tu* to a circle). As was customary at most periods in Chinese history, the lengths of cometary tails were expressed in terms of linear units, *chih* (here translated 'ft') and *chang* (equal to 10 *chih*). The angular equivalent of the *chih* was very roughly 1 deg, so that for a few days round closest approach to the Earth (10 Apr) the tail must have stretched across a considerable part of the sky. The comet was described as a *hui-hsing* ('broom star'), the usual Chinese term for a comet with a developed tail. The position of the head of Halley's Comet was measured relative to the 28 *hsiu* ('lunar mansions'). Each of these star groups— situated in the vicinity of the celestial equator— defined a range of R.A.

Finally, in the most ancient period (iii), cometary data from any part of the world are so scarce that effective use must be made of whatever material is still extant. Chinese cometary observations at this remote period tend to give minimal details, while European data are of negligible value. As was demonstrated by Stephenson *et al.* (1985), Babylonian astronomical texts, although in a very fragmentary condition, prove to be the most accurate source of information on P/Halley in ancient times. Historical records of consecutive apparitions in 87 BC (from both Babylon and China), 164 BC (Babylon alone) and 240 BC (China alone) have so far been identified. Before 240 BC, extant cometary records from any part of the world are of poor quality and at best give little more than the year of occurrence, so that Halley's Comet becomes lost in the mists of time.

3 ACCURACY OF CURRENT ORBITAL THEORIES

Commencing with the investigation of Brady &

Carpenter (1971), several long-term numerical integrations of the motion of Halley's Comet have been published in recent years. These include the solutions of Chang (1979), Yeomans & Kiang (1981), Brady (1982)—an extension of his original paper with Carpenter—Landgraf (1984), (1986). The more recent integration by Landgraf (1986) effectively supersedes his earlier work. In each case, the various authors computed sets of osculating orbital elements for every perihelion passage back to at least the 5th century BC. The rather ambitious investigations of Brady (1982) and Landgraf (1984), actually extended well into the 3rd millennium BC!

A fundamental difference between the various numerical integrations cited above is the degree of reliance on medieval and ancient observations of P/Halley. Each integration was initiated with an orbit determined from telescopic observations of the comet. However, both Chang (1979) and Brady (1982) made a single continuous integration of the comet's motion back to the appropriate terminal epoch; they did not incorporate any pre-telescopic observations in their solution. In contrast, Yeomans & Kiang (1981) and Landgraf (1986) constrained their integrations by using selected ancient and medieval Chinese observations of relatively high accuracy. The basis for the use of pre-telescopic data in this way may be outlined as follows.

In the historical past, Comet Halley has come within about 0.1 AU of the Earth on several occasions—see Fig. 1. By far the closest known encounter occurred in the spring of AD 837 when the comet passed by only about 0.035 AU away from our planet. Hence from time to time the motion of P/Halley has suffered significant perturbations by the Earth. The extent of these perturbations is difficult to compute precisely since a small change in the time of perihelion passage may seriously affect the comet's minimum distance from the Earth. An extreme example is provided by the apparition in AD 837. Calculations show that if Halley's Comet had been delayed by only four days, its nucleus would have come within three times the radius of the Moon's orbit. Under such circumstances, the effect of the Earth's gravitational field on the motion of the comet would have been severe. The possibility of making accurate calculations of P/Halley's previous orbit

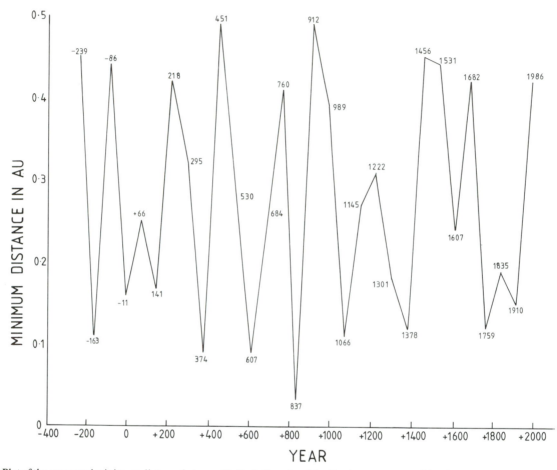

Fig. 1 — Plot of the computed minimum distance between Halley's Comet and the Earth at every apparition between 240 BC (-239) and AD 1986. Computations were made by the author, using the orbital elements of Yeomans & Kiang (1981). Distances are in AU (Astronomical Units, roughly equal to 150 million km). Note the unusually close approach in AD 837.

would have been precluded without the benefit of early observational data. In practice, the situation in AD 837 was less serious, but still resulted in marked perturbations of the comet's motion.

Even rather primitive observations can be used to investigate the reliability of modern computations of the past orbit of Comet Halley. This results from the rapid apparent motion of the comet when near the Earth; it often travels at more than a degree per hour (faster than the Moon). Hence only an approximate position among the constellations may be quite adequate. Especially in the case of a close flyby, the apparent motion of a comet across the celestial sphere is very sensitive to small changes in the time of perihelion passage (T). On the contrary, there is little

response to minor variations in the other orbital elements such as eccentricity and inclination. Hence in comparing theory with observation, it is only necessary to make a single parameter solution— solving for T alone. Observations of comparable accuracy to those made by the Chinese astronomers in AD 837 (quoted in section 2) enable dates of perihelion passage to be calculated to within about 0.1 day for several returns. This represents remarkably high precision, even by modern standards.

Deduction of a particular date of perihelion from observations can be most readily effected by interpolation. At the selected return, the course of the comet is first computed for a suitable range of T values, leading to a result for T which best matches observation. The dates of perihelion passage for P/Halley which can be deduced from observation in period (ii) are listed in Table 1. This table is a revised version of Table 3 of Stephenson & Yau (1985) and is based on an extensive analysis of Oriental observations of the comet between AD 1378 and 12 BC. At some returns during this period, the observations are not sufficiently accurate to enable a useful value of T to be deduced. However, the table is particularly complete for apparitions since AD 837. The relative accord between observation and theory for the various numerical integrations under discussion is shown in Fig. 2. This is based on Fig. 1 of Stephenson *et al.* (1985), but incorporates the new data of Table 1.

Table 1. Dates of perihelion deduced for Halley's Comet from observations made between 12 BC and AD 1378

Year	Observed T
12 BC	Oct 8.5 ± 0.3
AD 66	Jan 24.0 ± 2.0
141	Mar 23.3 ± 0.5
374	Feb 17.1 ± 0.5
530	Feb 27.0 ± 0.2
837	Feb 28.3 ± 0.1
989	Sep 9.0 or later
1066	Mar 23.2 ± 0.1
1145	Apr 21.3 ± 0.5
1222	Oct 1.1 ± 0.1
1301	Oct 24.8 ± 0.1
1378	Nov 10.8 ± 0.2

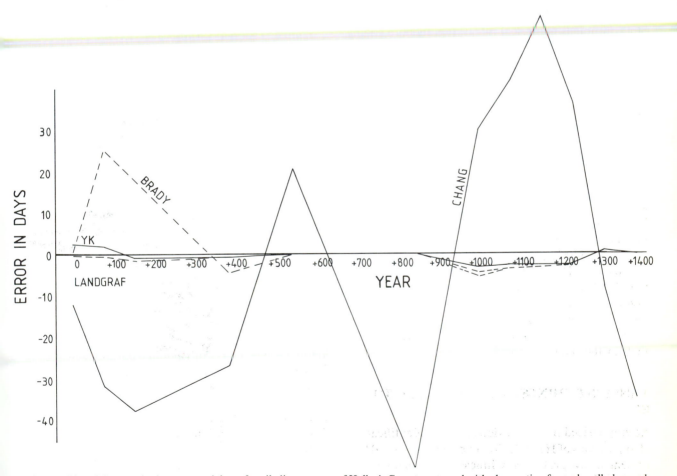

Fig. 2 — Plot of the error in the computed date of perihelion passage of Halley's Comet compared with observation for each well-observed apparition between 12 BC and AD 1378 (see Table 1). Computations are due to Chang (1979), Yeomans & Kiang (1981) [=YK], Brady (1982), and Landgraf (1986). The solution of Chang is clearly subject to large and fluctuating errors. (Note also the small but systematic discrepancies for the other solutions around AD 1000.)

Yeomans & Kiang (1981) used observations made in AD 837, 374, and 141 to constrain the computed motion of P/Halley, while Landgraf (1986) restricted his attention to the earliest of these observations. It is evident from Fig. 2 that both solutions are in excellent accord with observation throughout the period in question. By contrast, the results of both Chang (1979) and Brady (1982) are somewhat discordant. Chang's solution is in general of low precision, and the discrepancies between his results for T and those derived from observation exhibit marked variations. Although the T values computed by Brady (1982) are very accurate throughout the period since about AD 400, they are at variance with observation at most returns before that date. The explanation of these deviations probably lies—at least in part—in the use of a continuous numerical integration by both Chang and Brady. To achieve a reliable long-term solution, there seems little alternative to the incorporation of selected early observations.

As discussed by Stephenson et al. (1985), the values of T initially computed by Landgraf (1984) showed marked discord with those of Yeomans & Kiang (1981) in ancient times. However, the agreement between the revised solution of Landgraf (1986) and that of Yeomans and Kiang is impressive, despite the use of different techniques for modelling non-gravitational forces. As far back as 467 BC—where Landgraf's latest solution terminates—mutual discrepancies in T exceed a single day on only three occasions. The largest deviation is only about 4 days (in 164 BC). Hence it seems reasonable to conclude that if either of these numerical integrations is used to investigate the ancient history of Halley's Comet, limitations are more likely to be imposed by the deficiencies in the observations themselves rather than in the theory adopted.

4 ANCIENT CHINESE RECORDS: 12 BC TO 164 BC

As emphasized in the previous section, identification of apparitions of Halley's Comet during period (ii) has presented few problems. Chinese astronomers documented even the earliest apparition in this period—that of 12 BC—in considerable detail. This was a fairly close approach; at minimum distance; the

comet was about 0.16 AU from the Earth. The preserved text, in the 'Treatise on the Five Elements' (chapter 27) of the *Han-shu* ('History of the Former Han Dynasty'), is a summary of nearly two months of observation. It traces the comet's motion from Gemini to Scorpius, noting the change in direction after conjunction with the Sun. In the following translation by Stephenson & Yau (1985), dates are expressed in terms of the Julian Calender:

'On Aug 26 there was a 'Bushy Star' at *Tung-ching*; it was treading on *Wu-chu-hou*. It appeared to the north of *Ho-shu* and passed through *Hsuan-yuan* and *T'ai-wei*. Later it travelled at more than 6 deg daily. In the morning it appeared at the eastern direction. On the evening of the 13th day (Sep 7) it was seen at the western direction. It trespassed against *Tz'u-fei*, *Ch'ang-ch'iu*, *Pei-tou* and *Chen-hsing* (Saturn). Its 'swarming flames' again penetrated within *Tzu-wei*, with *Ta-huo* (Antares) right behind. It reached *T'ien-ho* (the Milky Way), sweeping the region of *Hou* and *Fei*. It moved southwards, crossing and trespassing against *Ta-huo* (Arcturus) and *She-t'i*. When it reached *T'ien-shih* it moved slowly at a regular pace. Its 'flames' entered *T'ien-shih*. After a further ten days it went towards the west. On the 56th day (Oct 20) it went out of sight together with *Ts'ang-lung*.'

In later periods Bushy Star (*hsing-po*) was the standard term to describe a comet without an obvious tail. However, Former Han terminology may not have been so well-defined; the comet of 12 BC clearly developed a significant appendage. In general, the motion described above agrees very well with the computed path of Halley's Comet in 12 BC on either the solution of Yeomans & Kiang (1981) or that of Landgraf (1986). However, as Kiang (1971) and Stephenson & Yau (1985) pointed out, the allusions to Saturn, Antares, and the Milky Way are unintelligible; neither the head nor the tail of the comet could have passed nearby. All other stars and star groups were close to the calculated course of P/Halley. Possible the discordant positions are the result of copying errors by the compilers of the *Han-shu*.

Both of the previous returns—in 87 BC and 164 BC—also occurred during the Former Han Dynasty (206 BC to AD 9) but here Chinese records prove to be extremely disappointing. The calculations of Yeomans & Kiang (1981) indicate dates of perihelion of 8 Aug in 87 BC and 12 Nov in 164 BC, the close apparition in 164 BC being unusually spectacular. These dates are closely confirmed by Landgraf (1986). No known reference to the 164 BC return survives in the history of China, while merely a brief Chinese account of the 87 BC apparition is extant. The latter is found only in the Imperial Annals (chapter 7) of the *Han-shu* rather than in one of the sections of the history specifically devoted to celestial phenomena. The text may be translated as follows:

'During the autumn, the 7th month (of the 2nd year of the Hou-yuan reign period), there was a Bushy Star at the eastern direction'.

Such brevity is quite characteristic of the few astronomical records found in the Imperial Annals. The date is equivalent to some time between 10 Aug and 8 Sep in 87 BC. Taken literally, the record would imply that the comet, following its customary retrograde path, had not yet reached conjunction with the Sun. However, it would be unwise to assign much weight to such a condensed secondary report.

The lack of useful Chinese descriptions of Halley's Comet in 87 BC and 164 BC may be due to the loss of many Han records (both astronomical and otherwise) before the official history of the dynasty was compiled (AD 68–76). In AD 23, the original Han capital of Ch'ang-an in Western China was destroyed by rebels. The devastation was so severe that soon afterwards a new capital was established in Eastern China. A study of the observations contained in the various sections of the *Han-shu* illustrates the sporadic survival of astronomical records other than eclipses of the Sun. Solar eclipses (which had special calendar importance) are systematically reported throughout the dynasty. Apart from the last few years of the dynasty, four distinct periods may be recognized. These are: (a) 206–163 BC, (b) 162–138 BC, (c) 137–68 BC, and (d) 67–1 BC. In periods (a) and (c) there are few extant observations (solar eclipses excepted), and generally dates are only roughly recorded—to the nearest month at best. Intervals (b) and (d) yield fairly frequent astronomical records, some of which are quite detailed; in most cases the exact date of the phenomenon is also noted. Presumably some of the original observations made by the court astronomers were still accessible from periods (a) and (c) when the

official history was put together, but at other times use had to be made of secondary material of lower reliability.

5 BABYLONIAN RECORDS: 87 BC AND 164 BC

Faced with the above difficulties, the author initiated a search of the Late Babylonian astronomical texts in the British Museum. This previously untapped source of ancient cometary observations yielded accounts of Halley's Comet in both 87 BC and 164 BC. Full details are given by Stephenson *et al.* (1985) and Stephenson & Walker (1985) and only a summary need be provided here. The Babylonian texts, which were accidentally discovered about a century ago, are largely in the form of astronomical diaries. They are inscribed on clay tablets using a cuneiform script which even today is understood by only a few scholars. The diaries are mainly concerned with lunar and planetary phenomena. They originally contained daily reports extending over several centuries— probably the entire interval from about 750 BC to AD 100. Although the extant tablets number more than 1000, most are in a very fragmentary condition, having been extensively damaged when they were dug up. As little as 5 per cent of the original vast corpus of material can now be accounted for. Sachs (1955) published drawings of many of the known texts, and later (1974) he produced a valuable commentary on their nature and content. At present, H. Hunger of the University of Vienna is in process of publishing transliterations and translations of this material.

Surviving pieces of Babylonian astronomical diaries mainly range in date from about 400 to 50 BC. Although a typical diary originally reported six or seven months of day-to-day observation, many preserved texts contain only fragmentary entries. In many examples, the date is broken away. Nevertheless, in such cases it is often possible to deduce the correct date by astronomical calculations based on the planetary and lunar data still intact. Much pioneering work in dating texts in this way was done by Sachs (1955). Detailed examples of the techniques used are given by Stephenson & Walker (1985). Hunger (personal communication) has been able to identify datable references to 9 comets (termed *sallammu*) among the extant diaries. The date range of these is rather restricted—from 234 to 87 BC. Obviously the loss of both earlier (possibly back to 690 BC) and later sightings of Comet Halley is to be much regretted. However, it is perhaps fortunate that two references to its apparition in 164 BC and one in 87 BC are still accessible.

Objects described as *sallammu* (the term cannot be satisfactorily translated) can be recognized as comets in a variety of ways. Although there are occasional references to a tail (*mishu*), their lengthy duration (typically of the order of a month) clearly rules out meteors. In addition, the recorded motion of *sallammu* deviates considerably from that of planets—which in any case were readily recognized by the Babylonian astronomers. Finally, the dates of *sallammu* reported in 234, 157, 138, 120, and 87 BC agree well with those of Broom Stars catalogued by Ho Peng Yoke (1962) from Chinese sources.

Cometary records are found in two sections of a Babylonian astronomical diary—the day-to-day entries and the monthly summaries of more important events. Judging from the preserved texts—all of which are badly damaged—only the following observations were reported: (a) discovery (b) heliacal setting (c) heliacal rising (d) any stationary point (e) last visibility. As an example of a typical entry, the following description of the comet of 138 BC may be cited:

'That month (the 2nd lunar month), the comet which had set in Libra reappeared on the night of the 20th in the west in the area of . . .'

This and similar descriptions are helpful in interpreting the Babylonian observations in 87 and 164 BC. It is interesting to note that the astronomers of Babylon, like their counterparts in China, were able to recognize the reappearance of a comet after conjunction with the Sun.

The apparition of Halley's Comet in 87 BC is described only on a single fragment of a tablet (only about 5 cm by 5 cm) in the British Museum (catalogue number: WA 41018). The context is much broken.

'On the 13th, the interval between moonrise and sunset was 8 degrees, measured. First part of the night, the comet . . . which in month IV day beyond day one cubit . . . between north and west its tail 4 cubits . . .'

This entry is followed by an allusion to a lunar eclipse. The cubit was normally a linear unit, but in astronomy its angular equivalent was about 2 or 2.5 deg. This can be readily demonstrated by calculating Babylonian records of planetary conjunctions; separations are expressed in terms of the same unit. Fragments of the entries for three lunar months are preserved. As shown by Stephenson & Walker (1985), the lunar and planetary data recorded on the tablet are sufficient to enable a unique date for the cometary entry to be established by calculation—24 Aug in 87 BC. Reference to the chronological tables of Parker & Dubberstein (1956) shows that this corresponds to the 13th day of month V (the Babylonian year began in March or April). Calculations using the date of perihelion computed by Yeomans & Kiang (1981) or Landgraf (1986)—respectively 6 Aug and 3 Aug—show that P/Halley would be close to last visibility (heliacal setting) in the west on 24 Aug. Hence this is probably the observation referred to on that date. The minimum separation between Halley's Comet and the Earth was fairly large—0.44 AU. Hence around the time of closest approach in month IV (13 Jul to 10 Aug), the comet would move at a fairly regular pace (anything from about 1 to 6 deg or roughly 1 cubit daily). Additionally, its tail would point roughly north-west in the first half of the same month. Hence computation is in satisfactory accord with observation. The Chinese record in section 4 is at variance both with the Babylonian text and with calculation, since it alleges that the comet was visible in the east at some time between 10 Aug and 8 Sep. As was noted by Kiang (1971), a minor scribal error in either month or direction may well be to blame here.

In 164 BC, two Babylonian accounts of P/Halley have survived, and although both texts are damaged, the descriptions overlap sufficiently to provide a fairly detailed report. A composite translation based on the entries on British Museum tablets WA 41462 and WA 41628 reads as follows:

'The comet which had previously appeared in the east in the path of Anu in the area of Pleiades and Taurus, to the west . . . and passed along in the path of Ea in the region of Sagittarius, 1 cubit in front of Jupiter, 3 cubits high towards the north . . .'

Once again, both year and month are missing from both tablets. Calculation of the numerous planetary and lunar data by Stephenson & Walker (1985) yields a date for the cometary sighting of some time between 21 Oct and 19 Nov in 164 BC—i.e. lunar month VIII.

The Babylonians divided the sky into three 'paths' (kaskal), each named after a deity. Enlil was the region north of about declination +17, Anu was the equatorial zone, while Ea was the region south of about declination −17. Calculations using the dates of perihelion deduced by Yeomans & Kiang (1981) and Landgraf (1986)—respectively 12 Nov and 8 Nov in 164 BC—indicate a likely discovery of P/Halley during late September in the constellation of Taurus. Weather permitting, the whole course of the comet, in a generally south-westerly direction, would be visible until the final disappearance in the evening sky about a month later. The accord between observation and theory is thus excellent; there can be no doubt about the identity of the comet. Halley's Comet would be

unusually bright in late September and early October; it passed within 0.11 AU of the Earth. Regarding the precise date of perihelion, the most critical observations are (a) discovery 'in the area of Pleiades and Taurus' and (b) the position '1 cubit in front (i.e. to the west) of Jupiter, 3 cubits high towards the north'. The date of (a) is not preserved, while it is only possible to conclude that observation (b) was made some time in month VIII 21 Oct to 19 Nov. The latter may represent last visibility, overcast weather possibly preventing observation when comet and planet appeared closest together. Calculations of the path of P/Halley for a range of T indicates the most likely date of perihelion based on these two independent observations between about 9 and 15 Nov in 164 BC—see Figs 3a and 3b, which are based on Figs 4a and 4b of Stephenson et al (1985). In the original paper,

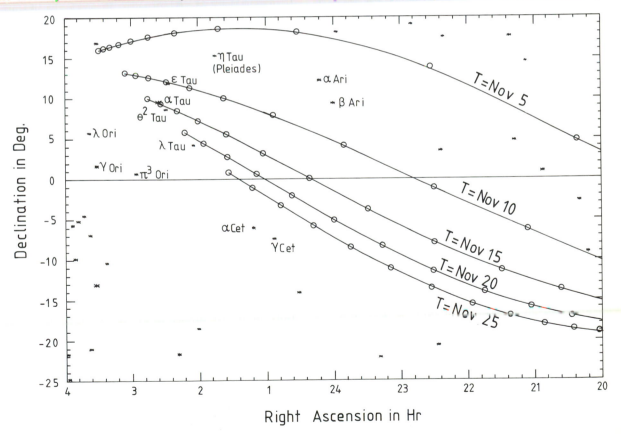

Fig. 3a — The path of P/Halley in the autumn of 164 BC among the stars in the Taurus region, computed for dates of perihelion passage (*T*) at 5-day intervals between 5 and 25 Nov. To correspond with the Babylonian observation that the comet was seen 'in the area of Pleiades and Taurus', a date of perihelion between about 7 and 15 Nov seems indicated.

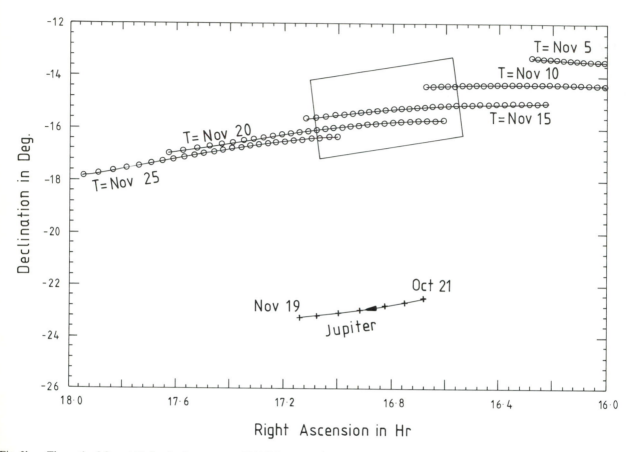

Fig. 3b — The path of Comet Halley in the autumn of 164 BC among the stars in the Sagittarius region, computed for dates of perihelion passage (*T*) at 5-day intervals between 5 and 25 Nov. The rectangular region above the path of Jupiter is the area estimated to be '1 cubit in front of Jupiter, 3 cubits high towards the north' during the lunar month 21 Oct to 19 Nov—the position reported by the Babylonian astronomers on some unknown date during the 8th lunar month.

Stephenson *et al.* were rather conservative in specifying the range of *T* as 9 to 24 Nov; a position of the comet roughly between eta and alpha Tau indicates *T* no later than about 15 Nov (depending to some extent on the precise date of discovery). The refined date range is in excellent agreement with the computations of both Yeomans & Kiang (1981) and Landgraf (1986).

6 THE APPARITION IN 240 BC AND EARLIER RETURNS

The most ancient identifiable return of Halley's Comet is that of 240 BC. It is unfortunate that no Babylonian diary is preserved for the critical year, so that we must rely on a brief Chinese account. This is contained in the Annals (chapter 6) of the *Shih-chi*

('Historical Record'), a work compiled by the Grand Historian Szu-ma Ch'ien around 100 BC. The original record may have been more detailed, but is no longer extant. Much of ancient Chinese history seems to have been destroyed in two literary disasters: the 'Burning of the Books' in 213 BC and the sacking of the capital of Hsien-yang only a few years later.

'In the 7th year (of Ch'in Shih-huang), a Broom Star first appeared at the northern direction. During the 5th month it was again seen at the western direction . . . The Broom Star was again seen at the western direction for 16 days'.

This record is devoid of any accurate positional information, but it represents all that is available. From the calendar tables of Tung Tso-pin (1960), the 5th lunar month was equivalent to 24 May–23 Jun in

240 BC. Yeomans & Kiang (1981) deduce a date of perihelion of 25 May, while Landgraf (1986) derives a date only one day earlier. On these T values, Halley's Comet would probably become visible in mid-May and would be seen before dawn in the eastern sky. By the end of May it would be visible in a north-north-east direction, roughly confirming the northerly aspect mentioned in the text. After conjunction with the Sun at the beginning of June, the comet would reappear in the western sky after dusk, but by the end of the month it would have faded considerably. Hence within the limitations of the text, the agreement between observation and modern computation is satisfactory, and once more we have an identifiable reference to P/Halley.

Both Yeomans & Kiang (1981) and Landgraf (1986) derive previous dates for the return of Comet Halley in 315, 391, and 466 BC. The Shih-chi, the only reliable Chinese history of the period, cites the appearance of comets (Broom Stars) in 296, 303, 305, 361, 467, 470, 481, 482, and 500 BC (as well as three earlier events). Hence we may conclude that there is no extant record of Halley's Comet in 315 and 391 BC. However, the reference in 467 BC has often attracted attention—e.g. Ho Peng Yoke (1962). If there were any observational details, it might be possible to give a definitive opinion on the suggested identity of this broom star with P/Halley. However, the Chronological Tables (chapter 15) of the Shi-chi merely record that at some unknown time 'in the 10th year of Li (ruler of the state of Ch'in) there was a Broom Star'. The stated year corresponds to 467 BC, whereas both Yeomans & Kiang (1981) and Landgraf (1986) deduce dates of perihelion in July of the following year. Analysis of several reports of solar eclipses in the Shih-chi around this time (in years corresponding to 443, 435, and 410 BC) shows that an error of a year in the recorded date is typical. The calculated dates of the eclipses are respectively 444, 436, and 409 BC. Hence the true year of the cometary sighting may well be 466 or 468 BC. An additional difficulty is caused by the fairly frequent occurrence of bright comets other than P/Halley. Statistics of oriental cometary sightings compiled by Ho Peng Yoke (1962) indicates an average frequency of one every five years or so during the past two millennia. Hence it may be concluded that although the Shih-chi report in 467 BC *may* relate

to Comet Halley, a decision is impossible.

Going back still further in time, the Shih-chi (chapter 15) notes Broom Stars in 516, 525, 532, and 613 BC. None of these can represent sightings of P/Halley, whose computed dates of return in this period are 540 and 616 BC. The 613 BC record deserves special mention as probably the earliest reliable sighting of a comet from anywhere in the world. The Ch'un-ch'iu ('Spring and Autumn Annals'), a chronicle of the period from 722 to 481 BC, provides a few descriptive details:

'In autumn, during the 7th month of the 14th year of Wen (ruler of the state of Lu), a Bushy Star entered Pei-tou'.

The above date corresponds to some time in August of 613 BC. Pei-tou is the well known Plough in Ursa Major. Before 613 BC, Ho Peng Yoke (1962), who made a systematic search of the available Chinese literature, could trace no more than three allusions to comets. All of these are of dubious reliability and are cited only in late works. Hence they have no bearing on the history of P/Halley.

CONCLUSION

The history of Halley's Comet is securely established as far back as 240 BC. However, there seems little likelihood of discovering still earlier sightings of the comet unless further important archives come to light. Astronomy cannot make progress in this direction without the benefit of archaeology. Pending such finds, the not infrequent occurrence of brilliant comets in recent decades is a reminder that caution needs to be exercised in interpreting the brief allusions to comets preserved from antiquity. The temptation to link such sightings with P/Halley without adequate evidence must be resisted.

REFERENCES

Brady, J.L. (1982) *J. Brit. Astron. Assoc.* **92** 209–215
Brady, J.L., & Carpenter, E. (1971) *Astron. J.* **76** 728–739
Chang, Y.C. (1979) *Chin. Astron.* **3** 120–131
Ho Peng Yoke (1962) *Vistas Astron.* **5** 127–225
Kanda, S. (1934) *Nihon Temmon Shiryo Soran.* Tokyo
Kanda, S. (1935) *Nihon Temmon Shiryo.* Tokyo
Kiang, T. (1971) *Mem. R. Astron. Soc.* **76** 27–66
Landgraf, W. (1984) *ESTEC* EP/14.7/6184
Landgraf, W. (1986) *Astron. & Astrophys.* **163** 246–260
Parker, R.A., & Dubberstein, W.H. (1956) *Babylonian chronology:*
    626 B.C.–A.D. 75. Brown Univ. Press, Providence, R.I
Pingre, A.G. (1783–4) *Cometographie.* Paris
Sachs. A.J. (1955) *Late Babylonian astronomical and related texts.*
    Brown Univ. Press, Providence, R.I
Sachs, A.J. (1974) *Phil. Trans. Roy. Soc.* **A 276** 43–50
Stephenson, F.R., & Walker, C.B.F. eds (1985) *Halley's Comet in*
    history. British Museum Publications, London
Stephenson, F.R., & Yau, K.K.C. (1985) *J. Brit. Interplan. Soc.* **38**
    195–216
Stephenson, F.R., Yau, K.K.C., & Hunger, H. (1985) *Nature* **314**
    587–592
Tung-tso-pin (1960) *Chronological tables of Chinese history.* Hong
    Kong
Yeomans, D.K., & Kiang, T. (1981) *Mon. Not. Roy. Astron. Soc.* **197**
    633–646

17

The effects of non-gravitational forces on the long-term orbital evolution of comets

Grzegorz Sitarski

1 INTRODUCTION

Nongravitational effects in cometary motions have been known since the famous discovery by Encke (1820) of a secular acceleration in the motion of the short-period comet now bearing his name. He found that the period of revolution of the comet was decreasing by about 0.1 day per period, although all the planetary perturbations had been taken into account. To explain that phenomenon Encke (1823) postulated that a resisting interplanetary medium produced the impeding force and influenced the comet's motion. Similar effects were then detected in the motion of other periodic comets. However, it also was found that some comets had experienced a secular deceleration, and this could not be explained by the resisting medium hypothesis. These investigations, performed with desk calculators and logarithmic tables, were concerned with determining the secular variations in the mean daily motions and with some other questions.

Modern studies of nongravitational effects in the motion of comets were begun and developed by Brian G. Marsden; the results were presented in a series of papers 'Comets and nongravitational forces' published by him and his co-workers in the *Astronomical Journal* (e.g. Marsden 1969, Marsden *et al.* 1973). Marsden recognized that a nongravitational force acting on a comet, whatever its cause, was expressed by its three components F_1, F_2, F_3 defined in orbital coordinates: F_1 directed along the comet's radius vector outward from the Sun, F_2 normal to the radius vector in a plane of the orbit, and F_3 normal to F_1 and F_2 and hence to the orbit plane. Marsden assumed that the nongravitational force had a constant direction in orbital coordinates, but that its value depended on the distance r of the comet from the Sun, that is (Marsden *et al.* 1973):

$$F_i = A_i g(r), \qquad (1)$$

and where A_i = const for i = 1,2,3, and the analytical form of the function $g(r)$ is known and has been deduced according to the results of studies connected with the icy model of cometary nuclei elaborated by Whipple (1950). Values of the nongravitational parameters A_i are determined from observations for an individual comet. Thus orbital computations were connected with the physical nature of nongravitational forces and with the properties of cometary nuclei.

To detect a nongravitational effect in the comet's motion and to determine reasonable values of the nongravitational parameters, it is necessary to make use of observations from at least three consecutive appearances of a periodic comet. Investigations of the nongravitational motion of comets, started by Marsden, were made on all the periodic comets. The

Catalogue of cometary orbits, prepared and published by Marsden (1986), contains orbital data concerning all the observed comets, and we can present the following statistics for the short-period comets:

135 comets have computed orbits,
85 comets were observed during more than one apparition,
46 comets have determined values of nongravitational parameters, and in several cases the nongravitational motion has been investigated in detail.

accounts for a secular acceleration or deceleration of the comet's motion. The normal component F_3 is correlated with F_2. A careful analysis of the orbital motion and recognition of nongravitational forces, especially of their evolution with time, allows some insight into the physical nature of the comet.

2 METHODS OF COMPUTATION

To determine the nongravitational motion of a comet, we have to compute positions of the comet according to an accepted mathematical model of motion and to

Table 1. Periodic comets observed during at least ten appearances

Comet	Actual period (in years)	Number of appearances	Interval of observation	Nongravitational effect
Encke	3.3	53	1786–1986	acceleration
Halley	76.0	30	−239†–1986	deceleration
Pons–Winnecke	6.4	19	1819–1983	accel. before 1900 decel. after 1900
Faye	7.3	18	1843–1984	decel. before 1950 accel. after 1950
Tempel 2	5.3	17	1873–1983	deceleration
Grigg–Skjellerup	5.1	15	1902–1987	acceleration
d'Arrest	6.4	14	1851–1982	deceleration
Wolf	8.2	13	1884–1984	deceleration
Brooks 2	6.9	12	1889–1980	acceleration
Kopff	6.4	12	1906–1983	decel. before 1936 accel. after 1936
Giacobini–Zinner	6.6	11	1900–1984	decel. before 1968 accel. after 1968
Borelly	6.8	10	1905–1981	acceleration
Finlay	7.0	10	1886–1981	deceleration
Tuttle	13.7	10	1790–1980	deceleration

† That is, BC

Table 1 lists 14 periodic comets observed during many returns to the Sun. For these comets the nongravitational motion could be examined in details, and in some cases it was possible to deduce a variability of nongravitational parameters with time. The Whipple model excellently explains a nature of nongravitational forces acting on the comet owing to the rocket effect of ejection of material of the comet nucleus. The radial component F_1 is always positive and an order of magnitude larger than F_2. The transverse component F_2 arises owing to the rotation of the comet's nucleus because there is a lag between the direction of maximum ejection and the subsolar point; F_2 may be negative or positive, and in fact

compare them with the observed positions of the comet. It is obvious that in the gravitational model of motion all the planetary perturbations and also subtle gravitational effects (e.g. relativistic effects) should be taken into consideration. The Marsden assumption that a nongravitational force is acting on the comet leads to the presence of additional terms in equations of the comet's motion.

The equations of nongravitational motion of the comet have the following vectorial form:

$$\ddot{\mathbf{r}} + k^2 \frac{\mathbf{r}}{r^3}\left[1 + \frac{3}{c^2}(\dot{\mathbf{r}}\cdot\dot{\mathbf{r}} - 2\dot{r}^2)\right] = \frac{\partial R}{\partial \mathbf{r}} + f(r, A) \quad (2)$$

where

r — radius vector of the comet,

$\dot{\mathbf{r}}$ — vector of its velocity, $\dot{r} = (\mathbf{r} \cdot \dot{\mathbf{r}})/r$,

k — Gaussian gravitation constant,

c — the speed of light,

R — the known planetary disturbing function,

f — nongravitational force as dependent on a distance r of the comet from the Sun and on the constant nongravitational parameter A, a value of which is determined from the observational equations. The solar term in equation (2) is modified by including the relativistic effects (Sitarski 1983).

Starting from some initial values of parameters of the comet's motion we can numerically integrate equation (2) and obtain values of the position **r** of the comet for observational moments. To improve the initial parameters of motion we can apply a general method of orbit improvement as developed by Kulikov (1950) and adapted by Sitarski (1971) to the correcting of cometary orbits. The deviation $\Delta\mathbf{r}$ from the true position of the comet, being a consequence of the inaccuracy of initial data, satisfies the following differential equation:

$$\Delta\ddot{\mathbf{r}} + k^2\frac{\Delta\mathbf{r}}{r^3} - 3k^2\frac{\mathbf{r}}{r^5}(\mathbf{r}\cdot\Delta\mathbf{r}) = \frac{\partial}{\partial\mathbf{r}}\left(\frac{\partial R}{\partial\mathbf{r}}\cdot\Delta\mathbf{r}\right)$$

$$+ \frac{\partial\mathbf{f}(\mathbf{r}\cdot\Delta\mathbf{r})}{\partial r}\frac{}{r} + \frac{\partial\mathbf{f}}{\partial A}\Delta A. \qquad (3)$$

which can be integrated by substituting

$$\Delta\mathbf{r} = \sum_{i=1}^{6} \mathbf{G}_i\Delta E_i + \mathbf{G}_7\Delta A \qquad (4)$$

where ΔE_i are arbitrary corrections to the six orbital elements E_i. Hence we obtain a set of differential equations for $\mathbf{G}_i$, $i=1,\ldots,7$, a numerical integration of which allows us to calculate values of differential coefficients in the observational equations (Sitarski 1971, 1981, 1984). Solving the observational equations by the least squares method, we can obtain the corrected values of orbital elements and of parameter A; they can be used as starting data for the next iteration.

The Marsden method of assuming the nongravitational force in the form (1) involves the following term **f** in equation (2):

$$\mathbf{f} = A_1 g(r)\frac{\mathbf{r}}{r} + A_2 g(r)\frac{r\dot{\mathbf{r}} - \dot{r}\mathbf{r}}{h} + A_3 g(r)\frac{\mathbf{h}}{h} \qquad (5)$$

where A_1, $A_2 A_3$ are constant, $\mathbf{h} = \mathbf{r} \times \dot{\mathbf{r}}$, and

$$g(r) = 0.1113\left(\frac{r}{2.8}\right)^{-2.215}\left[1 + \left(\frac{r}{2.8}\right)^{5.093}\right]^{-4.6142}$$

An alternative method of studying the nongravitational effects refers to the traditional searching of the secular variation in the mean daily motion. If instead of that variation we use an equivalent value of the daily change of the semi-major axis of the orbit $\dot{a} = da/dt$, this can be expressed in terms of the equations of motion in rectangular coordinates. We then put in equation (2) (Sitarski 1981)

$$\mathbf{f} = \frac{\dot{a}}{2a}\dot{\mathbf{r}} \quad \text{where} \quad \frac{1}{a} = \frac{2}{r} - \frac{\dot{\mathbf{r}}\cdot\dot{\mathbf{r}}}{k^2}; \qquad (6)$$

here $\dot{a}$ plays the role of the constant nongravitational parameter to be corrected along with the six orbital elements.

Using the relation (6) is the simplest way of including the nongravitational term in the equations of motion. A supposition that the nongravitational effect in the comet's motion manifests itself solely by a secular change of the semi-major axis of the orbit, corresponds to an assumption that some secular change in energy of the comet's motion exists, but there is no need to guess the source and nature of this change when looking for its value.

Preparation of observational material for the orbit improvement is a very important problem. There are convenient formulae allowing us to calculate in rectangular coordinates the necessary reductions, i.e. for precession, aberration, etc., concerning the positional observations (Bielicki & Ziolkowski 1976). There are also mathematically objective criteria for selecting and weighting the observations (Bielicki 1972). Recently, a new method was elaborated allowing us to use the instants of perihelion passages of a periodic comet as observational data for orbit improvement (Sitarski 1987a). The method was successfully applied to the study of variability of nongravitational effects in the motion of Comet Encke. It appeared to be the best procedure for making use of the full observational material of Comet Halley when improving its orbit and studying the nongravitational motion over two millennia (see the next section).

It must be emphasized that the numerical integration of equation (3) with the substitution (4) is the only way to obtain accurate values of differential coefficients in observational equations when investigating the comet's motion over long intervals of time. The old method of varying the appropriate parameters and repeatedly integrating the equations of motion to obtain the numerical values of partial derivatives, does not suffice, therefore the iterative process of orbit improvement may become divergent.

3 RECENT RESULTS FOR INDIVIDUAL COMETS

3.1 Comet P/d'Arrest

Comet d'Arrest was discovered in 1851 and observed during 14 returns to the Sun. A nongravitational deceleration in the comet's motion was found. Table 2 contains values of the nongravitational parameter A_2 as obtained by linkages of four or three consecutive apparitions of the comet.

Table 2. Values of the nongravitational parameter A_2 for the periodic comet d'Arrest (after Marsden 1986)

No.	Observation interval	Centre point	$10^8 A_2$
1	1851–1870	1860.06	+0.1038
2	1870–1897	1884.05	+0.0960
3	1890–1910	1899.61	+0.0937
4	1910–1943	1926.05	+0.0957
5	1923–1951	1939.29	+0.1014
6	1943–1964	1952.63	+0.0961
7	1963–1977	1970.27	+0.1201

According to the values of A_2 presented in Table 2, the secular deceleration in the motion of Comet d'Arrest may be considered as being steady with time over almost 130 years, therefore it seems to be possible to link all the observed apparitions by one system of orbital elements with constant values of A_1 and A_2.

We have made such an attempt at preliminary linkage. The initial orbital elements as well as the values of perihelion times used as observational data were taken from Marsden's (1986) Catalogue. Solving 14 observational equations, we corrected four parameters of motion: two orbital elements, the perihelion time T and perihelion distance q, and two nongravitational parameters A_1 and A_2. We obtained the following result:

Epoch: 1982 Sept. 28.0 ET,   Equinox: 1950.0

$$T = 1982 \text{ Sept. } 14.2656 \text{ ET} \qquad \omega = 1950.0$$
$$q = 1.2910866 \text{ AU} \qquad \Omega = 138.8598$$
$$e = 0.624811 \qquad i = 19.4301$$
$$10^8 A_1 = +0.4337 \pm 0.0148$$
$$10^8 A_2 = +0.09821 \pm 0.00049$$

There was indeed no problem in linking all the observed perihelion times. The preliminary solution represents quite well the observational data: the maximum difference between the observed and computed perihelion time amounts to 0.046 day.

Comet d'Arrest is an example of a comet with constant values of nongravitational parameters over the interval of its observed orbital evolution. An interpretation might be that the rotation axis of the comet's nucleus was stable over the 130-year observational interval. A physical model for the nucleus of Comet d'Arrest should take into account this fact.

3.2 Comet P/Wolf-Harrington

This comet was discovered in 1924, lost for observation, and rediscovered again in 1951; seven returns of the comet were observed during 1924–1985. A secular acceleration was detected in the comet's motion, and the method of $\dot{a}$ for nongravitational effects was applied to link the five apparitions in the interval 1951–1978 (Sitarski 1981).

Recently, Szutowicz (1987) has undertaken investigations of motion of the comet. She applied both methods, of $\dot{a}$ and of the Marsden's parameters A_1 and A_2, to study the nongravitational effects in the comet's motion. Fig. 1 shows the nongravitational parameter A_2 plotted against time. Abnormal behaviour of nongravitational effects is apparent, since A_2 shows a

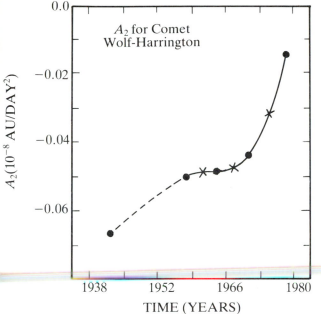

Fig. 1 — Plot of the nongravitational parameter A_2 versus time for Comet Wolf-Harrington. The dots mark values of A_2 obtained by linkages of three apparitions, and the crosses those by linkages of four apparitions.

jump in its value after 1978. Hence it would be very difficult to approximate A_2 reasonably by any continuous function of time in order to link all the appearances of the comet. To predict the next return, Szutowicz linked the last three apparitions of the comet and thus obtained the orbital elements and nongravitational parameters for the 1991 prediction.

The irregular variation of nongravitational forces is presumably related to a physical model of the nucleus of Comet Wolf-Harrington. The observed anomaly in the A_2 evolution with time might be explained by a sudden change of orientation of the spin axis of the comet's nucleus.

3.3 Comet P/Kopff

Since 1906 Comet Kopff has been observed during 12 returns to the Sun. The nongravitational motion of the comet was investigated by Yeomans (1974) who found that the nongravitational parameter A_2 has changed sign during the orbital evolution since the discovery of the comet (Table 3).

Table 3. Values of the nongravitational parameter A_2 for the periodic comet Kopff (after Marsden 1986)

No.	Observation interval	Centre point	$10^8 A_2$
1	1906–1933	1921.14	+0.0267
2	1919–1946	1932.61	+0.0078
3	1945–1965	1954.96	−0.0455
4	1958–1977	1967.59	−0.0843
5	1963–1977	1970.44	−0.0959
6	1970–1983	1977.18	−0.0912

The change of sign of A_2 can be explained by a precessional motion of the spin axis of the rotating nucleus of the comet. A physical model allowing solutions for the spin axis precession of Comet Kopff was elaborated by Sekanina (1984) who used the Yeomans' values of A_2 and A_1 to constrain the properties of the nucleus. i.e. its size, mass, etc. Recently, an attempt to refine Sekanina's model used also the values of the parameter A_3 corresponding to the normal component of the nongravitational force (Rickman *et al.* 1987). Determination of all three nongravitational parameters was made by using the observations of the last five apparitions of Comet Kopff in 1958–1983 (Table 4).

Interval of obs.	Centre point	$10^8 A_1$	$10^8 A_2$	$10^8 A_3$
1958 – 1970	1964.46	0.44 ± 0.11	−0.0793 ± 0.0022	−0.257 ± 0.046
1963 – 1977	1970.89	0.66 ± 0.07	−0.0947 ± 0.0019	−0.157 ± 0.024
1970 – 1983	1977.27	0.37 ± 0.05	−0.1020 ± 0.0031	−0.020 ± 0.035

From the data of Table 4 we can see that the absolute values of A_2 increase while A_3 tends to approach zero during the interval 1964–77. The results for A_2 confirm those of Yeomans (Table 3), and the results for A_3 seem to confirm the variability expected by a precessing spin axis hypothesis. The study of nongravitational parameter A_3 will be extended also to the earlier apparitions of Comet Kopff. One may presume that it would be able to approximate both A_2 and A_3 by some regular functions of time, and then to link all the observed apparitions of Comet Kopff. This seems to be possible in spite of considerable perturbations in the comet's motion caused by the close approach of the comet to Jupiter in 1954, since there are no evident changes in size and shape of the orbit due to that approach.

3.4 Comet P/Encke

Among the known periodic comets, Comet Encke has the shortest period of revolution, and its returns to the Sun have already been observed 53 times. This comet was the first periodic comet in whose motion a secular acceleration has been discovered (Encke 1820). It was also the first object for which it was found that the nongravitational acceleration in the comet's motion has been diminishing with time. A study of the variability of nongravitational effects in the motion of Comet Encke was made by Marsden & Sekanina (1974), who determined values of the Marsden's nongravitational parameters A_1 and A_2 and found that the evolution of A_2 over the period of almost two hundred years could be represented by a sinewave.

An attempt to approximate the regular changes of nongravitational effects of Comet Encke by some function of time, and to link all the observed apparitions of the comet by one system of orbital elements, has been undertaken by Sitarski (1987a). The investigations were made by using two methods: by determination of $\dot{a}$ or of the Marsden parameters A_1 and A_2. It was found that in both cases the corresponding nongravitational parameters were similar functions of time, and $\dot{a}(t)$, as well as $A_2(t)$, could be approximated by a sinusoidal function.

A refined approximation of $\dot{a}(t)$ by an asymmetric sinusoid was obtained recently (Sitarski 1987b). The approximating function for $\dot{a}(t)$ was assumed to be as follows:

$$\dot{a}(t) = a_1(1 + a_2 \sin(a_3 + a_4 t + a_5 t^2))]. \qquad (7)$$

To improve the orbit and to determine values of the parameters $a_1, \ldots, a_5$, 116 positional observations made during 1970–86; and 48 perihelion times observed in 1876–1967, taken from the Marsden's (1986) catalogue, were used as observational data.

Then 11 dynamical parameters of motion were iteratively corrected on solving 277 observational equations by the least squares method; the following final solution was obtained:

Epoch: 1984 April 10.0 ET, Equinox: 1950.0
$T = 1984$ March 27.6831 ET $\omega = 185°.9962$
$q = 0.3410011$ AU $\Omega = 334.1812$
$e = 0.8463325$ $i = 11.9272$
$a_1 = (-5.6984 \pm 0.0274) \times 10^{-8}$ AU/day
$a_2 = -1.1266 \pm 0.0107$
$a_3 = (-5.6369 \pm 0.4259) \times 10^{-2}$
$a_4 = (+4.0785 \pm 0.0241) \times 10^{-5}$
$a_5 = (-3.3563 \pm 0.0363) \times 10^{-10}$

In formula (7) $t = 0$ for the epoch: 1899 Dec. 12.0 ET $=$ JD 2415000.5, and t is counted in days. The presented model of nongravitational motion successfully links all the observed perihelion times of Comet Encke: the greatest deviation of the computed perihelion time from the observed time does not exceed 0.025 day.

The time-dependence for $\dot{a}(t)$ given by (7) is plotted in Fig. 2 for comparison with the empirical curve for $A_2(t)$ found by Marsden & Sekanina (1974). The striking similarity of the two curves suggests that a variation of nongravitational effects can be successfully investigated by the simple method of $\dot{a}$, and then the main features of the found variability might be assigned to A_2 for the natural interpretation connected with physical properties of the comet's nucleus. It must be questioned, however, whether the assumed time-dependence for $\dot{a}(t)$ may be extrapolated beyond the observation interval to investigate the long-term nongravitational motion of Comet Encke, e.g. before its discovery.

However, the accepted sinusoidal function of time allows us to avoid any abnormal values of nongravitational parameters which could appear during long-term integration, and it seems that the most important need is to start integrating the equations of motion with good initial data obtained on a basis of all the apparitions of the comet. Recently, the sinusoidal time-dependence was admitted for the transverse component of the nongravitational force, the orbit of Comet Encke was again improved, and the equations

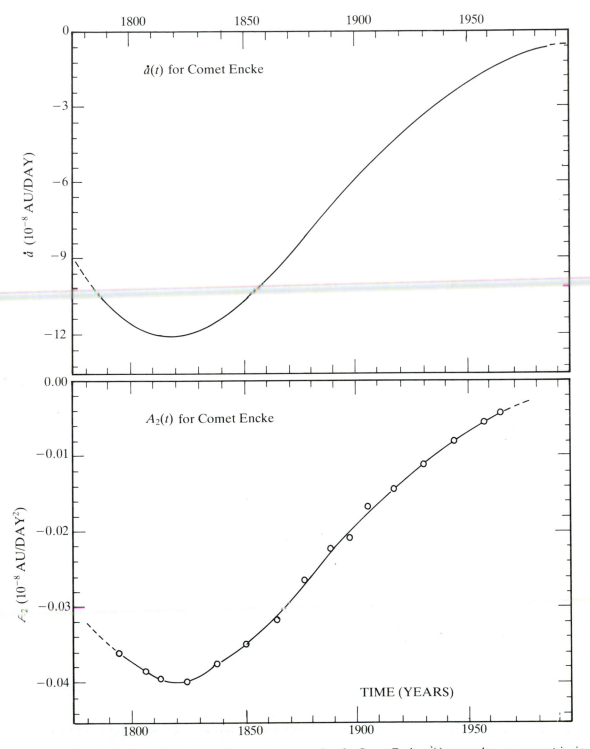

Fig. 2 — Plot of the corresponding nongravitational parameters versus time for Comet Encke: $\dot{a}(t)$ assumed as an asymmetric sinusoid. $A_2(t)$ drawn according to the discrete values of A_2 (after Marsden & Sekanina 1974).

of motion were integrated backwards to 1201 to predict 177 perihelion time of the comet before its discovery in 1786 (Sitarski 1988).

3.5 Comet P/Halley

There is little doubt that the famous Comet Halley has been observed since 240 BC during its 30 returns to the Sun. The motion of the comet over the whole two-millennium observation interval has been investigated by various authors. However, only two studies were concerned with the nongravitational effects examined by the modern Marsden method. Yeomans & Kiang (1981) used the nongravitational parameters A_1 and A_2 as constant values with time, while Landgraf (1984) used A_2 as being some linear function of time. In both studies the orbit was improved by the least squares method, using the positional observations from only the last few apparitions. Then the equations of motion were integrated backward in a time-span of thousands years; however, to obtain good agreement of computed perihelion times with the observed times, the authors had to make some subjective changes in the orbital elements for AD 837 when the comet closely approached the Earth to within 0.03 AU. Although the computational procedures in both studies, and the fit to the observed perihelion times, were similar, the results of integration were quite different when comparing perihelion times far from the observational interval (see Table 6).

Investigations based on a new method, enabling the use of the full observational material since 240 BC to orbit improvement, indicated a source of the above-mentioned discordance (Sitarski & Ziolkowski 1986). It appeared that the secular change in the nongravitational parameter A_2, included in the process of orbit improvement, resulted in different values of improved orbital elements, and this caused considerably different results of prediction of the past perihelion times. Hence it was clear that evolution of non-gravitational effects played an important role in the motion of Comet Halley, and should be carefully examined when we study the long-term motion of the comet.

Here we are presenting a very recent preliminary result of our own investigations of the nongravitational motion of Comet Halley. We used the method of orbit improvement successfully applied in the case of Comet Encke (Sitarski 1987a) and also tested for Halley's Comet (Sitarski & Ziolkowski 1986). The method allows us to look for variability of nongravitational effects with time, using the instants of perihelion passages as observational data. It is especially useful in the case of Comet Halley, since for 24 early appearances of the comet there are inaccurate observations known from historical records, from which the dates of perihelion passages of the comet have been deduced (Kiang 1972, Hasegawa 1979, Yeomans & Kiang 1981).

To find a variability of nongravitational effects in the motion of Comet Halley over two millennia, we determined values of a daily change $\dot{a}$ of the semi-major axis of the orbit by successive linkages of four or five consecutive perihelion times. Thus we obtained the values of $\dot{a}$ for the 24 dates (Table 5). Graphical presentation of these values versus the time, showed that the simplest reasonable approximation for $\dot{a}(t)$ would be parabolic:

$$\dot{a}(t) = a_0(1 + a_1 t + a_2 t^2). \qquad (8)$$

Table 5. Values of the nongravitational parameter $\dot{a}$ obtained by the linkages of four or five consecutive perihelion times of Comet Halley; $\dot{a}(t)$ was approximated by a parabolic function

No.	Observation interval	Centre point	$10^8 \dot{a}$ determ.	Mean error	$10^8 \dot{a}(t)$ adjusted
1	1835–1986	1910.76	+6.3822	0.0043	+6.2293
2	1759–1986	1872.87	+6.4864	0.0778	+6.2869
3	1682–1910	1797.02	+6.5053	0.0727	+6.3891
4	1607–1835	1721.40	+6.3586	0.0103	+6.4733
5	1531–1759	1645.36	+5.7047	0.1591	+6.5403
6	1378–1682	1531.52	+5.7343	0.1670	+6.6074
7	1301–1607	1455.35	+6.3448	0.3905	+6.6301
8	1222–1531	1378.34	+7.3128	0.4144	+6.6350
9	1145–1456	1301.07	+6.8704	0.5164	+6.6216
10	1066–1378	1223.02	+6.1383	0.3667	+6.5894
11	989–1301	1145.19	+6.8052	0.4492	+6.5387
12	837–1222	1052.26	+7.1915	0.2305	+6.4538
13	760–1145	959.79	+6.8506	0.1104	+6.3430
14	684–1066	867.68	+5.9845	2.5845	+6.2066
15	607–989	775.08	+3.7464	2.8481	+6.0432
16	530–837	684.08	+4.3012	1.8481	+5.8570
17	451–760	606.94	+2.8825	3.1172	+5.6792
18	374–684	529.69	+3.8698	2.8103	+5.4829
19	295–607	451.80	+5.6872	2.4451	+5.2664
20	218–530	374.04	+5.9184	3.0008	+5.0317
21	141–451	296.14	+4.6268	1.3000	+4.7779
22	66–374	219.06	+6.5921	1.8775	+4.5085
23	−11†–295	142.18	+3.4647	1.3639	+4.2217
24	−86†–218	66.04	+2.9879	1.1373	+3.9318

† BC

The determined values of $\dot{a}$ were weighted according to their mean errors, and then the parabolic function (8) was adjusted by the least squares method; the following values of parameters in (8) were obtained:

$$a_0 = (+6.1470 \pm 0.0376) \times 10^{-8} \text{ AU/day}$$
$$a_1 = (+7.7125 \pm 0.1376) \times 10^{-7} \qquad (9)$$
$$a_2 = (-1.8714 \pm 0.1335) \times 10^{-12}$$

In relation (8) t is expressed in days, and $t = 0$ for the epoch: 832 Feb. 25.0 ET = JD 2025000.5.

Table 6.   Prediction of 20 perihelion times of Comet Halley for the years BC according to the different authors

No.	Yeomans & Kiang	Landgraf	Sitarski
1	12.78	12.77	12.77
2	87.60	87.59	87.59
3	164.87	164.83	164.81
4	240.40	240.29	240.22
5	315.69	315.37	315.12
6	391.70	391.32	392.95
7	466.55	466.28	467.92
8	540.36	542.96	543.28
9	616.57	618.72	620.79
10	690.06	692.02	695.92
11	763.59	769.09	771.78
12	836.35	846.38	846.40
13	911.38	924.14	923.58
14	986.92	1002.78	1000.76
15	1059.92	1082.97	1075.02
16	1129.25	1159.45	1152.49
17	1198.36	1237.27	1227.27
18	1266.68	1316.32	1304.41
19	1334.65	1394.78	1380.36
20	1404.79	1473.18	1458.49

To improve the orbit of Comet Halley we selected the 250 best positional observations from 1835–1986 and joined to them 26 observed perihelion times from the interval 87 BC—AD 1835 (the perihelion times in 164 BC and 240 BC are too uncertain). The nongravitational effects were taken into account by including $\dot{a}(t)$ in the form (8) to the equations of motion. We corrected the six orbital elements and the parameter a_0, accepting the constant shape of parabola (8) determined by the constant parameters a_1 and a_2 given in (9). Processing of the orbit improvement required many iterations, and finally we obtained the following result:

Epoch: 1835 Nov. 18.0 ET, Equinox: 1950.0
$T = 1835$ Nov. 16.4367 ET $\omega = 110°.6908$
$q = 0.5866008$ AU $\Omega = 56.8157$
$e = 0.9673963$ $i = 162.2783$
$10^8 a_0 = (+6.0332 \pm 0.0013)$ AU/day

Starting from the above elements the equations of motion were integrated backwards to 1500 BC by the recurrent power series method (Sitarski 1984). To avoid a dangerous extrapolation of $\dot{a}(t)$ far from the observation interval, the constant value $\dot{a} = +2.7134 \times 10^{-8}$ AU/day was kept for the integration before 240 BC. The perihelion times for the years BC obtained during the integration are given in the last column of Table 6.

The three considered solutions for the long-term motion of Comet Halley, including our recent preliminary result, well represent the observed perihelion times during 240 BC–AD 1986 (see the paper by Ziolkowski in this volume). However, a comparison of the predicted perihelion times before 240 BC seems to confirm the earlier supposition that a secular variation in nongravitational parameters, taken into account in the process of orbit improvement, does significantly influence the corrected elements, and hence the result of extrapolation of the solution strongly depends on this variation.

The data in Table 6 also show that our results are much closer to the Landgraf's results than to Yeomans–Kiang. Therefore it seems that the secular variation of nongravitational effects in the motion of Comet Halley is indeed detectable, and one cannot really retain an assumption that the comet's nongravitational forces remained constant with time over two millennia.

4 FINAL REMARKS

Cometary investigations have shown that nongravitational forces play an important role in the orbital evolution of a periodic comet. Marsden's remarkable idea to include the nongravitational terms in the equations of motion, based on the Whipple model of the comet's nucleus, allowed us to relate the nongravitational forces with the physical properties of cometary nuclei. It has appeared that the results of investigation of the long-term motion of a comet could be a rich source of information for studying the physical nature of the comet; this was confirmed by

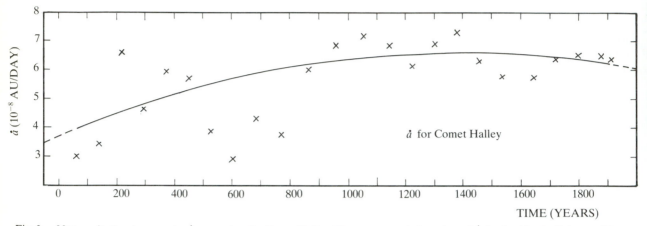

Fig. 3 — Nongravitational parameter $\dot{a}$ versus time for Comet Halley. The crosses mark the values of $\dot{a}$ obtained by the linkages of four or five consecutive perihelion times of the comet; the curve presents the parabolic approximation of $\dot{a}(t)$.

the observational results of the space missions to Halley's Comet.

We have looked at a number of comets for which the orbital evolution has been studied in details. Variation of nongravitational parameters with time was observed for some comets. In the case of regular changes it was possible to approximate the evolution of nongravitational parameters by a continuous function of time, and to find one system of orbital elements for the nongravitational model of the comet's motion over the whole observation interval. The regular changes of nongravitational force of a comet can be explained by a precessional motion of the spin axis of the rotating nucleus, whereas erratic behaviour of the rotation axis in a comet's dying stages would be expected to produce some wild changes in the nongravitational parameters. Therefore, investigations of variations of nongravitational parameters may throw some light on the origin of comets. Hence the recent results presented seem to start a new phase of investigations of comets and their nongravitational forces, initiated twenty years ago by Brian G. Marsden.

Comet Halley is a special case among periodic comets. It was observed during 30 appearances over two millennia; however, accurate positional observations were made during only the last few apparitions, and for more than twenty observed returns very uncertain positions have been deduced from ancient records. Recently we found a method for making use of all the observed apparitions of the comet. The positional observations as well as the past moments of perihelion passages will be carefully examined to

assess their use for orbit improvement in a homogeneous way. The preliminary results allow us to hope that in the near future the new revised theory of long-term motion of Comet Halley will be elaborated.

REFERENCES

Bielicki, M. (1972) In: *The motion, evolution of orbits, and origin of comets*, D. Reidel, Dordrecht, 112
Bielicki, M., & Ziolkowski, K. (1976) *Acta Astron.* **26** 371
Encke, J.F. (1820) *Berliner Astron. Jarbuch für 1823* 211
Encke, J.F. (1823) *Berliner Astron. Jarbuch für 1826* 124
Hasegawa, I. (1979) *Publ. astr. Soc. Japan* **31** 257
Kiang, T. (1972) *Mem. R. astr. Soc.* **76** 27
Kulikov, D.V. (1950) *Bull. I.T.A.* **IV** 380
Landgraf, W. (1984) *ESTEC EP/14.7/6184 Final Report.*
Marsden, B.G. (1969) *Astron. J.* **74** 720
Marsden, B.G. (1986) *Catalogue of cometary orbits*, fifth edition, IAU Central Bureau for Astronomical Telegrams (Cambridge, Massachusetts)
Marsden, B.G., & Sekanina, Z. (1974) *Astron. J.* **79** 413
Marsden, B.G., Sekanina, Z., & Yeomans, D.K. (1973) *Astron. J.* **83** 64
Rickman, H., Sitarski, G., & Todorovic-Juchniewicz, B. (1987) *Astron. Astrophys.* **188** 206
Sekanina, Z. (1984) *Astron. J.* **89** 1537
Sitarski, G. (1971) *Acta Astron.* **21** 87
Sitarski, G. (1981) *Acta Astron.* **31** 471
Sitarski, G. (1983) *Acta Astron.* **33** 295
Sitarski, G. (1984) *Acta Astron.* **34** 53
Sitarski, G. (1987a) *Acta Astron.* **37** 99
Sitarski, G. (1987b) *Proc. Symposium on the Diversity and Similarity of Comets*, Brussels 6–9 April 1987, ESA SP-278 ESTEC (Noordwijk-Holland), 751
Sitarski, G. (1988) *Acta Astron.* **38** 269
Sitarski, G., & Ziolkowski, K. (1986) Proc. 20th Eslab Symposium on the Exploration of Halley's Comet, Heidelberg 27–31 October (1986) ESA SP-250. 299
Szutowicz, S. (1987) *Acta Astron.* **37** 179.
Whipple, F.L. (1950) *Astrophys. J.* **111** 375
Yeomans, D.K. (1974) Publ. Astron. Soc. Pacific **86**. 125.
Yeomans, D.K., & Kiang, T. (1981) *Mon. Not. R. astr. Soc.* **197** 633

18

Investigating the motion of Comet Halley over several millennia

Donald K. Yeomans

1 THE HISTORY OF COMET HALLEY UNTIL THE 1909–1911 APPARITION

1.1 The prediction of future perihelion passage times

Since 240 BC, Chinese observers have documented a nearly unbroken record of scientifically useful observations of Comet Halley (Ho Peng Yoke 1962, Ho Peng Yoke and Ang Tian-Se 1970). After the probable 240 BC apparition, only the 164 BC return went unrecorded by the Chinese, and with the exception of occasional Babylonian, Korean and Japanese sightings, useful Comet Halley observations made outside China were virtually nonexistent for over a millennium thereafter. Beginning with the cometary observations of the Florentine physician and astronomer, Paolo Toscanelli (1397–1482), quantitative and accurate cometary positions became available throughout the West (Celoria 1893). However, the necessary theory for representing a comet's motion was not available until the publication of Isaac Newton's *Principia* in 1687. Newton (1687) outlined a semi-analytic orbit determination theory, and used the comet of 1680 as an example. While Newton never applied the method to another comet, Edmond Halley began what he termed 'a prodigious deal of calculation' and applied Newton's method to determine the parabolic orbits for two dozen well observed comets (Halley 1705). Struck by the similarity in the orbital elements for the comets observed in 1531, 1607, and

1682, Halley suggested that these three apparitions were due to the same comet, and that it might be expected again in 1758. Halley's subsequent calculations indicated that a close Jupiter approach in 1681 would cause an increase in the length of the next period. Halley then revised his earlier prediction and suggested in a publication appearing after his death (Halley 1749) that the comet that was to bear his name would return again in late 1758 or early 1759.

To refine Halley's prediction, Clairaut (1758) used a modified version of his analytic lunar theory to compute the perturbations on the comet's orbital period due to the effects of Jupiter and Saturn over the interval 1531–1759. Noting that calculations over the intervals 1531–1607 and 1607–1682 predicted the 1682 perihelion passage time to within one month, Clairaut stated that his mid-April 1759 prediction should be good to a similar accuracy. The actual time of perihelion passage in 1759 was March 13.1 (Unless otherwise stated, all times are given in UT). Beginning with Clairaut's work in 1758, all subsequent work to 1910 on the perturbed motion of Comet Halley was based upon the variation of elements technique (Lagrange 1783). The various works differed only in how many perturbing planets were included, how many orbital elements were allowed to vary, and how many times per revolution the reference ellipse was rectified by adding the perturbations in elements. Until after the 1909–1911 apparition, no attempt was made to link the observations of two or more apparitions into one orbital solution.

In anticipating the 1835 return, Damoiseau (1820) computed the perturbative effects of Jupiter, Saturn, and Uranus on Comet Halley over the interval 1682–1835. Since the actual time of perihelion passage in 1835 was November 16.4, Damoiseau's initial prediction of November 17.15 was remarkable. However, Damoiseau (1829) later added the perturbations due to the Earth and revised his prediction to November 4.81. De Pontécoulant considered the perturbative effects of Jupiter, Saturn, and Uranus over the interval 1682–1835 as well as the Earth's perturbative effects near the 1759 time of perihelion passage. His predictions for the 1835 perihelion passage times were successively, November 7.5, November 13.1, November 10.8, and finally November 12.9 (de Pontécoulant 1830, 1834, 1835). The most complete work leading up to the 1835 return was undertaken by O.A. Rosenberger. After a complete reduction of available observations, Rosenberger recomputed an orbit for the 1759 and 1682 apparitions (Rosenberger 1830a, 1830b). Rosenberger (1834, 1835) computed the effect on all the orbital elements from the perturbations of the seven known planets over the 1682–1835 interval. Assuming the comet's motion was unaffected by a resisting medium, Rosenberger's prediction for the 1835 perihelion passage time was November 12.0. Lehmann (1835) also investigated the motion of Comet Halley over the 1607–1835 interval, taking into account the perturbative effects of Jupiter, Saturn, and Uranus. However, his perihelion passage prediction was late by more that 10 days.

In an effort to anticipate the next apparition of Comet Halley, de Pontécoulant (1864) took into account the perturbative effects of Jupiter, Saturn, and Uranus before predicting May 24.36, 1910 as the next time of perihelion passage. The actual time of perihelion passage turned out to be April 20.18. Cowell & Crommelin began their work with preliminary calculations to see if de Pontécoulant's prediction was approximately correct (Cowell & Crommelin 1907a, 1907b, 1907c, 1908c). Their computations used

the variation-of-elements technique, included perturbations by all the planets from Venus to Neptune (except Mars), and predicted a return to perihelion on April 8.5. Cowell & Crommelin (1910) then began a new study of the comet's motion by using numerical integration whereby the perturbed rectangular coordinates are obtained directly at each time step. This time they computed the perturbations from Venus through Neptune and used a time step that varied from 2 to 256 days. They predicted a 1910 perihelion passage time of April 17.11. The 1909 recovery of the comet required that their prediction be corrected by 3 days, and they then revised their work by reducing the time steps by one half, carrying an additional decimal place, and correcting certain errors in the previous work (Cowell & Crommelin 1910). Their post recovery prediction was then revised to April 17.51, and they concluded that at least 2 days of the remaining discordance was due to causes other than errors in the calculations or errors in the planetary positions and masses. We note here that the best predictions for the 1835 perihelion passage time by Rosenberger and de Pontécoulant as well as the 1910 prediction by Cowell & Crommelin were too early by 4.4, 3.5, and 2.7 days respectively. As pointed out in section 2, this is just what one would expect, since none of these predictions included the effects of the so-called nongravitational forces.

1.2 The identification of early Comet Halley apparitions

Until the 20th century, all attempts at identifying ancient apparitions of Comet Halley were done by either determining orbits directly from the observations or by stepping back in time at roughly 76 year intervals and testing the observations with an approximate orbit of Comet Halley. Pingré (1783–84) confirmed the suspicion of Halley (1705) by showing that the comet of 1456 was an earlier apparition of Comet Halley. Biot (1843) pointed out that an orbit by Burckhardt (1804) for the comet of 989 closely resembled that of Comet Halley, and Laugier (1843, 1846) correctly identified as Comet Halley the comets seen by the Chinese in 451, 760, and in the Autumn of 1378. Laugier (1842) also noted that four of the five parabolic orbital elements for the comet seen in 1301 were close to those of Comet Halley. By stepping backward in time at roughly 76–77 year intervals and analyzing European and Chinese observations, Hind (1850) attempted to identify Comet Halley apparitions from 11 BC to AD 1301. Approximate perihelion passage times were often determined directly from the observations, and an identification was suggested if Halley-like orbital elements could satisfy existing observations. Although many of Hind's identifications were correct, he was seriously in error for his suggested perihelion passage times in AD 1223, 912, 837, 608, 373 and 11 BC.

Using a variation of elements technique, Cowell & Crommelin (1907d) began the first effort to actually integrate the comet's equations of motion backward in time. They assumed that the orbital eccentricity and inclination were constant with time and the argument of perihelion and the longitude of the ascending node changed uniformly with time their rates being deduced from the values computed over the 1531–1910 interval. By using Hind's (1850) times of perihelion passage or by computing new values from the observations, they deduced preliminary values of the orbital semi-major axes for the perturbation calculations. The motion of the comet was accurately carried back to 1301 by taking into account first order perturbations in the comet's period from the effects of Venus, Earth, Jupiter, Saturn, Uranus, and Neptune. Using successively more approximate perturbation methods, Cowell & Crommelin (1907d, 1908a-e) carried the motion of the comet back to 239 BC. At this stage, their integration was in error by nearly 1.5 years in the perihelion passage time, and they adopted a time of 15 May, 240 BC, not from their integration, but rather from their consideration of the observations themselves. After a complete and careful analysis of the European and Chinese observations, Kiang (1972) used the variation of elements technique to investigate the motion of Comet Halley over the 240 BC—AD 1682 interval. By determining the time of perihelion passage time directly from the observations and considering the perturbations from all nine planets on the other orbital elements, Kiang traced the motion of Comet Halley for nearly two millennia. Hasegawa (1979) also empirically determined perihelion passage times for Comet Halley. For each apparition from AD 1378 to 240 BC, he computed several ephemerides, using Kiang's (1972) orbital elements, except for the perihelion passage times

which were chosen to make the best fit with the observations. Attempts to represent the ancient observations of Comet Halley by using the numerical integration of the comet's gravitational and nongravitational accelerations are presented in the next section.

2 NONGRAVITATIONAL FORCES AND COMET HALLEY

Beginning with the work of Bessel (1836a, 1836b), it became clear that the motion of Comet Halley was influenced by more than the solar and planetary gravitational accelerations. Michielsen (1968) pointed out that perihelion passage time predictions that had been based upon strictly gravitational perturbation calculations required a correction of $+4.4$ days over the past several revolutions. Kiang (1972) determined a mean correction of $+4.1$ days. In an attempt to account for this 4 day discrepancy between the actual period of Comet Halley and that computed using perturbations from the known planets, some unorthodox solutions have been proposed. Brady (1972) suggested the influence of a massive trans-Plutonian planet, and Rasmusen (1967) adjusted the ratio of the Sun:Jupiter mass ratio from the accepted value of 1047 to 1051. Both of these suggested solutions must be rejected, because they would produce effects on the motion of the known planets that are not supported by observation. Rasmusen (1981) derived a 1986 perihelion date of February 5.46 from a fit to the observations in 1835 and 1910, and then added $+3.96$ days to yield a 1986 perihelion passage time prediction of February 9.42. Brady & Carpenter (1967) first suggested a 1986 perihelion passage time of Feb. 5.37 based upon a 'trial and error' fit to the observations during the 1835 and 1910 returns. Brady & Carpenter (1971) then introduced an empirical secular term in the radial component of the comet's equations of motion. Although this device had the unrealistic effect of decreasing the solar gravity with time, it did allow an accurate 1986 perihelion passage time prediction of Feb. 9.39. It is now clear that the actual 1986 perihelion passage time (Feb. 9.46) was accurately predicted by both Rasmusen (1981) and Brady & Carpenter (1971). However, if the orbit of the comet is to be accurately computed throughout a particular apparition, or if the comet's motion is to be traced back to ancient times, the mathematical model used to represent the obvious nongravitational forces must be based upon a realistic physical model and not upon empirical mathematical devices.

In introducing the icy conglomerate model for a cometary nucleus, Whipple (1950, 1951) recognized that comets may undergo substantial perturbations due to reactive forces or rocket-like effects acting upon the cometary nucleus itself. In an effort to accurately represent the motions of many short periodic comets, Marsden (1968, 1969) began to model the nongravitational forces with a radial and transverse term in the comet's equations of motion. Marsden et al. (1973) modified the nongravitational force terms to represent the vaporization flux of water ice as a function of heliocentric distance. The cometary equations of motion are written;

$$\frac{d^2 r}{dt^2} = -\mu\frac{r}{r^3} + \frac{\partial R}{\partial \vec{r}} + A_1 g(r)\hat{r} + A_2 g(r)\hat{T}$$

where $g(r) = \alpha(r/r_o)^{-m}(1 + (r/r_o)^n)^{-k}$

The acceleration is given in astronomical units/(ephemeris day)², μ is the product of the gravitational constant and the solar mass, while R is the planetary disturbing function. The scale distance r_0 is the heliocentric distance where reradiation of solar energy begins to dominate the use of this energy for vaporizing the comet's nuclear ices. For water ice, r_0 = 2.808 AU and the normalizing constant α = 0.111262. The exponents m, n, k equal 2.15, 5.093, and 4.6142 respectively. The nongravitational acceleration is represented by a radial term, $A_1 g(r)$, and a transverse term, $A_2 g(r)$, in the equations of motion. If the comet's nucleus were not rotating, the outgassing would always be preferentially toward the Sun, and the resulting nongravitational acceleration would act only in the antisolar direction. However, the rotation of the nucleus, coupled with a thermal lag angle between the nucleus subsolar point and the point on the nucleus where there is maximum outgassing, introduces a transverse acceleration component in either the direction of the comet's motion or contrary to it—depending upon the nucleus rotation direction. The radial unit vector ($\hat{r}$) is defined outward along the Sun–comet vector, while the transverse unit vector ($\hat{T}$)

is directed normal to $\hat{r}$ in the orbit plane and in the direction of the comet's motion. An acceleration component normal to the orbit plane is certainly present for most comets, but its periodic nature makes detection difficult in these computations because we are solving for an average nongravitational acceleration effect over three or more apparitions. While the nongravitational acceleration term $g(r)$ was originally established for water ice, Marsden et al. (1973) have shown that if the Bond albedo in the visible range equals the infrared albedo, then the scale distance r_o is inversely proportional to the square of the vaporization heat of the volatile substance.

Using observations of Comet Halley over the 1607–1911 interval, Yeomans (1977) used a least squares differential correction process to solve for the six initial orbital elements and the two nongravitational parameters A_1 and A_2. Different values for the scaling distance were tried, with the result that r_0 = 2.808 AU was the optimum input value. This suggests that the outgassing causing the nongravitational forces acting on Comet Halley are consistent with the vaporization of water ice. This result is a general one for nearly all comets for which nongravitational force parameters have been determined. The positive sign for the determined value of A_2 for Comet Halley indicates that the comet's nucleus is rotating in a direct sense—in the same direction as the orbital motion. Yeomans (1977) integrated the motion of Comet Halley back to 837, and forward to predict a perihelion passage time of 1986 Feb. 9.66. Using improved computer software, Yeomans (1984) revised his perihelion passage time prediction to Feb. 9.49.

3 INTEGRATING THE MOTION OF COMET HALLEY BACKWARD IN TIME

Brady & Carpenter (1971) were the first to apply direct numerical integration to the study of Comet Halley's ancient apparitions. Using their empirical secular term to represent the nongravitational effect, they initiated their integration with an orbit that was determined from the 1682 through 1911 observations and integrated the comet's motion back to 240 BC in one continuous run. Because their integration was tied to no observational data prior to 1682, their early perihelion dates diverged from the dates Kiang (1972) had determined directly from the Chinese observations. Using Brady & Carpenter's (1971) orbit for Comet Halley, Chang (1979) integrated the comet's motion back to 1057 BC. However, this integration was not based upon any observations prior to 1909, nor were nongravitational effects taken into account.

Yeomans & Kiang (1981) began their investigation of Comet Halley's past motion with an orbit based upon the 1759, 1682, and 1607 observations, and numerically integrated the comet's motion back to 1404 BC. Planetary and nongravitational perturbations were taken into account. In nine cases, the perihelion passage times calculated by Kiang (1972) from Chinese observations were redetermined, and the unusually accurate observed perihelion times in AD 837, 374, and 141 were used to constrain the computed motion of the comet. The dynamic model, including terms for nongravitational effects, successfully represented all the existing Chinese observations of Comet Halley. This model assumed the comet's nongravitational forces remained constant with time; hence it seems that the comet's spin axis has remained stable, without precessional motion, for more than two millennia. In this regard, the recently discovered Babylonian observations in 87 BC and 164 BC were also found to be in agreement with the times of perihelion passage computed by Yeomans & Kiang (1981), further suggesting that no time dependence is warranted (Stephenson et al. 1985). Recently, Landgraf (1986a) revised his earlier integration of Comet Halley's motion backward in time (Landgraf 1984), bringing the comet's motion in line with the long-term integration of Yeomans & Kiang (1981). While Landgraf's integration did utilize a small time dependence in the nongravitational forces (as did the 1988 work by Sitarski outlined in this volume), the earlier work of Yeomans & Kiang (1981) succeeded without them. In this regard, we should also note that an attempt to include a time dependence in the nongravitational effects over the observed interval 1682–1982 produced a poor prediction for the 1986 perihelion passage time that was in error by −0.073 day (Landgraf 1986b). Thus it appears that the comet's nongravitational effects are nearly time independent, and that the spin axis has remained stable, without substantial precessing, for more than two millennia. Over the same interval, the comet's

average ability to outgas could not have changed substantially. This latter conclusion is consistent with a nearly constant intrinsic brightness over roughly the same interval of time (Broughton 1979).

Although the motion of Comet Halley is quite well determined for a period exceeding two millennia, the comet's dynamic behaviour cannot be extrapolated, either forward or backward, for long periods. Chang (1979) integrated the comet's motion back to 1057 BC, Yeomans & Kiang (1981) integrated back to 1404 BC, Landgraf (1984) went back as far as 2317 BC, and Brady (1982) went back even further to 2647 BC. However, only the integrations of Yeomans & Kiang and that of Landgraf agree back to 87 BC and all four integrations diverge wildly from one another prior to 87 BC; by 700 BC, the differences between the predicted times of perihelion passage are best measured in years. As pointed out by Stephenson et al. (1985), only the integration of Yeomans & Kiang (1981) agrees with the recently recovered Babylonian observations of 87 BC and 164 BC. Although Landgraf's more recent work (1986a) presents a backward integration to 466 BC that is very similar to that of Yeomans & Kiang (1981), one must still question whether the motion of Comet Halley can be extrapolated more than a few perihelion passages beyond the period of observational data. Chirikov & Vecheslavov (1989) caution that the exponential divergence of close trajectories (chaos) occurs rather quickly for Comet Halley's dynamic system. They point out that the perturbations of Jupiter and Saturn on the highly elliptical orbit of Comet Halley render the comet's motion chaotic within less than 30 returns to perihelion. The strong instability of a chaotic trajectory restricts its extrapolation to a relatively short time interval—irrespective of the model's accuracy in fitting the observations used to create the model.

Table 1 presents the orbital elements of Comet Halley from AD 2134 back to 466 BC, a period over which there should be no problem with the extrapolation accuracy.

Table 1. Orbital elements for Comet Halley

Year	T (E.T.)	q (AU)	e	Arg.	Node	I	Epoch
2134	2134 Mar. 27.82435	0.5932150	0.9666411	113.98820	60.59200	161.74573	Mar. 15
2061	2061 Jul. 28.71115	0.5927814	0.9665754	112.03301	58.67602	161.96220	Aug. 4
1986	1986 Feb. 9.45894	0.5871029	0.9672755	111.84653	58.14339	162.23925	Feb. 19
1910 II	1910 Apr. 20.17862	0.5872086	0.9673028	111.71769	57.84547	162.21559	May 9
1835 III	1835 Nov. 16.43965	0.5865638	0.9673947	110.68489	56.80119	162.25570	Nov. 18
1759 I	1759 Mar. 13.06333	0.5844878	0.9675306	110.70817	56.54679	162.38297	Mar. 21
1682	1682 Sep. 15.29064	0.5826372	0.9677987	109.24246	54.86670	162.27537	Aug. 31
1607	1607 Oct. 27.53437	0.5836877	0.9673329	107.57718	53.06780	162.91222	Oct. 24
1531	1531 Aug. 26.23846	0.5811975	0.9677499	106.95724	52.34044	162.91385	Aug. 14
1456	1456 Jun. 9.63257	0.5797014	0.9769974	105.81647	51.15021	162.88607	June. 28
1378	1378 Nov. 10. 68724	0.5762013	0.9683723	105.27668	50.30348	163.10897	Nov. 5
1301	1301 Oct. 25.58194	0.5727097	0.9689307	104.48199	49.43575	163.07179	Nov. 9
1222	1222 Sep. 28.82294	0.5742108	0.968844	103.83087	48.58845	163.18782	Oct. 15
1145	1145 Apr. 18.56090	0.5747921	0.9687853	103.68573	48.33830	162.22004	Apr. 2
1066	1066 Mar. 20.93405	0.5744956	0.9688655	102.45543	46.90873	163.10814	Mar. 8
989	989 Sep. 5.68757	0.5819144	0.9678887	101.46581	45.84533	163.39474	Aug. 19
912	912 Jul. 18.67429	0.5801559	0.9680692	100.75913	44.93122	163.30679	Jul. 14
837	837 Feb. 28.27000	0.5823182	0.9678055	100.08403	44.21516	163.44258	Mar. 10
760	760 May 20.67126	0.5818368	0.9678541	99.98016	43.97218	163.43860	Jun. 2
684	684 Oct. 2.76682	0.5795841	0.9681495	99.13197	43.08465	163.41338	Sep. 29
607	607 Mar. 15.47581	0.5808315	0.9680396	98.78209	42.54593	163.47190	Mar. 18
530	530 Sep. 27.12998	0.5755915	0.9687113	97.56504	41.26006	163.38977	Oct. 8
451	451 Jun. 28.24911	0.5737438	0.9689123	97.01122	40.49602	163.47468	Jun. 25
374	374 Feb. 16.34230	0.5771940	0.9685857	96.49409	39.86451	163.53760	Mar. 1
295	295 Apr. 20.39842	0.5759148	0.9687528	95.22565	38.39767	163.36268	Apr. 25
218	218 May 17.72347	0.5814660	0.9679755	94.13158	37.19436	164.56891	Apr. 29
141	141 Mar. 22.43405	0.5831377	0.9678439	93.67835	36.50620	163.43259	Mar. 24
66	66 Jan. 25.96014	0.5851046	0.9675458	92.63672	35.41600	163.57158	Feb. 6
12 BC	−11 Oct. 10.84852	0.5871999	0.9673664	95.54339	35.19064	163.58392	Oct. 8
87 BC	−86 Aug. 6.46171	0.5856047	0.967679	90.76383	33.30553	163.33505	Aug. 23
164 BC	−163 Nov. 12.56604	0.5845470	0.9676686	89.09882	31.35152	163.69946	Nov. 15
240 BC	−239 May 25.11796	0.5853647	0.9678571	88.09919	30.09811	163.46207	Jun. 7
315 BC	−314 Sep. 8.52367	0.5874295	0.9673085	86.86997	28.83174	163.59479	Sep. 29
391 BC	−390 Sep. 14.36897	0.5880489	0.9672597	86.80007	28.61248	163.59935	Sep. 28
466 BC	−465 Jul. 18.23879	0.5902897	0.9671425	85.23459	26.86861	163.25878	Jul. 4

Notes: Following the year, the orbital elements are respectively the time of perihelion passage (T) in ephemeris time, the perihelion distance in AU (q), eccentricity (e), the argument of perihelion (Arg.), the longitude of the ascending node (Node), and the inclination (I). The last column gives the epoch of osculation where the decimal of a day is 0.0 (ephemeris time). The angular elements are given in degrees and referred to the ecliptic (equinox 1950.0). Prior to 1582, the Julian calendar is used.

The orbital elements for 1835 through 2134 are based upon a fit of 7469 observations over the interval 21 Aug. 1835 through January 9, 1989. It is designated International Halley Watch (IHW) orbit No. 61, and the appropriate nongravitational parameters are $+0.0388 \times 10^{-8}$ and $+0.0155 \times 10^{-8}$ AU/(ephemeris day)^2. The 1607, 1682, and 1759 elements are from a fit of 228 observations over the interval 28 Sept. 1607 through 3 June 1759. Elements prior to 1759 were taken from Yeomans & Kiang (1981).

ACKNOWLEDGEMENTS

This research was carried out by the Jet Propulsion Laboratory, California Institute of Technology, under contract with the National Aeronautics and Space Administration.

REFERENCES

Bessel, F.W. (1836a) *Astron. Nachr.* **13** 185–232
Bessel, F.W. (1836b) *Astron. Nachr.* **13** 345–350
Biot, E.C. (1843) *Conn. des temps pour l'an 1846*, Additions 69–84
Brady, J.L. (1972) *Publ. Astron. Soc. Pac.* **84** 314–322
Brady, J.L., & Carpenter, E. (1967) *Astron. J.* **72** 365–369
Brady, J.L., & Carpenter, E. (1971) *Astron. J.* **76** 728–739
Brady, J.L. (1982) *J. Brit. Astr. Assoc.* **92** 209–215
Broughton, R.P. (1979) *J.R. Astr. Soc. Canada* **73** 24–36
Burckhardt, J.C. (1804) *Monat. Corres.* **10** 162–167
Celoria, G. (1893) In: Uzielli, G., *La vita di Toscanelli*, Rome; reprinted 1921, Pubbl. Oss. astron. di Brera, no. 55
Chang, Y.C. (1979) *Chin. astron.* **3** 120–131
Chirikov, B.V., & Vecheslavov, V.V. (1989) *Astron. & Astrophys.* **221** 146–154
Clairaut, A.C. (1758) *J. des scavans* (Jan. 1759) **41** 80–96
Cowell, P.H., & Crommelin, A.C.D. (1907a) *Mon. not. R. Astron. Soc.* **67** 174
Cowell, P.H., & Crommelin, A.C.D. (1907b) *Mon. not. R. Astron. Soc.* **67** 386–411, 521
Cowell, P.H., & Crommelin, A.C.D. (1907c) *Mon. not. R. Astron. Soc.* **67** 511–521
Cowell, P.H., & Crommelin, A.C.D. (1907d) *Mon. not. R. Astron. Soc.* **68** 111–125
Cowell, P.H., & Crommelin, A.C.D. (1908a) *Mon. not. R. Astron. Soc.* **68** 173–179
Cowell, P.H., & Crommelin, A.C.D. (1908b) *Mon. not. R. Astron. Soc.* **68** 375–378
Cowell, P.H., & Crommelin, A.C.D. (1908c) *Mon. not. R. Astron. Soc.* **68** 379–395
Cowell, P.H., & Crommelin, A.C.D. (1908d) *Mon. not. R. Astron. Soc.* **68** 510–514
Cowell, P.H., & Crommelin, A.C.D. (1908e) *Mon. not. R. Astron. Soc.* **68** 665–670
Cowell, P.H., & Crommelin, A.C.D. (1910) *Publ. Astron. Gesellschaft*, no. 23
Damoiseau, T. de (1820) *Mem. R. Accad. delle scienze di Torino* **24** 1–76
Damoiseau, T. de (1829) *Conn. des temps pour l'an 1832*, Additions, 25–34
Halley, E. (1705) *Astronomiae cometicae synopsis*, Oxford
Halley, E. (1749) *Tabulae astronomicae*, London
Hasegawa, I. (1979) *Publ. Astron. Soc. Japan* **31** 257–270, 829
Hind, J.R. (1850) *Mon. not. R. Astron. Soc.* **10** 51–58

Ho Peng Yoke (1962) *Vistas astron.* **5** 127–225
Ho Peng Yoke & Ang Tian-Se (1970) *Oriens extremus* **17** 63–99
Kiang, T. (1972) *Mem. R. Astron. Soc.* **76** 27–66
Lagrange, J.L. (1783) *Nouv. mem. Acad. royale Berlin* 161–223
Landgraf, W. (1984) *ESTEC EP/14.7/6184*
Landgraf, W. (1986a) *Astron. & Astrophys.* **163** 246–260
Landgraf, W. (1986b) *Quart. J. Roy. Astr. Soc.* **27** 118
Laugier, P.A.E. (1842) *Comptes rendus* **15** 949–951
Laugier, P.A.E. (1843) *Comptes rendus* **16** 1003–1006
Laugier, P.A.E. (1846) *Comptes rendus* **23** 183–189
Lehmann, J.W.H. (1835) *Astron. Nachr.* **12** 369–400
Marsden, B.G. (1968) *Astron. J.* **73** 367–379
Marsden, B.G. (1969) *Astron. J.* **74** 720–734
Marsden, B.G., Sekanina, Z., & Yeomans, D.K. (1973) *Astron. J.* **78** 211–225
Michielsen, H.F. (1968) *J. Spacecr. Rockets* **5** 328–334
Newton, I. (1687) *Philosophiae naturalis principia mathematica*, bk. 3, London
Pingré, A.G. (1783–84) *Cometographie*, Paris.
Pontécoulant, P.G.L. de (1830) *Conn. des temps pour l'an 1833*, Additions, 104–113
Pontécoulant, P.G.L. de (1834) *Conn. des temps pour l'an 1837*, Additions, 102–104
Pontécoulant, P.G.L. de (1835) *Mem. presentes pars divers savants*, ser. 2 **6** 875–947
Pontécoulant, P.G.L. de (1864) *Comptes rendus* **58** 825–828,915
Rasmusen, H.Q. (1967) *Publ. og mindre medd. fra Kobenhavns Observatorium* No. 194
Rasmusen, H.Q. (1981) *Fourth expected return of Comet Halley: elements and ephemerides, 1981 to 1985* (dated July)
Rosenberger, O.A. (1830a) *Astron. Nachr.* **8** 221–250
Rosenberger, O.A. (1830b) *Astron. Nachr.* **9** 53–68
Rosenberger, O.A. (1834) *Astron. Nachr.* **11** 157–180
Rosenberger, O.A. (1835) *Astron. Nachr.* **12** 187–194
Sitarski, G. (1988) *Acta Astron.* **38** 253–268
Stephenson, F.R., Yau, K.K.C., & Hunger, H. (1985) *Nature* **314** 587–592
Whipple, F.L. (1950) *Astrophys. J.* **111** 375–394
Whipple, F.L. (1951) *Astrophys. J.* **113** 464–474
Yeomans, D.K. (1977) *Astron. J.* **82** 435–440
Yeomans, D.K., & Kiang, T. (1981) *Mon. not. R. Astron. Soc.* **197** 633–646
Yeomans, D.K. (1984) In: *Cometary astrometry, proceedings of a workshop held at the European Southern Observatory Headquarters*, JPL Publication 84–92

19

Comparison of old and present-day methods and results of investigations into the long-term motion of Comet Halley

Krzysztof Ziolkowski

1 INTRODUCTION

Our reliable present-day knowledge of Comet Halley's motion reaches back to 240 BC. The existing models of this motion have been verified by available descriptions of direct observations retrieved from old records for all apparitions of the comet over much the same period. Attempts, so far, to extrapolate backward in time beyond the observational period, cannot, however, be regarded as reliable, because of discrepancies between the results reported by different authors, despite their quite reasonable approaches to the solution of the problem. Therefore a thorough study of the hitherto used methods of investigating the long-term motion of Halley's Comet as well as comparison of the results obtained, seems to be not only interesting from the historical viewpoint but also important for further researches and discussions on this puzzling matter.

The history of the comet's motion is described in a general outline, in the main, a single parameter: the time of perihelion passage. Determination of this parameter for successive apparitions of Halley's Comet is the main purpose of the present paper. As a starting point the famous paper by Halley (1705) can be considered, where the perihelion time and other orbital elements in 1531, 1607, and 1682 are given and a supposition formulated that an earlier perihelion passage took place in 1456, and that the next would occur in 1758. The correctness of the last supposition is commonly known. The justness of Halley's suggestion concerning the 1456 apparition has been demonstrated by Pingré (1783).

2 FIRST ATTEMPTS

Initially, the investigations of Comet Halley's motion in the past were closely associated with search for and identification of observations made during previous apparitions, as described in old records. The investigation method used was very simple. From the known time of perihelion passage the mean orbital period of the comet was subtracted, and in the vicinity of a date so obtained searches for some chronicler's records which may be interpreted as those concerning the comet were carried out. If a successful retrieval result was obtained, then a previously adopted perihelion time was correspondingly corrected, obtaining in this way a new, more reliable value for determining perihelion passage.

For example, following this procedure, Laugier (1843, 1846) demonstrated that the comet observed in China in the autumn 1378, as well as in 760 and 451, was Comet Halley. He did not, however, manage to do this for the comet which appeared in 1152 (Halley's Comet had passed its perihelion in 1145). The most thorough and systematic investigations of this kind were carried out by Hind (1850). Proceeding in this way he managed to retrieve the moments of perihelion passage for Halley's Comet as far back as 11 BC. And though some of his identifications appeared later to be erroneous, owing to his work the motion of Halley's Comet through nearly two millennia had been made known by the middle of the nineteenth century in general outline (for more details see the paper by Yeomans in this volume).

Analysis of these data enabled Angström (1862) to reveal some periodicity of changes in the orbital period of Comet Halley. While assuming that the main causes of such irregularities are perturbations by the large planets Jupiter and Saturn only, he stated that these periodicities can be approximately described by the sum of two periodic functions having amplitudes of about 1.5 and 2.2 years and periods of about 2650 and 780 years respectively. He calculated that the arguments of trigonometric functions which would give such periods should have the forms:

$$13\, n_\mathrm{C} - 2\, n_\mathrm{J} \text{ and } n_\mathrm{J} + n_\mathrm{S} - 9\, n_\mathrm{C}$$

where n_C, n_J, and n_S denote mean yearly motions of the comet, Jupiter, and Saturn respectively. This enabled him in turn to derive a simple formula for calculation of the time of perihelion passage T_N (in years) for each Nth apparition:

$$T_\mathrm{N} = 913.97 + 76.93\, N + 1.46 \sin(10°49\, N + 19°2) + {} \\ + 2.16 \sin(34°93\, N + 233°6),$$

where $N = 0, \pm 1, \pm 2, \ldots$ This gave $N = -12$ for the 11 BC apparition, $N = 0$ for the 912 apparition, and $N = +12$ for the 1835 apparition. The first summand is an initial epoch which has been assumed by Angström to be the arithmetic mean of the known times of perihelion passage within the interval 11 BC to AD 1835. The numerical factor of the second summand incorporating N is a mean orbital period of Comet Halley, found to be 1/24 of the interval between the times of perihelion passage in 11 BC and AD 1835. The next two summands are those describing the perturbations by Jupiter and Saturn. The mean residual of the times of perihelion passage calculated by means of the above formula amounts to ± 0.48 year, whereas maximum deviation does not exceed 1 year. It turned out, however, that the time of

perihelion passage in 1910 so calculated differs from actuality by as much as 2.7 years, and for the year 1986 apparition by more than 5 years. This formula is, therefore, unsuitable for making extrapolations beyond the limits of the interval for which it has been deduced.

3 SEMI-ANALYTICAL APPROACH

The methods of investigating the long-term motion of Halley's Comet as devised in the middle of the 19th century were followed, in general, in further work on that problem until the middle of 20th century. They were, in fact, refined during that time, with rectification of some errors made by earlier workers and improvement in reliability of results, but no essential qualitative changes have been made therein during the present century.

Cowell & Crommelin (1907, 1908a, b, c, d), instead of sequentially subtracting the mean value of the orbital period of the comet, have calculated planetary perturbations on the comet's motion by the variation of elements technique. While being aware, however, since exact calculation of planetary perturbations from all planets for an interval of nearly two millennia is not feasible within a reasonable time, they were forced to make some simplifying assumptions (see Yeomans, this volume). In this way they determined not only the times of perihelion passage by Comet Halley but also its remaining orbital elements from AD 1910 to 240 BC. Though further investigations have demonstrated that the error in their results obtained for the earlier times of perihelion passage was as large as 1.5 year, their unparalleled, in those days, success has been regarded as an outstanding scientific achievement, and for many years thereafter numerous different investigations not only of Comet Halley but of comets in general followed the methods of Cowell and Crommelin.

Viliev (1917) extended the computations made by Cowell and Crommelin back beyond the era of observations, and calculated the perturbations of Comet Halley's motion for the period AD 451 to 622 BC. It should be noted that the main purpose of his investigations was an attempt to demonstrate that the biblical vision of the prophet Jeremiah (1, 13–14† was an observation of Halley's Comet in 623/622 BC.

The Angström concept to describe the changes in orbital period of the comet by an interpolation formula was developed by Kamieński (1949, 1951, 1957). He proposed the following general formula for the time of perihelion passage T_N for each Nth apparition:

$$T_N = T_0 + N\tau + A_1 \sin [(13\, n_C - 2\, n_J)\, N\tau + P_1] + $$
$$+ A_2 \sin [(n_J + n_S - 9\, n_C)\, N\tau + P_2] + $$
$$+ A_3 \sin [(13\, n_C - 2\, n_J)\, N\tau + P_3] + $$
$$+ A_4 \sin [3(8\, n_C - 3\, n_S)\, N\tau + P_4],$$

where T_0 is the initial epoch, τ is mean orbital period of the comet, and n_C, n_J and n_S are mean yearly motions of the comet, Jupiter, and Saturn respectively. Kamieński based his work on 29 times of perihelion passage by Halley's Comet as calculated by Cowell & Crommelin for the period from 240 BC to AD 1910. With the moment of the AD 837 apparition as starting point, he computed by the least squares method the corrections for provisionally adopted values of mean orbital period as well as for amplitudes A_1, A_2, A_3, A_4, and arguments P_1, P_2, P_3, P_4. The correction for the initial epoch appeared to be so small as to be negligible. In this way Kamieński has determined that:

$$\tau = 76.903 \pm 0.005 \text{ years},$$

and finally:

$$T_N = 837.0 + 76.903\, N + 1.60 \sin(12°13\, N + 9°8) + $$
$$+ 2.17 \sin(33°77\, N + 188°4) + $$
$$+ 0.10 \sin(12°13\, N + 270°0) + $$
$$+ 0.23 \sin(181°45\, N + 90°0),$$

where $N = 0, \pm 1, \pm 2, \ldots$, wherein $N = -14$ for the 240 BC apparition, $N = 0$ for the AD 837 apparition, and $N = +14$ for the 1910 apparition. This formula represents a single time of perihelion passage with a mean error of ± 0.49 year.

Proceeding in a similar way, Kamieński derived an independent formula for calculation of the duration of the time interval between two successive perihelion passages by Comet Halley:

†13. And the word of the LORD came unto me the second time, saying, What seest thou? And I said, I see a seething pot, and the face thereof was toward the north.
14. Then the LORD said unto me, Out of the north an evil shall break forth upon all the inhabitants of the land.   Authorised Version.

$$T_{N+1} - T_N = 76.903 + 0.43 \cos(12°13\, N + 10°4) + $$
$$+ 1.18 \cos(33°77\, N + 184·0) + $$
$$+ 0.13 \sin(38°39\, N - 29·7) - $$
$$- 0.75 \sin 181.45\, N.$$

The consistency between times of perihelion passages as calculated by the two formulae has proved the correctness of the calculations. Furthermore, attempts to extrapolate back beyond the observational era gave results fairly consistent with those obtained by Viliev (1917). This encouraged Kamieński to make a thorough study of the motion of the comet as far back in time as the 24th century BC. The earliest chronicler's records go back just into that century (Hasegawa 1980).

To check, however, the results of those calculations, as well as to make them more reliable, Kamieński has repeated them while using a quite different method, i.e. the cyclic method of finding orientational positions of planets in very distant times which he had developed (Kamieński 1947). This method is based on the commensurability of the revolutions of the comet, of Jupiter, and of Saturn round the Sun. Although the orbital period of the comet varies and fluctuates within the limits of a few years, it can, however, be assumed that during the long period of many revolutions of the comet these inequalities are being smoothed out to some extent. As Kamieński pointed out, the situation is the same as that of the period of eclipses of Saros (6585 days), where the inequalities of lunar months are almost entirely smoothed out. Kamieński has found four such cycles:

$$23\, C \cong 149\, J \cong 60\, S \cong 1768.1 \text{ years},$$
$$25\, C \cong 162\, J \cong 65\, S \cong 1921.5 \text{ years},$$
$$48\, C \cong 311\, J \cong 125\, S \cong 3690.0 \text{ years},$$
$$54\, C \cong 350\, J \cong 141\, S \cong 4152.5 \text{ years},$$

where C, J, and S are mean orbital periods of the comet, Jupiter, and Saturn respectively. On the basis of the results of Cowell & Crommelin as well as on those of Viliev, and using the first two of these cycles, Kamieński (1956) obtained fair consistency between the times of perihelion passage by Halley's Comet as calculated by using the cyclic method, and those obtained from the formulae given above. This confirmed him in the conviction as to the rightness of this procedure and the feasibility of rough reproduction of the comet's motion in the very distant past (See Table 1).

While considering all four cycles, Kamieński (1961) calculated the times of perihelion passage as far back in time as the middle of the tenth millennium BC. Counting back from the present, this encompasses 150 revolutions of the comet around the Sun. It is difficult to judge reliability of these unique results, which are unparalleled in the history of cometary investigation. Their originator was fully aware of the fact that only observations can be considered as ultimate proof. Because of the absence of such proof he approached the problem in quite another way. Namely, while assuming the genuineness of the times of perihelion passage that he had determined, and assuming the probability that some descriptions of unusual phenomena accompanying famous historical events can reasonably be attributed to apparition of Halley's Comet, he determined with relatively high accuracy the dates of corresponding events. Since some of them are precisely dated by historians, their consistency with the calculated dates may to some extent justify the correctness of the assumptions made. In this manner Kamieński (1965) analysed thoroughly, among other events, King David's foundation of the First Temple in Jerusalem (1010 BC), the fall of Troy (1150 BC), the destruction of Sodom and Gomorrah (1757 BC), the birth of Abraham (1856 BC), and even the biblical flood (3850 BC) and the catastrophe of legendary Atlantis (9542 BC). Of course, the credulous acceptance of the results of such study as ascertained facts would be as unjustified as definite rejection of them. They constitute an interesting contribution to the chronology of ancient history, and may be helpful in visualizing Comet Halley's motion in the very distant past.

4 NUMERICAL INTEGRATION OF THE EQUATIONS OF THE COMET'S MOTION

Today's great advances in science and technology make possible a qualitatively new approach to the investigation of the long-term motion of Comet Halley. The main contributing factors are:
- the use of highly advanced computational equipment and numerical methods;
- the availability of new and more precise observational data;

- deeper and more thorough understanding of various subtle effects (especially nongravitational forces) influencing the comet's motion.

The modelling of the motion of Comet Halley based on the integration of appropriate differential equations has been made possible by the use of computer techniques. Consequently, better and better methods of numerical integration of the equations of motion are being developed which ensure appropriate accuracy, minimize cumulative errors, and speed up highly labour-consuming computations (e.g. the recurrent power series method (Sitarski 1979)). To find the best initial conditions for integration, improved methods of differential improvement of orbits are being developed (e.g. Sitarski 1983, Yeomans 1984). Simultaneously, attempts are being made to verify and more precisely determine the observational data on the comet. In the tradition of Hind (1850) and Cowell & Crommelin (1907, 1908), Kiang, analyzing ancient Far Eastern and other records, and calculating planetary perturbations by the variation of elements method, has determined new times of perihelion passage by Halley's Comet from 240 BC onward (Kiang 1972, Yeomans & Kiang 1981). The work of Kiang and others is described in more details by Stephenson in this volume. Many positional observations from recent apparitions have been re-elaborated in order to refer them to modern star catalogues (e.g. Morley 1983, Bielicki et al. 1984). Finally, a thorough investigation of nongravitational effects on the comet's motion (described in details by Sitarski in this volume) has created a suitable theoretical basis for modelling the evolution of Comet Halley's orbit through the ages.

The first successful attempt to link by one system of elements several apparitions of the comet was made by Brady & Carpenter (1967, 1971). Taking into account an empirical term in the equations of motion giving the effect of secular decrease in the solar attraction with time, they have linked the four apparitions in 1682, 1759, 1835, and 1910. Although they applied a 'trial and error' fit to the observations, the result was then used by Brady (1982) for the numerical integration of the equations of Comet Halley's motion as far back as 2647 BC.

Using a least squares differential improvement process and applying Marsden's model for the nongravitational forces as rocket-like effects acting on the cometary nucleus owing to the vaporization flux of water ice, Yeomans (1977) has found a set of orbits linking three or four apparitions within the 1607–1910 interval. Yeomans & Kiang (1981) have shown that the orbit based on the 1607, 1682, and 1759 observations is the best one to use for numerical integration far back into the past. They have integrated Comet Halley's motion back to 1404 BC. But to fit the accurately observed perihelion passage times in AD 837, 374, and 141 (during these apparitions the comet passed very near to Earth) they introduced some empirical adjustments to the computed perihelion time and eccentricity in AD 837. Some details of this work and its main results are cited in the review by Yeomans in this volume.

A similar approach to the problem can be found in the paper by Landgraf (1984). He also improved the orbit of Comet Halley by the least squares method, but used the positional observations and normal places from the last six apparitions between 1607 and 1984. An important distinction associated with the nongravitational effects must be noted. Although Yeomans & Kiang (1981) and Landgraf (1984) used the same model, Yeomans & Kiang assumed that the comet's nongravitational force remained constant with time, while Landgraf considered the transverse component of that force to be some linear function of time. After making some subjective changes in the orbital elements for AD 837 also, Landgraf extended the process of numerical integration back to 2317 BC.

An entirely new procedure for orbit improvement (Sitarski 1987) opens a new approach to investigations of the long-term motion of Comet Halley. This method, using the times of perihelion passages as observational data for the least squares correction of the orbit, enables us to use, in a uniform way, both modern positional observations, as well as ancient observations for which the perihelion times have been deduced. Thus, all the observational material can be used to link, by one system of orbital elements, all the apparitions of Comet Halley since 240 BC. In particular, Sitarski (see his review in this volume), using a secular change in the semi-major axis of the orbit as a model of nongravitational effects, has found a parabolic time dependence in the nongravitational forces over the 87 BC–AD 1986 period. Taking into

account the best 250 observations from the 1835–1986 period and the observed perihelion times since 87 BC, he obtained one system of dynamical parameters of the comet, which he used as input data for numerical integration back to 1457 BC.

5 COMPARISON AND DISCUSSION OF RESULTS

The results of the first attempts to chart Comet Halley's motion during the last two millennia, summarized in Table 1, seem to be interesting only

Table 1. Results of the first efforts to determine the times of past perihelion passages of Halley's Comet (in years)

Hind	Ångström	Cowell, Crommelin	Viliev	Kamienski
		1910.3		1910.3
1835.9	1836.0	1835.9		1837.1
1759.2	1759.1	1759.2		1760.8
1682.7	1683.0	1682.8		1684.4
1607.8	1607.4	1607.8		1609.3
1531.6	1531.8	1531.6		1532.4
1456.4	1455.8	1456.4		1456.7
1378.8	1379.1	1378.9		1378.7
1301.8	1301.6	1301.8		1301.8
1223.5	1223.4	1222.7		1222.8
1145.3	1145.1	1145.3		1145.3
1066.2	1067.0	1066.2		1066.5
989.7	989.5	989.7		989.8
912.2	912.7	912.5		912.3
837.3	836.6	837.1		837.0
760.4	760.7	760.4		760.2
684.8	684.5	684.9		684.8
608.8	607.8	607.2		607.1
530.8	530.4	530.9		530.3
451.5	452.3	451.5	451.5	451.3
373.8	373.9	473.8	374.1	373.8
292.2	295.7	295.3	295.3	295.1
218.3	218.0	218.3	218.3	218.5
141.2	141.3	141.2	141.2	141.1
66.1	65.2	66.1	66.0	65.9
−10.2	−10.4	−10.2	−10.2	−10.6
		−85.4	−85.4	−85.3
		−161.6	−161.7	−161.7
		−238.6	−238.0	−238.1
			−312.7	−313.1
			−389.7	−390.0
			−465.7	−465.8
			−543.9	−543.7
			−621.0	−620.6
				...
				...
				...
				−9540.6

Table 2 Comparison of the times of past perihelion passages of Halley's Comet deduced from observations (K – Kiang 1972, Yeomans & Kiang 1981) and computed (B – Brady & Carpenter 1971, Y – Yeomans & Kiang 1981, L – Landgraf 1984, S – Sitarski 1987)

K			K − B	K − Y	K − L	K − S
y	m	d	d	d	d	d
1910	4	20.18	0.0	0.0	0.0	0.0
1835	11	16.44	0.0	0.0	0.0	0.0
1759	3	13.05	0.0	0.0	0.0	+0.8
1682	9	15.27	0.0	0.0	0.0	+0.7
1607	10	27.56	+0.3	0.0	0.0	+2.4
1531	8	25.8	−0.3	−0.4	−0.5	+2.0
1456	6	9.1	−0.4	−0.5	−0.4	+0.9
1378	11	9.02	−2.4	−1.7	−1.6	−0.7
1301	10	24.53	−2.4	−1.0	−0.7	−0.8
1222	·9	30.8	+1.2	+2.0	+2.2	+1.1
1145	4	21.25	+2.9	+2.7	+3.2	+0.7
1066	3	23.5	+3.5	+2.6	+3.4	+0.9
989	9	9.0	+5.5	+3.3	+4.9	+1.5
912	7	9.5	−7.6	−9.2	−7.5	−9.6
837	2	28.27	−0.1	0.0	−0.1	−0.1
760	5	22.5	+0.2	+1.8	+1.9	+1.9
684	9	28.5	−8.7	−4.3	−2.9	−4.0
607	3	13.0	−5.7	−2.5	−0.6	−2.1
530	9	26.7	−0.7	−0.4	+1.1	−0.6
451	6	24.5	−1.8	−3.7	−2.7	−3.5
374	2	17.4	+4.3	+1.1	+2.1	+2.0
295	4	20.5	−2.5	+0.1	−0.1	+0.4
218	5	17.5	−10.6	−0.2	−0.2	−0.2
141	3	22.35	−19.4	−0.1	+1.2	−0.2
66	1	26.5	−25.0	+0.5	+4.6	+1.0
−11	10	5.5	−3.6	−5.3	−0.5	−3.4
−86	8	2.5	+22.6	−4.0	−1.0	−0.9
−163	11	17†	+147.6	+4.4	+17.9	+24.7
−239	3	30.5	+118.9	−55.6	−17.0	+7.9

† The centre point of the interval obtained by Stephenson *et al.* (1985).

from a historical viewpoint. Further consideration will therefore be concentrated on results obtained during the second half of the present century.

Conclusive for prediction of Comet Halley's motion before the observational era is, first of all, the model of motion adopted for investigation, as well as the initial values of orbital elements. Both factors are verified by observations, therefore in Table 2 the residuals O-C of the times of perihelion passage by the comet are compared, taking as observational data the corresponding values obtained by Kiang. It is not difficult to note that, except for the data reported by Brady & Carpenter, all remaining data give satisfactory consistency with observations. It should be noted, however, that the integrations made by Brady & Carpenter gave a fairly good description of the comet's motion back to AD 295, despite the fact that to take into account the nongravitational effect a simple mathematical trick instead of some physical model has been used; and a not particularly precise, in terms of mathematical accuracy, orbit improvement process has been adopted (trial and error technique). On the other hand the recent results obtained by Sitarski seems to excel all others primarily because the initial orbital elements for the integration process and the variable with time parameters of nongravitational effects have been obtained by linking practically all observed apparitions of the comet.

It is worth stressing that the important role of the time of perihelion passage in 240 BC is determined on the basis of observations. While assuming its correctness (which seems not to be a well grounded assumption) it can be stated that the models with time dependence of nongravitational forces (Landgraf, Sitarski) better fit this apparition than those with constant nongravitational effects (Yeomans & Kiang). The same conclusion arises from the work of Sitarski & Ziolkowski (1987). The time of perihelion passage in 164 BC, as given in Table 2, cannot be considered to be equivalent to other observations, since Stephenson *et al.* (1985) have estimated that the perihelion passage in that year would happen only between 9 and 26 November; but it is interesting that the centre point is in good agreement with the results of Yeomans & Kiang only.

The model of Comet Halley's motion by Yeomans and Kiang was used by Sitarski & Ziolkowski (1986) to link, by one system of orbital elements, all apparitions of the comet since 11 BC and to integrate thereafter its equations of motion back beyond the

Table 3. Comparison of the times of past perihelion passages of Halley's Comet (in years) predicted by Kamieński (K), Yeomans & Kiang (Y), Brady (B), Landgraf (L), and Sitarski (S)

K	K − Y	K − B	K − L	K − S
−238.1	+0.5	+1.0	+0.6	+1.1
−313.1	+0.2	+2.1	+0.5	+1.1
−390.0	−0.7	+1.7	−0.3	+0.4
−465.8	−1.3	+0.7	−1.1	−0.6
−543.7	−5.1	−1.0	−3.7	−2.2
−620.6	−6.2	−2.4	−4.3	−2.8
−700.1	−11.2	−7.9	−9.1	−7.0
−777.2	−15.8	−9.6	−9.3	−8.0
−855.6	−21.0	−12.6	−11.0	−11.0
−931.8	−22.3	−15.6	−8.9	−10.4
−1008.9	−24.8	−17.1	−8.7	−10.7
−1084.3	−27.2	−16.8	−4.3	−10.3
−1161.3	−36.6	−19.6	−3.8	−10.8
−1237.2	−40.6	−19.3	−1.5	−11.5
−1315.4	−51.1	−21.5	−0.7	−13.8
−1392.3	−59.9	−24.5	−0.1	−13.7
−1471.3	−69.1	−27.8	+0.5	−14.7
−1548.6		−29.2	+0.8	
−1626.8		−34.7	+0.2	
−1703.1		−34.8	+1.3	
−1780.0		−35.7	+1.2	
−1855.2		−37.1	+2.0	
−1931.6		−37.8	+2.6	
−2007.3		−37.7	+1.7	
−2082.7		−39.5	+3.1	
−2159.3		−40.3	+1.9	
−2234.6		−39.7	+3.4	
−2311.9		−43.0	+3.2	
−2388.2		−42.7		
−2468.2		−46.9		
−2545.3		−50.1		
−2623.7		−53.2		
−2699.9		−54.2		

observational era. Nearly identical results obtained by using quite different methods is indicative not only of their correctness (although some empirical adjustments had to be made to the orbital elements during the process of numerical integration by Yeomans & Kiang) but also of consistency of models of the Solar System used by the authors to calculate planetary perturbations. This fact leads one also to accept as reliable the essential results of Sitarski & Ziokowski (1986) who give reasons for the large difference between results obtained by Yeomans & Kiang and those obtained by Landgraf (see Table 3). It appears that the reasons for such discrepancies are secular changes in nongravitational effects. This surprising conclusion is confirmed by the fact that when Landgraf (1986) has drastically reduced the value of a parameter characterizing the linear variability with time of nongravitational effects, he also obtained results almost identical with those of Yeomans & Kiang.

The data summarized in Table 3 show the very good agreement between results of computations made by Kamieński and by Landgraf. Thus we have a model of motion which confirms earlier calculations made in a simple manner and by primitive methods. Is this only fortuitous? It is, as yet, impossible to answer this question. It is worth stressing that the procedure used by Kamieński has the advantage that, over long time spans, it smooths out small inaccuracies and compensates for the absence of a model of cometary and planetary motions. In other words, it minimizes the effects of those factors which, in the process of numerical integration of the equations of motion, play a particularly important role because of systematic increase of their values. It seems that in the light of the most recent work and new possibilities of making full use of observational data, as well as better understanding of the mechanism of nongravitational effects, the amazing agreement of Kamieński's and Landgraf's results can throw much light on the true motion of Comet Halley, and can indicate the direction of future research.

6 CONCLUDING REMARKS

In the present review we have omitted work concerning the long-term evolution of the orbit of Halley's Comet (e.g. Babadzhanov & Obrubov 1982, Hajduk 1983, Kiang 1974, Kozai 1979). It must be stressed, however, that a critical analysis of their results should by no means be disregarded in future investigations of the motion of the comet. Moreover, it is impossible to neglect the consistency of some present-day results and those of Kamieński's researches made many years ago. Kamieński's work is scattered in many publications, some of which are hard to find. We hope that, by dealing with it at some length, we may save it from being forgotten.

Revision of earlier calculations of perihelion passage, based on old observations, is well worthwhile, as is further search for ancient records of observations of Halley's Comet. Existing achievements have been based on theories of cometary motion which have been since verified and made more accurate. The selection and weighting of observational material of all kinds deserves thorough study.

It seems paradoxical that, in the investigation of the long-term motion of Halley's Comet, we have not made, in fact, any essential progress beyond what was already known a hundred years ago. The accuracy and reliability of results obtained at that time have been, indeed, improved, but any attempts to go further back than 240 BC lead only to more or less probable hypotheses. And, despite some doubt as to whether it is possible to determine the motion of the comet in the yet more distant past, comparative analysis of the methods used, and results obtained, not only encourages us to make further investigations but also seems to indicate some possible ways to the solution of the problem.

REFERENCES

Angström, A.J. (1862) *Nova Acta Reg. Soc. Sc. Upsal Ser. 3*, 1–10

Babadzhanov, P.B., & Obrubov, J.V. (1982) *Bull. Inst. Astrofiz. Akad. Nauk Tadzhik. SSR*, No. 71, 11–15

Bielicki, M., Sitarski, G., & Ziolkowski, K. (1984) In: *Cometary astrometry* eds Yeomans, D.K., West, R.M., Harrington, R.S., Marsden, B.G., NASA/JPL Publ. 84–82, 203–206

Brady, J.L. (1982) *J. Br. astr. Ass.* **92** 209–215

Brady, J.L., & Carpenter, E. (1967), *Astr. J.* **72** 365–369

Brady, J.L., & Carpenter, E. (1971) *Astr. J.* **76** 728–739

Cowell, P.H., & Crommelin, A.C.D. (1907) *Mon. Not. R. astr. Soc.* **68** 111–125

Cowell, P.H., & Crommelin, A.C.D. (1908a) *Mon. Not. R. astr. Soc.* **68** 173–179

Cowell, P.H., & Crommelin, A.C.D. (1908b) *Mon. Not. R. astr. Soc.* **68** 375–378

Cowell, P.H., & Crommelin, A.C.D. (1908c) *Mon. Not. R. astr. Soc.* **68** 510–514

Cowell, P.H., & Crommelin, A.C.D. (1908d) *Mon. Not. R. astr. Soc.* **68** 665–670

Hajduk, A. (1983) In: *Asteroids, comets, meteors* eds Lagerkvist, C.-I., & Rickman, H., Uppsala, 425–429

Halley, E. (1705) *Phil. Trans. R. Soc. London* **24** 1882–1899

Hasegawa, I. (1980) *Vistas in Astronomy* **24** 59–102

Hind, J.R. (1850) *Mon. Not. R. astr. Soc.* **10** 51–58

Kamieński, M. (1947) *Bull. Acad. Pol. Sci. Ser. A* 69–101

Kamieński, M. (1949) *Bull. Acad. Pol. Sci. Ser. A* 101–140

Kamieński, M. (1951) *Bull. Acad. Pol. Sci. Ser. A* 33–38

Kamieński, M. (1956) *Acta Astr.* **6** 3–23

Kamieński, M. (1957) *Acta Astr.* **7** 111–118

Kamieński, M. (1961) *Acta Astr.* **11** 223–229

Kamieński, M. (1965) *Bull. Soc. Amis Sc. Poznań, Ser. B*, **18** 117–135

Kiang, T. (1972) *Mem. R. astr. Soc.* **76** 27–66

Kiang, T. (1974) In: *Asteroids, comets, meteoric matter*, Bucuresti, 171–173

Kozai, Y. (1979) In: *Dynamics of the solar system* ed. Duncombe, R.L., Dordrecht, 231–237

Landgraf, W. (1984) *ESTEC EP/14.7/6184 Final Report*.

Landgraf, W. (1986) *Astron. Astrophys.* **163** 246–260

Laugier, P.A.E. (1843) *Comp. Rend. Acad. Sci. Paris* **15** 949–951

Laugier, P.A.E. (1846) *Comp. Rend. Acad. Sci. Paris* **23** 183–189

Morley, T.A. (1983) *Giotto Study Note* No. 46

Pingré, A.G. (1783) *Cometographie* **1** Paris.

Sitarski, G. (1979) *Acta Astr.* **29** 401–411

Sitarski, G. (1983) In: *Asteroids, comets, meteors* eds Lagerkvist, C.-I., & Rickman, H., Uppsala, 167–170

Sitarski, G. (1987) *Acta Astr.* **37** 99–113

Sitarski, G., & Ziolkowski, K. (1986) In: *Exploration of Halley's Comet*, ESA SP-250 **III** 299–301

Sitarski, G., & Ziolkowski, K. (1987) *Astron. Astrophys.* **187** 896–898

Stephenson, F.R., Yau, K.K.C., Hunger, H. (1985) *Nature*, **314**, 587–592

Viliev, M. (1917) *Izv. Russ. Obshch. Lub. Miroved.* **6** 215–219

Yeomans, D.K. (1977) *Astron. J.* **82** 435–440

Yeomans, D.K. (1984) In: *Cometary Astrometry* eds Yeomans, D.K., West, R.M., Harrington, R.S., Marsden, B.G., NASA/JPL Publ. 84–82, 167–175

Yeomans, D.K., Kiang, T. (1981) *Mon. Not. R. astr. Soc.* **197** 633–646

20

Physical processing of cometary nuclei since their formation

Paul R. Weissman

1 INTRODUCTION

It has become almost a matter of faith among solar system astronomers that 'comets are the best obtainable source of original solar nebula material.' Thus, there is a temptation to quickly apply many of the new findings from Comet Halley's 1986 apparition to a description of the primordial solar nebula at the time the comets were forming. But Comet Halley, like every other solar system body we have studied with flyby and orbiter spacecraft, is an evolved body. Over its 4.5 Gyr history it has been subjected to a variety of physical processes which have modified it, admittedly much less than the larger planets and satellites, but still in very significant ways.

To fully understand and interpret the Halley results it is necessary to consider the comet's complete physical and dynamical history, and the processes which have probably acted to modify it from its original pristine state. This is necessarily a statistical exercise, considering those 'likely' processes which we expect from our current understanding of the solar system, the Oort cloud, and the galaxy. There is no way that we can foresee unique, low probability events in Comet Halley's history that have modified it in some significant fashion that is different from that generally experienced by the majority of other comets. Nor can we say that Comet Halley is not a unique body, formed (or captured) in some unique way that again does not represent the majority of the cometary

population. However, we will see that much of the evidence points to Halley being a quite typical comet.

A comet's physical processing history can be broken into three distinct periods: (1) formation, presumably coincident with the formation of the Sun and planetary system; (2) storage in the Oort cloud at large solar distances for most of the history of the solar system; and (3) re-entry into the planetary system and the observable region, $r < 5\,\mathrm{AU}$. Each of these will be discussed below. In addition, the uncertainty associated with the various dynamical paths by which comets may evolve from the Oort cloud to short-period orbits, and the possible dynamical history of short-period comets, will be examined.

Exactly what constitutes a 'pristine' cometary nucleus has different meanings to different investigators. Even the interstellar medium prior to the comet's formation might not be considered truly 'pristine' because of the complex chemistry that apparently occurs when volatile ices condense on interstellar grains and are irradiated by UV photons and cosmic rays (Greenberg & D'Hendecourt 1985). Given that the interstellar medium we observe today has already undergone that same processing, we will assume that condition as our starting point. It will be seen that the degree of modification as the comet forms and evolves can usually only be defined in a relative sense. The object of this paper is to survey the possible physical processes and to rank them in relative importance, leaving the detailed quantitative evaluation to future work.

2 COMETARY FORMATION

Hypotheses of cometary origin fall into two major categories: primordial hypotheses in which comets formed at the same time, and as part of the formation of the Sun and planets; and episodic hypotheses in which comets are formed or captured at essentially random times, often as a result of catastrophic processes, and possibly repeatedly over the history of the solar system (Weissman 1985a). In general, the episodic hypotheses have not gained very wide acceptance, and will not be discussed further here.

All primordial hypotheses agree on one basic fact: comets formed far from the Sun in the cooler parts of the solar nebula. They disagree in just how far away the formation zone actually was. The suggestions range from the Uranus–Neptune zone at 20 to 30 AU from the Sun, to neighbouring fragments of the protosolar nebula, at distances of 10^3 AU or more.

Formation of comets among the outer planets was first suggested by Kuiper (1951) who noted that water ice would not condense any closer to the Sun than about Jupiter's orbit. But dynamical studies (Safronov 1969, Fernandez & Ip 1981) have shown that Jupiter and Saturn tend to eject the majority of the icy planetesimals in their zones to interstellar space, whereas Uranus and Neptune with their smaller masses and larger semimajor axes are more likely to place a sizeable fraction of planetesimals into distant bound orbits with dimensions of the order of 10^4 AU or more, the region of the Oort cloud. Temperatures in the Uranus–Neptune zone are expected to have been 100 K or less, allowing volatile ices such as H_2O, CO_2, NH_3, and CH_4 to condense, and others to become trapped as clathrate hydrates in the icy matrix.

According to the accepted scenario (Greenberg *et al.* 1984) ice and silicate grains in the infalling solar nebula would descend to the equatorial plane of the nebula, forming an accretion disc. Agglomeration of grains would lead to the growth of larger particles, both during the fall toward the nebula plane, and while circulating in the accretion disc (Weidenschilling 1980), leading to growth of initial icy-conglomerate objects as large as 10 metres. When the density of

material in the plane reached a sufficient value, gravitational instabilities (Goldreich & Ward 1973) would lead to fragmentation of the disc and collapse into planetesimals several kilometres in dimension. These were presumably the protocomets. The total initial mass of planetesimals in the Uranus–Neptune zone is estimated to be of the order of 10 times the combined present masses of those two planets. The protocomets were ejected to the Oort cloud or to interstellar space by the growing protoplanetary cores.

The degree to which interstellar material is modified as it is brought together by the processes described above is highly uncertain. Infalling nebula material in the Uranus–Neptune zone may be moving inward with a velocity of a few km s^{-1}, and is probably slowed and shocked as it reaches the denser parts of the nebula around the accretion disc. This could raise the temperature of the material significantly, vaporizing icy particles and driving volatiles off of silicate grains. The degree of heating is also affected by the opacity of the accretion disc, how well it traps the heat generated at the shock boundary.

The grains would then rapidly cool as they began to agglomerate and settle toward the nebula plane. The interparticle velocities at this point would be very low, and material would be brought together with relatively little compaction (Goldreich & Ward 1973). This would lead to a highly porous, low density, loosely bound structure composed of a heterogeneous mixture of ice and dust grains covering a wide range of particle sizes (Donn & Rahe 1982, Greenberg 1986).

The breakup of the accretion disc and collapse into planetesimals due to gravitational instabilities would also be a relatively mild process, with interparticle velocities of the order of only a few m s^{-1} at most. The total gravitational potential energy of a 10 km diameter icy conglomerate planetesimal is not enough to raise the average temperature of the material by even one Kelvin. Some local compaction and melting might be expected at interfaces where larger particles come together, probably resulting in a 'welding' of the particles into a loosely bound agglomerate of icy chunks covering a fairly wide size range.

The exact structure of the cometary nucleus created in this fashion is still a matter of considerable debate. Fig. 1 shows four suggested models for cometary nuclei (Whipple 1950, Donn *et al.* 1985, Weissman

a)

b)

c)

d)

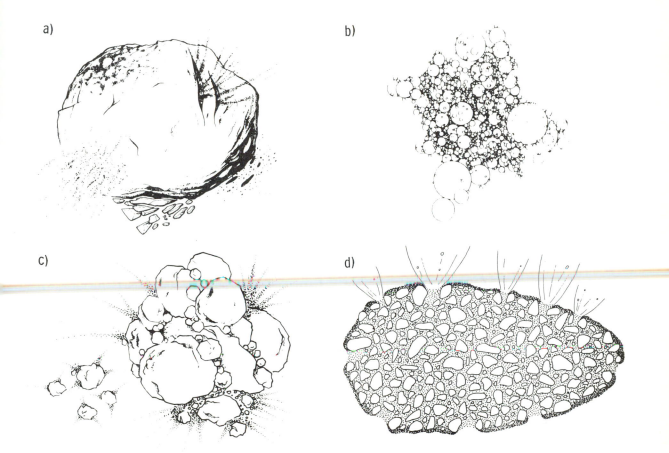

Fig. 1 — Four suggested models for the structure of cometary nuclei: (a) the icy conglomerate model (Whipple 1950); (b) the fractal model (Donn *et al.* 1985), (c) the primordial rubble pile (Weissman 1986a); and (d) the icy-glue model (Gombosi & Houpis 1986). All but (d) were suggested prior to the Halley spacecraft encounters in 1986.

1986a, Gombosi & Houpis 1986). Common features of all the models are the irregular shape of the nucleus, the heterogeneous mixture of icy and non-volatile materials, and the existence of substantial voids within the nucleus. Differences exist over the exact degree and scale of the heterogeneity, and the degree to which the initial agglomeration of particles has been modified into a single well-compacted body. The initial nucleus structure will be important to its subsequent evolution, in particular the way energy is deposited in or liberated from the nucleus interior, and its survivability against collisional destruction.

Mixed in with the icy conglomerate material of the Uranus–Neptune zone will be some material formed closer to the Sun and then dynamically ejected by the growing proto-Jupiter and proto-Saturn. Some contamination is even possible with material from the terrestrial planets zone, but it is likely that Jupiter will be a strong dynamical filter that ejects most of that material hyperbolically, rather than just into the Uranus–Neptune zone (material from the terrestrial planets zone approaches Jupiter with a high relative velocity as compared with the low velocity of material already in the Jupiter–Saturn zone, and thus has a much higher probability of dynamic ejection). Nevertheless, any initial chemical differentiation in the solar nebula due to the radial temperature gradient will be blurred by dynamic exchange between different protoplanetary accretion zones.

Once the nuclei form they will continue to circulate

in their orbits with low relative velocities. Collisions should initially be highly efficient in growing larger bodies. But as the larger accretion cores grow they will also serve to perturb the orbits of the remaining small icy planetesimals, increasing their relative velocities and decreasing the accretion efficiency. Higher velocity collisions will result in more destructive collisions, creating large amounts of debris which will quickly be swept up again by the planetesimals. Collisions which are not totally disruptive will result in crushing and compaction of the initial nucleus structure. The nuclei will thus probably go through a period of competing accretionary and destructive forces.

This will end when the runaway growth of one of the planetary embryos becomes sufficient to start ejecting comets to Oort cloud distances, of the order of 10 000 to 50 000 AU. At those distances stellar perturbations will be sufficient over one orbital period to detach the protocomet's perihelion from Uranus and/or Neptune and to make it a semi-permanent resident of the Oort cloud. In fact, Duncan et al. (1987) have recently shown that the comets can begin to have their perihelia raised out of the planetary region by stellar and galactic perturbations when their aphelia reach approximately 6000 AU. As material was either ejected or incorporated into the protoplanetary cores, the density of planetesimals remaining in the planetary zone dropped dramatically, and collisional evolution rapidly declined.

Alternative hypotheses place the formation zone for the cometary nuclei farther from the Sun. Among the attractive features of such hypotheses are the facts that the nebula material will undergo less heating prior to being incorporated into the icy planetesimals, and that subsequent evolution will also probably be milder with lower collision rates and collision velocities between planetesimals and less acceleration of velocities, since no large protoplanets grew beyond the Uranus–Neptune zone (as far as we know, at the present time). The principal drawback of these hypotheses involves the difficulty of bringing 10 km sized icy planetesimals together at large solar distances.

One suggestion (Cameron 1962, 1978) is that the nebula accretion disc did not stop at the orbit of Neptune, but rather extended out several hundred AU

or more. The lack of planets beyond Neptune (Pluto is probably better thought of as the largest icy planetesimal to grow in its zone, but not a true planet) was not a result of the nebula disc running out of material, but rather of it running out of sufficient time to build a giant planet. Accretionary times are dependent on the orbital period as well as the density of material, both of which decrease as one moves outward from the Sun.

Observational support for this concept has come from the IRAS discovery of dust shells around some young main sequence stars in the solar neighborhood (Aumann et al. 1984). A visual light photograph of one of these discs edge-on (Smith & Terrile 1984) has shown a flat disc extending up to 800 AU from the star, with a thickness of about 50 AU. Estimates of the mass of these discs range from a minimum of about 15 Earth masses, to possibly 200 to 300 Earth masses if an asteroidal size distribution can be assumed.

This, and other evidence (Weissman 1985b), has led to the suggestion that there exists a massive inner Oort cloud, interior to the distant spherical cloud that is the source of the long-period comets, and with perhaps 5 to 10 times as many comets. The inner cloud originated as this extended accretion disc, and the protocomets in it have since been pumped up by stellar perturbations until the inner and outer clouds have merged. In fact, by this point in the solar system's history, the original outer Oort cloud of comets has probably been lost owing to perturbations from stars and giant molecular clouds in the galaxy. The outer cloud has been replenished, however, by comets from the inner Oort cloud disc. The inner Oort cloud may also serve as a source for the short-period comets (see section 5).

An entirely different formation process involves differential radiation pressure on distant nebula fragments, perhaps 5×10^3 AU from the forming proto-Sun (Whipple & Lecar 1976, Hills 1982, Cameron 1985). Because of the opacity of the nebula fragments, they will feel a net radiation pressure from the proto-Sun, forcing material together. At some point the density of the fragment will rise sufficiently for self-gravity to take over and for the fragment to collapse to form the protocomet. In this manner comets could be formed in isolation and would be very cold, with virtually no processing of material from its pristine state.

The suggestion has also been made that distant nebula fragments may form their own, less massive accretion discs, in which comets would form within a relatively benign environment (Cameron 1973). Or perhaps the comets formed in neighbouring nebula fragments around other stars in the same open cluster in which the Sun formed. Initial relative velocities between the stars might be 1 km s^{-1} or less, low enough to allow capture of comets from the other stars (Donn 1976).

There is no good way at present to discriminate between all these various hypotheses. Each has its particular strong points and advocates, and each its weak areas and detractors. Comets may be a direct consequence of planetary formation in the outer solar system, or they may be a largely unrelated by-product of far more distant nebula processes. Or a fraction of

the present day comets may have formed through each of the suggested processes, and what we observe is a truly heterogeneous mixture of material from different formation sites and with different degrees of physical processing.

Some attempts have been made to look for 'cosmic thermometers' in comets that might indicate the heliocentric distance of their formation zones, or isotopic anomalies that could be interpreted as evidence for formation elsewhere in the galaxy. In the case of the former, observations of S_2 in comets has been put forward as evidence for a low temperature formation, $T < 25$ K (A'Hearn & Feldman 1985).

As for the latter, measurements of the $^{12}C/^{13}C$ ratio in comets (including Halley) using groundbased techniques have yielded values around 100, close to the terrestrial value of 89 (Vanysek & Rahe 1978,

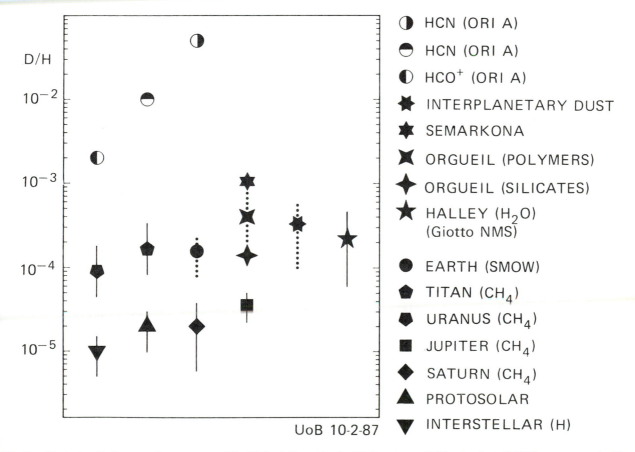

UoB 10-2-87

Legend:
- ◑ HCN (ORI A)
- ◒ HCN (ORI A)
- ◐ HCO$^+$ (ORI A)
- ✷ INTERPLANETARY DUST
- ✶ SEMARKONA
- ✦ ORGUEIL (POLYMERS)
- ✚ ORGUEIL (SILICATES)
- ★ HALLEY (H_2O) (Giotto NMS)
- ● EARTH (SMOW)
- ⬟ TITAN (CH_4)
- ⬟ URANUS (CH_4)
- ■ JUPITER (CH_4)
- ◆ SATURN (CH_4)
- ▲ PROTOSOLAR
- ▼ INTERSTELLAR (H)

Fig. 2 — Deuterium/hydrogen ratios as measured for Halley's Comet by the Giotto spacecraft (Eberhardt *et al.* 1987), as compared with other solar system and galactic reservoirs.

Wehinger *et al.* 1986). The spacecraft encounters with Halley have yielded a measurement of the D/H ratio between 0.6×10^{-4} and 4.8×10^{-4} (Eberhardt *et al.* 1987). This value is shown in Fig. 2 and is compared to D/H values for various other solar system and galactic reservoirs. It is seen that the cometary value is enriched over that for the interstellar medium and the giant planets, but matches the range for terrestrial seawater and the atmospheres of Titan and Uranus, each of which has been suggested to have come from cometary bombardment. In addition, Eberhardt *et al.* found that the $^{18}O/^{16}O$ ratio in Halley was 0.0023 ± 0.0006, in good agreement with the measured solar system value of 0.00205. All this suggests that the comets (or at the very least, Comet Halley) formed from a cosmo-chemical reservoir that was very similar, if not identical, to the rest of the solar system.

Following their formation, however it might occur, the cometary nuclei will undergo warming due to long and short-lived radionuclides mixed in with the other non-volatiles. A study of the internal temperatures of icy satellites and planetesimals in the outer solar system, assuming a chondritic composition diluted 3:1 by ices and with K, U, and Th as radioactive heat sources (and no solar heating), found that the internal temperatures of objects less than 300 km in radius would be less than 25 K (Lewis 1971). That study assumed a thermal conductivity typical of solid water ice, considerably higher than the conductivities suggested for cometary materials. However, assuming more reasonable conductivities would still result in very little internal heating for bodies less than 30 km in radius, about four times the size of the nucleus of Comet Halley.

However, if short-lived radionuclides are present, then the results can be very different. ^{26}Al with a half-life of 7.4×10^5 yr has been suggested as an early solar system heat source, and evidence for it has been found in inclusions in carbonaceous chondrites. If similar concentrations were incorporated into the comets, then, assuming a low thermal conductivity, it would be possible to melt the interiors of nuclei greater than 3 to 6 km in radius (Wallis 1980). This could lead to an interesting nucleus structure, melted and refrozen at the centre, and least modified in the near surface layers. Such a comet might go through interesting changes in behaviour as it aged and the outer layers

sublimed away.

On the other hand, it is entirely possible that the cometary formation process took long enough that short-lived radionuclides were exhausted before they could be buried in large, kilometre sized bodies. Planetesimal formation times, particularly in the outer solar system, are not very well defined, and might well have exceeded the lifetime of heat sources such as ^{26}Al. Support for this latter possibility comes from Jessberger & Kissel (1987) who found no evidence of anomalous ^{26}Mg (the ^{26}Al decay product) in Comet Halley dust grains. Thus, the initial Halley results suggest that comets were not melted by short-lived radionuclides.

One stage in the cometary formation process that has received very little attention to date is the effect of the Sun's T Tauri phase on the comets. This period is characterized by the development of a very strong stellar wind and substantial mass outflow, presumably clearing away the remnants of the solar nebula and ending the planetary formation process. The T Tauri stage is expected to begin within about 10^6 years of the initial nebula collapse, and thus favours hypotheses which lead to rapid giant planet formation. However, that time scale is not consistent with the very long accretion and clearing times for planetesimals in the Uranus–Neptune zone. Thus, the comets may still be relatively close to the Sun and will suffer considerable near surface modification as a result of the intense solar wind. Even if comets formed farther from the Sun, the T Tauri stage may still result in significant physical modification. As noted above, this is an area that certainly requires further study.

3 STORAGE IN THE OORT CLOUD

At first glance, the Oort cloud would appear to present a relatively benign environment. The average spacing between comets is of the order of 15 AU or more in the outer, dynamically active cloud, and about 1 AU in the inner Oort cloud. Expected comet–comet collision rates are virtually zero in the outer cloud and less than 0.1% over the history of the solar system in the inner cloud. The maximum surface temperature on cometary nuclei 100 AU from the Sun is only 40 K; comets at greater distances would be even colder.

Three physical processes which act to modify the

outer layers of cometary nuclei in the Oort cloud have been identified: irradiation by galactic cosmic rays and solar protons, heating by random passing stars, and interaction with the interstellar material through which the solar system is passing. The last process is a dual one, consisting of accretion of interstellar atoms and molecules, and erosion by hypervelocity impacts of interstellar grains.

Irradiation of volatile ices and ice–dust mixtures by high-energy particles has been studied in the laboratory (Shulman 1972, Lanzerotti *et al.* 1978, Moore *et al.* 1983) and has yielded a number of interesting results. The interaction breaks chemical bonds, producing volatile free radicals, some of which recombine to form a dense, dark polymer which is far less volatile than the original material. Polymerization occurs particularly if hydrocarbons are present, something that was clearly shown to be true for Halley. At the nucleus surface the high energy particle bombardment results in a net erosion by sputtering and probable escape of the more volatile species, leaving behind a low-volatility residue. At depth the volatiles are retained.

The depth to which the particles penetrate the nucleus surface depends strongly on the energy of the particles and the density of the nucleus surface layers. Low-energy solar protons of a few to a few hundred keV penetrate only the first millimetre of surface or less; cosmic ray protons with typical energies of 2 MeV penetrate about one metre (both depths assume a cometary density of about 1 g cm^{-3}).

This raises the interesting possibility that comets might already begin the process of developing non-volatile crusts while they are still in the Oort cloud and before they ever enter the planetary region. The polymerized hydrocarbons in the near surface layers would act as a 'cometary glue' to help begin the process of sealing off the nucleus surface against sublimation.

Random passing stars will occasionally penetrate the Oort cloud, resulting in heating of comets close to their paths. For the majority of stellar encounters, the comets within about 500 AU of the passing stars will be dynamically ejected from the Oort cloud (Weissman 1980), and thus it is very unlikely that the heated comets will ever be seen entering the planetary region. However, encounters with very massive and luminous

OB giants will heat comets out to a radius well beyond the ejection radius. Stern (1987) estimates that all comets in the Oort cloud have been heated to at least 27 K, and that 20 to 40% have experienced at least one episode of heating to 50 K. This warming will be over periods of several thousand years or more and thus should be sufficient to warm the entire nucleus, not just the immediate surface layers. As a result, the more volatile ices will have been mobilized within the nucleus, and some fraction can be expected to be depleted from near surface layers.

The cometary interaction with the interstellar medium is a duel between two competing processes: accretion of interstellar atoms and molecules, and erosion of the nucleus surface by hypervelocity impacts of interstellar dust grains. The former is estimated to add a layer of 10–100 micrometres thickness to the nucleus surface (O'Dell 1971) over a period of 4.5×10^9 years. It had been speculated (Whipple 1978) that this layer of additional interstellar volatiles accounted for the anomalous brightness of dynamically new long-period comets on their first pass through the planetary region. But the erosion process resulting from micro-cratering by the dust grains is estimated to remove a layer 0.5–8 cm in thickness over the same period (Stern 1986). Thus, the net change appears to be negative, comets losing mass to the interstellar medium, not gaining mass from it.

On the other hand, recent laboratory experiments simulating capture of cometary dust grains during high-speed spacecraft flybys have suggested that impacts into a very low density, 'fairy castle' surface structure may lead to intact capture of the particles, with little net mass loss (Tsou *et al.* 1984). The experiments have generally been done for relatively modest velocities, of the order of a few km s^{-1}. Whether or not the same effect would be true for the 20 km s^{-1} or more velocity expected for interstellar grains is open to some doubt. But it is possible that continued experiments might show that the erosion is not as significant as described above, and that in fact there is a net mass gain in the surface layers of cometary nuclei (Ostro *et al.* 1986).

The depth of erosion (whether positive or negative) is still considerably less than the depth of modification by galactic cosmic rays, so the polymerization process dominates while the comets are in the Oort cloud.

Free radicals produced by the cosmic ray bombardment and trapped in the near surface layers may contribute to the anomalous brightness of new comets at large heliocentric distances on their first pass through the planetary system.

It is likely that there are other, very slow-acting processes which might serve to modify the comets during their very long residence time in the Oort cloud. For example, creep of icy materials in the weak cometary gravitational field may lead to a compacting and densification of the nucleus. Thus, although the comet may be injected into the Oort cloud as a low-density, fractal structured object, it may return after 4.5 Gyr as a single, well-compacted nucleus with a density more typical of ordinary water ice. Conceivably, other unsuspected slow-acting processes could also serve to modify the nucleus in yet unknown ways over the same period.

4 RETURN OF COMETS TO THE PLANETARY REGION

When cometary nuclei return to the planetary region as short-period comets, their physical evolution is dominated by the heating they receive from direct solar radiation. Other processes such as irradiation by solar wind protons and impacts by interplanetary dust particles will also intensify with decreasing solar distance, but they will not compare in either mass removal rate or depth of penetration with the changes brought about by solar heating. The effect of the heating will manifest itself in a number of ways.

First will be the sublimation of volatile ices at the nucleus surface, resulting in development of an extended cometary atmosphere, the coma. The evolving gases will carry with them solid grains of fine dust (and perhaps ice), creating the dust coma around the nucleus. Larger grains of non-volatiles that are not carried off will begin to form a lag deposit on the nucleus surface, accumulating to form a crust that will either seal the nucleus off against further mass loss, or thermally insulate the nucleus ices interior to it, or both. Violent rupture of the crust due to the pressure of evolving gases below it may result in visible outbursts or even disruption of the nucleus.

Heating of amorphous water ice within the nucleus above about 140 K will result in an exothermic transition to crystalline ice, also possibly resulting in

visible outbursts. Thermal stresses on the nucleus caused by substantial temperature gradients within the ice may also result in cracking and exposure of 'fresh' ices, and possibly outburst or disruption phenomena.

More subtle effects will include the migration of more volatile molecules, both outward through the still frozen water ice matrix, and inward towards cooler regions of the nucleus where they will recondense. Also, as mass is lost from the rotating nucleus its moments of inertia will change and the nucleus will precess, changing the orientation of the rotation pole and the balance of solar insolation across the nucleus surface, with perhaps additional interesting implications.

For the purpose of analyzing the thermal history of cometary nuclei, the important factor will be the heating of the cometary nucleus at depth. The 'thermal skin depth' at the nucleus surface is defined as

$$d = (KP/\pi\rho C)^{1/2} \qquad (1)$$

where K is the conductivity, P is the period (length of day for diurnal skin depth, orbital period for orbital skin depth), ρ is the density, and C is the specific heat. The thermal skin depth is the depth over which a temperature perturbation at the surface will decrease by a factor of $1/e$. For conductivities typical of solid water ice, d is 20 cm for a rotation period of 24 hours, or 9.2 metres for an orbital period of 6 years. However, conductivity decreases sharply for porous, low-density structures as are suspected for cometary nuclei, and actual values of d may be considerably smaller. Most measured values of surface conductivity in the solar system, including the icy Galilean satellites, are extremely low. If comets are similar, then a more realistic estimate of the orbital thermal skin depth may be about 1 metre.

Internal temperature profiles for cometary nuclei are a complex function of the comets' orbital parameters, and the thermal properties of the comet surface materials. For example, the variation of temperature with depth for a typical 1 km short-period comet nucleus, ignoring the diurnal temperature cycle and sublimation, and assuming solid crystalline ice, is shown in Fig. 3 for four points

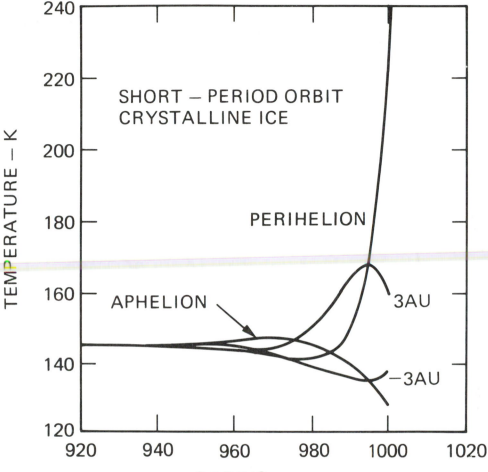

Fig. 3 — Temperature profiles in the near surface layers of a hypothetical 1 km radius crystalline ice comet in a short-period orbit, for four points around the orbit.

around the comet's orbit (Herman & Weissman 1987). Although the surface layer undergoes extreme temperature variations, the temperature several orbital thermal skin depths below the surface is virtually constant. Below this depth the nucleus has been heated to its average orbital temperature, but not appreciably higher.

Temperature profiles in a real case will not be as neat as shown in Fig. 4. Because of the typically low values of conductivity, it requires many returns for a comet to be heated to its equilibrium internal temperature. But the comet's orbit is likely to change significantly over that same period. Since the changes are essentially random, some will result in additional internal heating, while others will result in a net cooling for the cometary nucleus. The final result might be a fairly complex interweaving of warm and cooler layers.

Nevertheless, the equilibrium internal temperature of cometary nuclei can be approximated to within $\pm 10\%$ by the mean temperature for a nucleus in a circular orbit with the same semimajor axis

$$T_m = 280 \ (1 - A)^{1/4}/a^{1/2} \ \text{K} \qquad (2)$$

where A is the Bond albedo of the nucleus surface and a is the semimajor axis in AU. For comets in non-circular orbits the problem cannot be solved analyti-

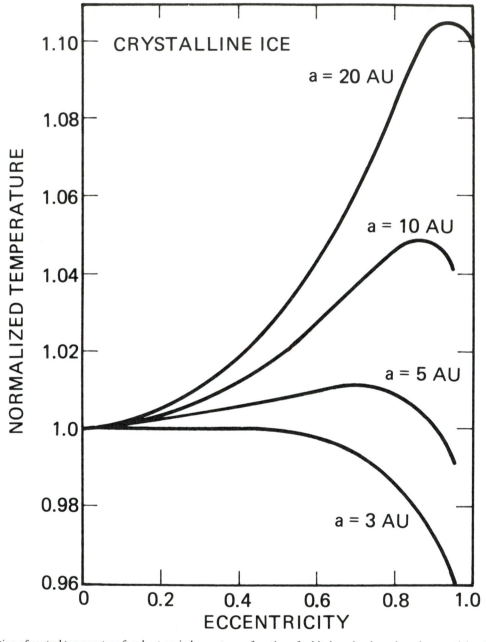

Fig. 4 — Variation of central temperature for short-period comets as a function of orbital semimajor axis and eccentricity. The normalized temperature is that for a comet in a circular orbit: see text.

cally. The increased solar heating near perihelion and increased time for cooling near aphelion, as well as the temperature dependence of the thermal conductivity, lead to a complex behaviour of the central temperature as a function of semimajor axis and eccentricity. That behaviour is illustrated in Fig. 4 for the case of a 1 km crystalline ice nucleus (Herman & Weissman 1987) where the 'normalized temperature' is the calculated central temperature divided by the mean temperature from equation (2).

The final central temperature which the nucleus reaches is also a function of the magnitude of the thermal conductivity, but is not a function of the nucleus radius. However, larger nuclei could take hundreds or even thousands of orbits to reach equilibrium, longer than the time constant for appreciable changes in the comet's orbit due to Jupiter perturbations.

As a result of this internal warming, volatile ices with sublimation points below the central temperature will diffuse through the icy matrix and possibly escape the nucleus, unless they are trapped as clathrate hydrates. The near surface layers will certainly be depleted in many of the more volatile ices. As a result, gas production rates for some volatile species may show a temporal dependence: as the comet approaches perihelion, the more volatile ices may not appear until the thermal wave has penetrated deep enough into the nucleus to reach undepleted layers. In addition, some deeper layers in the nucleus may be enriched in certain volatiles which were sublimated from the warm near surface layers, and were able to diffuse to, and recondense in, the cooler interior.

Calculated central temperatures for eight well known short-period comets, most of which are under consideration as targets for comet rendezvous and/or sample return missions, are given in Table 1. The quantity T_c is the equilibrium central temperature for the comet's present orbit, assuming normal water ice sublimation over the entire nucleus surface. T_c' is the same but assuming no sublimation, a condition the comet would evolve to as it built up a non-volatile crust covering most of the nucleus surface.

Note that with the exception of P/Halley, all the short-period comets in Table 1 have a central temperature above 140 K, the transition temperature for amorphous to crystalline water ice. For a low albedo nucleus, equation (2) gives a central temperature of 140 K at a semimajor axis of 4.0 AU, or an orbital period of 8.0 years. Of 135 known periodic comets (Marsden 1986), 82 have periods less than that value. Thus, those comets can be expected to have completely converted to crystalline ice in their interiors.

For comets with longer orbital periods, a sizeable fraction of the nucleus interior might still be expected to be amorphous ice (provided that the comet has not been significantly closer to the Sun in the past). In fact, the amorphous ice may be located only several orbital thermal skin depths beneath the surface (Herman & Podolak 1985). Recent work by Prialnik & Bar-Nun (1987) has suggested the following scenario for long-period comets. New comets approaching the planetary region for the first time will be entirely amorphous ice, and will undergo an amorphous–crystalline phase transition at 5 AU inbound as their surface temperatures first rise above 140 K. This transition may account for the anomalous brightness of dynamically new comets at large heliocentric distance inbound. In addition, the exothermic reaction may serve to blow off any cosmic-ray produced crust on the nucleus surface, exposing the fresh ice underneath while producing the large, slow-moving dust comae often observed for new, large perihelia comets.

On subsequent returns, the amorphous–crystalline phase transition point beneath the surface does not change until the surface layer of crystalline ice drops below about 15 metres in thickness, allowing the perihelion thermal wave to penetrate to that interface. Because of the thermal lag time, that transition will

Table 1 — Calculated central temperatures for eight short-period comets.

Comet	q (AU)	P (years)	T_c (K)	T_c' (K)
Encke	0341	3.30	171.0	183.4
Honda–Mrkos–Pajdusakova	0.581	5.28	149.8	154.9
Tempel 2	1.381	5.29	150.8	156.9
Wild 2	1.491	6.17	144.7	148.9
D'Arrest	1.291	6.38	141.9	146.1
Kopff	1.572	6.43	143.4	147.0
Churyumov–Gerasimenko	1.298	6.59	140.8	145.2
Halley	0.587	76.10	67.8	68.7

not take place until the comet is at about 5 to 8 AU outbound. Succeeding episodes of amorphous ice transition will not occur again for many orbits until the surface crystalline ice layer once again sublimates down to its minimum, 15 metre thickness. Successive transitions reach deeper into the nucleus, and eventually the entire cometary interior will be transformed to crystalline ice.

The sublimation of ices at the nucleus surface will lead to the growth of a non-volatile lag deposit of larger particles, too heavy to be carried off the nucleus by the evolving gas pressure. This deposit will combine with any polymerized hydrocarbon residue from cosmic ray and solar proton bombardment to form an insulating crust on the nucleus surface. Calculations have shown that a layer only one or two centimetres thick would probably be sufficient to insulate the ices below and greatly reduce the sublimation rate (Brin & Mendis 1979, Fanale & Salvail 1984). Estimates of the number of returns required to form the crust range from about one to 20. Various theories have proposed that the crust is either porous, allowing continued gas diffusion through it to the surface, or sealed, resulting in a buildup of pressure beneath it that might cause violent rupture events.

The results of the spacecraft flybys of Comet Halley have produced strong evidence for the existence of an insulating crust on the nucleus surface (Sagdeev et al. 1986, Reitsema 1986), but at the same time have presented us with several apparent paradoxes. The fraction of active area on the nucleus of the comet has been estimated at 10–30%, based on spacecraft imaging. However, it is not clear why such a small fraction of the nucleus is active. Virtually all predictions for Halley were that the nucleus surface would be crust free, the crust having been blown away by the high gas production.

Given that 70–90% of the nucleus surface was crusted over, why was it not 100%? Did the few active areas serve as pressure release points for the entire nucleus, implying a highly porous nucleus structure? Did the active areas change as the comet moved along its orbit, or from one perihelion passage to another? Earth-based observations of the comet suggested that the active areas did turn on and off irregularly, and Giotto imaging suggested at least one surface struc-

ture that may be a crusted-over, former active area. On the other hand, some observers alleged a link between active areas observed in 1985–86, and locations of active areas derived from 1910 observations (Sekanina & Larson 1986), a very tenuous possibility which is exceedingly difficult to prove.

A further question is whether or not the allegedly inactive areas on the Halley nucleus are indeed inactive? Weissman (1986b) found that the observed gas production rates for Halley implied that 30% of the sunlit nucleus surface area was active on the average, more than that estimated from Giotto imaging. Clearly, the sources of the dust jets appear to be confined to relatively small areas. But is it possible that gas from sublimating ices is still diffusing through the dark porous crust in other areas, while being unable to carry entrained dust with it? The low resolution of the Halley images does not permit an answer to this question.

Another piece of this very complex puzzle comes from the estimates of nongravitational forces in the motion of Comet Halley. The highly elongated nucleus seen in the spacecraft imaging, with its irregular and asymmetrically distributed active areas, would be expected to precess rapidly. But the fitted nongravitational forces on the comet have been virtually constant over the past 2100 years (Yeomans & Kiang 1981). The transverse component of the nongravitational force is sensitive to the orientation of the comet's rotation pole, the rotation period, and the distribution and number of active areas on the nucleus. How then can it be so constant when the nucleus appears to be so dynamic and ever-changing?

What this all must lead to is a complete rethinking of the physics and chemistry of the crust formation and removal process. Present models can simply not explain the observations of the Halley nucleus. Further analyses of the Halley spacecraft data will yield a refined definition of the problem. But extensive laboratory work and theoretical modelling, as well as in situ observations of an evolving cometary nucleus over a sizeable fraction of its orbit, are needed to completely understand crust formation.

An area that has not yet been extensively studied is densification. Smoluchowski et al. (1984) showed that pores in ices will migrate under the influence of even a very weak thermal gradient, resulting in possible densification of materials. Although the migration rate may be very slow, the long residence times of comets in the Oort cloud, and even their lifetimes in the planetary region, may be sufficient for the comet to evolve substantially from its original state. However, Smoluchowski *et al.* showed that migration rates in water ice were insignificant below about 125 K, and thus only short-period comets which have been resident in the planetary region for quite some time might be expected to show the effects of densification.

Another area of considerable interest is that of thermo-mechanical stresses on cometary nuclei (Kührt 1984, Green 1986). The dust jets in the Halley spacecraft images appear to be highly collimated, and appear to occur along linear features (Smith *et al.* 1986). This suggests crevices or faults opened up by stresses on the nuclei, or perhaps freshly exposed voids in a fractal nucleus structure. These crevices may penetrate sufficiently deep in the nucleus to expose relatively unmodified volatile rich layers, several thermal skin depths below the crusted surface. This, in turn, may lead to cometary outbursts and possibly even nucleus disruption events.

5 DYNAMICAL EVOLUTION OF LONG- AND SHORT-PERIOD COMETS

As noted above, a comet's thermal evolution once it returns to the planetary region is very much a function of its orbital history. There are a variety of dynamical paths that a long-period (LP) comet might take from the inner or outer Oort clouds to become a visible short-period (SP) comet. The relative efficiencies of the different paths vary by orders of magnitude, but because of the very large number of comets in the Oort cloud reservoir, any of the major paths (discussed below) have a finite probability of producing a particular SP comet.

Put another way, it is not possible to integrate the orbit of a known SP comet backwards in time to learn how it evolved into the planetary system. Small errors in the initial orbit determination or in the masses of the perturbing planets, unpredictable nongravitational forces from jetting of volatiles on the nucleus surface, and other unmodelled forces lead to a monotonic growth of uncertainty as the orbit is integrated backward. Close encounters with Jupiter,

the major force in evolving cometary orbits in the planetary region, also lead to major increases in this uncertainty.

Thus, the problem is limited to probabilistic arguments based on the relative efficiency of the various dynamical paths. Three have been identified: (1) repeated perturbation of dynamically new LP comets from the Oort cloud passing through the observable region; (2) repeated perturbation of dynamically new LP comets with perihelia in the Jupiter–Saturn zone, with a final dumping of comets into the terrestrial planets zone; and (3) perturbation of inner Oort cloud comets into Uranus–Neptune crossing orbits, with subsequent dumping into the Jupiter–Saturn zone, and finally the inner planets zone.

Early studies (Newton 1893) showed that single perturbations due to a close Jupiter encounter by a LP comet gave a probability of capture to a SP orbit of 2.7×10^{-6}, too small to account for the observed number of SP comets in the planetary region. Repeated passages through the planetary region were required, with the comets performing a random walk in orbital energy. This process requires some 10^2 to 10^3 passages through the planetary region, and results in a capture efficiency of the order of 10^{-4} to 10^{-3} (Weissman 1979, Fernandez & Ip 1983). It also leads to a preferred capture of comets in direct, low inclination orbits, since those comets experience the largest perturbation due to close Jupiter encounters.

A problem with this scenario is the large number of returns required for LP comets to evolve to SP orbits. Suggested lifetimes for SP comets (Kresák 1981) are of the order of 400 to 1000 returns, with comets either disintegrating or evolving to inactive asteroidal objects after that time. However, an analysis of Comet Halley's likely dynamical history suggests a much longer lifetime, at least for that one comet. Weissman (1987) estimated a lifetime of between 1800 and 2500 returns for Halley. Nevertheless, there is a feeling among cometary scientists that a means of evolving comets to SP orbits while preserving them physically is required.

The suggested solution is to consider new LP comets from the Oort cloud with perihelia near Jupiter, 4 to 6 AU (Everhart 1972). Such comets, particularly those in direct orbits, would evolve to SP orbits much faster than comets with perihelia in the terrestrial planets zone, owing to the larger Jupiter perturbations. At the same time, the comets would not be heated sufficiently for any substantial amount of water ice to sublimate, and for any significant growth of non-volatile crusts. After the semimajor axes of the orbits had been reduced to some moderate value, close Jupiter encounters would reduce the comets' perihelia to values in the terrestrial planets zone where they could be observed as SP comets. It is estimated that LP comets will evolve to observed SP orbits with periods less than 13 years in about 30 returns. This dynamical path would also yield a greater fraction of SP comets in direct orbits, in better agreement with the observed ratio.

Variations on this hypothesis include multiple perturbations by all the Jovian planets (Lowery 1973, Everhart 1978), with comets successively being passed from Neptune to Uranus to Saturn and finally to Jupiter. Because it is much more massive than the other outer planets, and because it is closer to the Sun, Jupiter's perturbations dominate a comet's motion once the comet comes under that planet's influence.

The most recent suggestion for the source of the SP comets has been the inner Oort cloud just beyond the orbit of Neptune (Fernandez 1980). A variety of evidence has suggested the existence of an extended ring or belt of comets beyond the planetary region and extending out as far as, and merging with, the distant spherical Oort cloud surrounding the solar system (Weissman 1985b). Stellar perturbations, or perturbations by large planetesimals circulating in the comet ring, would occasionally reduce the perihelia of comets to Neptune crossing orbits. The comets would then be passed inwards until Jupiter perturbations reduced their perihelia to values in the terrestrial planets zone. The advantage of this scenario is that the comets begin their evolution in considerably smaller orbits than LP comets from the outer Oort cloud. As a result, this dynamical path is estimated to be approximately two orders of magnitude more efficient than evolution of LP comets in the outer planets zone. However, such estimates are highly uncertain because of unknown parameters such as the number and orbit distributions of comets in the inner Oort cloud.

Subsequent to its being deposited in a SP orbit, a comet may continue to random walk in both perihelion distance and semimajor axis as a result of planetary perturbations. For example, Monte Carlo studies of the evolution of meteoroids in the planetary region (Arnold 1965) showed that 82% of all meteors which struck the Earth had at one time or another been within Mercury's orbit. To first order, those studies are equally applicable to the random walk of SP comet orbits.

Similar evidence can be found among the observed SP comets. Since its discovery in 1902, periodic comet Grigg–Skjellerup has evolved from an initial perihelion distance of 0.753 AU, to 1.003 AU in 1967, and 0.989 AU currently (Marsden 1986). Even more dramatic is the case of comet P/Lexell. It was discovered in 1770 when it approached to within 0.015 AU of the Earth, with an orbit perihelion of 0.67 AU. Integration of the orbit backwards in time (Kazimirchak–Polonskaya 1972) showed that it had passed within 0.02 AU of Jupiter in 1767, and had previously had a perihelion of about 3 AU. After two passages around the Sun in its new orbit, the comet re-encountered Jupiter in 1779 at a distance of 0.0015 AU (half the radius of Io's orbit) and was perturbed to a new orbit with a perihelion beyond Jupiter's orbit.

Thus, for any observed SP comet, it is impossible to say where in the solar nebula it formed, and where it has been since the time of formation. A comet could have spent a part of its past in the inner solar system, then have been returned to the Oort cloud, and subsequently evolved back into the planetary region a second time. The fact that a comet's current perihelion is relatively large is no proof that the comet has not been much closer to the Sun in the not too distant past. Similarly, comets which presently have small perihelia may be recently arrived at those close solar distances, and may not yet be heated very deep within the nucleus.

A possible example of the latter is P/Wild 2. The comet was deposited in its present orbit with a perihelion of 1.49 AU by a 0.0061 AU encounter with Jupiter in 1974 (Carusi et al. 1985). Prior to that time the comet had a perihelion near Jupiter's orbit. Integration of the orbit further backwards shows it being captured from a near-parabolic orbit about 400 years ago. However, as described above, such integrations are not reliable. They provide a representative example of possible cometary motion, rather than a detailed reconstruction of it.

6 DISCUSSION

The sections above have described a number of physical processes which have, or may have, modified cometary nuclei since their formation in the primordial solar nebula, 4.5 Gyr ago. Many of these processes are restricted to the nucleus surface or to near surface layers. But others affect the entire nucleus, and in at least one case (short-lived radionuclides) affects the centre of the nucleus the most.

The evidence appears to be that comets have received only modest heating over their histories, so as to drive off a fraction of the more volatile ices, but so as to leave the non-volatile constituents relatively unmodified. The two qualifications on that statement are the possible early melting of cometary interiors by short-lived radionuclides, though the effects of this are hidden deep below the nucleus surface, and the polymerization of hydrocarbons and other materials in the upper few metres of the nucleus surface, though the polymerized layer may have since sublimated away.

The greatest degree of processing has apparently occurred at and near the nucleus surface, as a result primarily of solar heating, and secondarily, cosmic ray bombardment. To find relatively unprocessed materials requires penetrating beneath those processed layers. In the case of solar heating that would probably require a depth of several metres. For cosmic ray bombardment the required depth would be comparable, but it is likely that the modified layer may have already been completely lost owing to sublimation and crust blow-off. On the other hand it may have provided the basis for later crust growth and thus could still be there.

Observational evidence provides considerable support for the belief that comets still retain a very large fraction of their original volatile and non-volatile constituents. The most abundant molecules detected in interstellar clouds have also been detected in cometary spectra (Irvine *et al.* 1985). Compositional measurements of solid grains during the Halley flybys show a mix of high temperature refractory grains and pure hydrocarbon grains, as well as more complex grains with a heterogeneous composition (Kissel *et al.* 1986a, 1986b). Recovered interplanetary dust particles, which are believed to come from comets, have a botryoidal or fractal structure of sub-micrometre grains, much like Fig. 1b, with a composition similar to the most primitive, undifferentiated meteorites (Fraundorf *et al.* 1982). In some cases the observations cited here are for 'fresh' long-period comets from the Oort cloud, but in other cases the evidence has also been found in short-period comets, particularly when observers have looked hard enough.

The 1985–86 apparition of Comet Halley has provided a wealth of new information on comets, that is continuing to yield new insights on their nature and origin. But it has also produced many new questions and new problems which require further study. It is clear that to answer these questions will require the next logical step: a comet rendezvous mission that will explore the cometary nucleus in detail and over an extended period. No other spacecraft mission holds such great promise of telling us about the origin of these unique bodies, and about the origin of the solar system itself.

ACKNOWLEDGEMENT

This work was supported by the NASA Planetary Geology and Geophysics Program, and was performed at the Jet Propulsion Laboratory under contract with the National Aeronautics and Space Administration.

REFERENCES

A'Hearn, M.F., & Feldman, P.D. (1985) S_2: a clue to the origin of cometary ice. In: *Ices in the solar system*, eds Klinger, J. *et al.* D. Reidel, Dordrecht, 463–472

Arnold, J.R. (1965) The origin of meteorites as small bodies. *Astrophys. J.* **141** 1536–1556

Aumann, H.H., *et al.* (1984) Discovery of a shell around alpha Lyrae. *Astrophys. J.* **278** L23–L27

Brin, G.D., & Mendis, D.A. (1979) Dust release and mantle development in comets. *Astrophys. J.* **229** 402–408

Cameron, A.G.W. (1962) The formation of the Sun and planets. *Icarus* **1** 13–69

Cameron, A.G.W. (1973) Accumulation processes in the primitive solar nebula. *Icarus* **18** 407–450

Cameron, A.G.W. (1978) The primitive solar accretion disc and the formation of the planets. In: *The Origin of the Solar System*, ed. Dermott, S.F., John Wiley & Sons, New York, 49–75

Cameron, A.G.W. (1985) Formation and evolution of the primitive solar nebula. In: *Protostars and planets II*, eds Black, D.C. and Matthews, M.S. Univ. Arizona Press, 1073–1099

Carusi, A. *et al.* (1985) First results of the integration of motion of short-period comets over 800 years. In: *Dynamics of comets: their origin and evolution*, eds. Carusi, A., & Valsecchi, G.B., D. Reidel, Dordrecht, 319–340

Donn, B. (1976) Comets, interstellar clouds, and star clusters. In: *The study of comets*, NASA SP-393, 663–672

Donn, B., & Rahe, J. (1982) Structure and origin of cometary nuclei. In: *Comets*, ed. Wilkening, L.L. Univ. Arizona Press, Tucson, 203–226

Donn, B., *et al.* (1985) On the structure of the cometary nucleus. *Bull. Amer. Astron. Soc.* **17** 520

Duncan, M., *et al.* (1987). The formation and extent of the solar system comet cloud. *Astron. J.* **94** 1330–1338

Eberhardt, P., *et al.* (1987) The D/H and $^{18}O/^{16}O$ isotopic ratios in Comet Halley. *Lunar & Planet. Sci. Conf.* **18** 252–253

Everhart, E. (1972) The origin of short-period comets. *Astrophys. Let.* **10** 131–135

Everhart, E. (1978) The evolution of comet orbits as perturbed by Uranus and Neptune. In: *Asteroids, comets, meteors: interrelations, evolution and origins*, ed. Delsemme, A.H., Univ. Toledo Press, 99–104

Fanale, F.P., & Salvail, J.R. (1984) An idealized short-period comet model: surface insolation, H_2O flux, dust flux, and mantle evolution. *Icarus* **60** 476–511

Fernandez, J.A. (1980) On the existence of a comet belt beyond Neptune. *Mon. Not. Roy. Astron. Soc.* **192** 481–491

Fernandez, J.A., & Ip, W.-H. (1981) Dynamical evolution of a cometary swarm in the outer planetary region. *Icarus* **47** 470–479

Fernandez, J.A., & Ip, W.-H. (1983) Dynamical origin of the short-period comets. In: *Asteroids, comets, meteors*, eds. Lagerkvist, C.-I., & Rickman, H., Uppsala Univ., 387–390

Fraundorf, P., *et al.* (1982) Laboratory studies of interplanetary dust. In: *Comets*, ed. Wilkening, L.L. Univ. Arizona Press, Tucson, 383–409

Goldreich, P., & Ward, W.R. (1973) The formation of planetesimals. *Astrophys. J.* **183** 1051–1061

Gombosi, T.I., & Houpis, H.L. (1986) The icy-glue model of the cometary nucleus. *Nature* **324** 43–44

Green, J.R. (1986) Stress, fracture, and outburst in cometary nuclei. *Bull. Amer. Astron. Soc.* **18** 800

Greenberg, R. *et al.* (1984) From icy planetesimals to outer planets and comets. *Icarus* **59** 87–113

Greenberg, J.M. (1986) Fluffy comets. In: *Asteroids, comets, meteors II*, eds Lagerkvist, C.-I., *et al.* Uppsala Univ. 221–223

Greenberg, J.M., & D'Hendecourt, L.B. (1985) Evolution of ices from interstellar space to the solar system. In: *Ices in the solar system*, eds Klinger, J., *et al.* D. Reidel, Dordrecht, 185–204

Herman, G., & Podolak, M. (1985) Numerical simulation of comet nuclei I. Water ice comets. *Icarus* **61** 252–266

Herman, G., & Weissman, P.R. (1987) Internal temperatures of cometary nuclei. *Icarus* **69** 314–328

Hills, J.G. (1982) The formation of comets by radiation pressure in the outer protosun. *Astron. J.* **87** 906–910

Irvine, W.M., *et al.* (1985) The chemical state of dense interstellar clouds: an overview. In: *Protostars and planets II*, eds Black, D.C., & Matthews, M.S., Univ. Arizona Press, Tucson, 579–620

Jessberger, E.K., & Kissel, J. (1987) Bits and pieces from Halley's Comet. *Lunar & Planet. Sci. Conf.* **18** 466–467

Kazimirchak-Polonskaya, E.I. (1972) The major planets as powerful transformers of cometary orbits. In: *The motion, evolution of orbits, and origin of comets*, eds Chebotarev, G.A., *et al.*, D. Reidel, Dordrecht, 373–397

Kissel, J., Sagdeev, R.Z., Bertaux, J.L., Angarov, V.N., Audouze, J., Blamont, J.E., Büchler, K., Evlanov, E.N., Fechtig, H., Fomenkova, M.N., von Hoerner, H., Inogamov, N.A., Khromov, V.N., Knabe, W., Krueger, F.R., Langevin, Y., Leonas, V.B., Levanseur-Regourd, A.C., Managadze, G.G., Podkolzin, S.N., Shapiro, V.D., Tabaldyev, S.R., & Zubkov, B. (1986a) Composition of Comet Halley dust particles from Vega observations. *Nature* **321** 280–282

Kissel, J., Brownlee, D., Büchler, K., Clark, B.C., Fechtig, H., Grün, E.G., Hornung, K., Igenbergs, E.B., Jessberger, E.K., Krueger, F.R., Kuczers, H., McDonnell, J.A.M., Morfill, G.M., Rahe, J., Schwehm, G.H., Sekanina, Z., Utterback, N.G., Völk, H.J., & Zook, H.A. (1986b) Composition of Comet Halley dust particles from Giotto observations. *Nature* **321** 336–337

Kresák, L. (1981) The lifetimes and disappearance of periodic comets. *Bull. Astron. Inst. Czech.* **2** 321–339

Kührt, E. (1984) Temperatures profiles and thermal stress on cometary nuclei. *Icarus* **60** 512–521

Kuiper, G.P. (1951) On the origin of the solar system. In: *Astrophysics*, ed. Hynek, J.A., McGraw Hill, New York, 357–424

Lanzerotti, L.J., *et al.* (1978) Low energy cosmic ray erosion of ice grains in interplanetary and interstellar media. *Nature* **272** 431–433

Lewis, J.S. (1971) Satellites of the outer planets: their physical and chemical nature. *Icarus* **15** 174–185

Lowery, B.E. (1973) The effect of multiple encounters on short-period comet orbits. *Astron J.* **78** 428–437

Marsden, B.G. (1986) *Catalogue of cometary orbits*, 5th edition, Smithsonian Astrophys. Obs., Cambridge, 102 pp.

Moore, M.H. *et al.* (1983) Studies of proton irradiated cometary type ice mixtures. *Icarus* **54** 338–405

Newton, H.A. (1893) On the capture of comets by planets, especially their capture by Jupiter. *Mem. Natl. Acad. Sci. USA* **6** 7–23

O'Dell, C.R. (1971) A new model for cometary nuclei. *Icarus* **19** 137–146

Ostro, S.J., *et al.* (1986) Impact cavities in underdense regoliths? *Amer. Meteor. Soc.* (abstract), submitted

Prialnik, D., & Bar-Nun, A. (1987) On the evolution and activity of cometary nuclei. *Astrophys. J.* **313** 893–905

Reitsema, H.J. (1986) Nucleus morphology of Comet Halley. In: *20th ESLAB Symposium on the Exploration of Halley's Comet*, ESA SP-250, **2** 351–354

Safronov, V.S. (1969) *Evolution of the protoplanetary cloud and formation of the Earth and planets*, Moscow, Nauka Press

Sagdeev, R.Z., *et al.* (1986) TV experiment of the Vega mission: Photometry of the nucleus and the inner coma. In: *20th ESLAB Symposium on the Exploration of Halley's Comet*, ESA SP-250, **2** 317–326

Sekanina, Z., & Larson, S.M. (1986) Dust jets in Comet Halley observed by Giotto and from the ground. *Nature* **321** 357–361

Shulman, L.M. (1972) The chemical composition of cometary nuclei. In: *The motion, evolution of orbits, and origin of comets*, eds Chebotarev, G.A., *et al.* D. Reidel, Dordrecht, 263–270

Smith, B.A., & Terrile, R.J. (1984) A circumstellar disc around beta Pictoris. *Science* **226** 1421–1424

Smith, B.A. *et al.* (1986) The spatial distribution of dust jets seen at Vega-2 fly-by. In: *20th ESLAB Symposium on the Exploration of Halley's Comet*, ESA SP-250, **2** 327–332

Smoluchowski, R. *et al.* (1984) Evolution of density in solar system ices. *Earth, Moon & planets* **30** 281–288

Stern, S.A. (1986) The effects of mechanical interaction between the interstellar medium and comets. *Icarus* **68** 276–283

Stern, S.A. (1987) Two important mechanisms contributing to cometary evolution in the Oort cloud. *Lunar & Planet. Sci. Conf.* **18** 951 (abstract)

Tsou, P.D. *et al.* (1984) Experiments on intact capture of hypervelocity particles. *Lunar & Planet. Sci. Conf.* **15** 866–867 (abstract)

Vanysek, V., & Rahe, J. (1978) The $^{12}C/^{13}C$ isotope ratio in comets, stars, and interstellar matter. *Moon & Planets* **18** 441–446

Wallis, M.K. (1980) Radiogenic melting of primordial comet interiors. *Nature* **284** 431–433

Wehinger, P.A. *et al.* (1986) Comet P/Halley: Echelle spectra of the coma. *Bull. Amer. Astron. Soc.* **18** 812

Weidenschilling, S.J. (1980) Dust to planetesimals: settling and coagulation in the solar nebula. *Icarus* **44** 172–189

Weissman, P.R. (1979) Physical and dynamical evolution of long-period comets. In: *Dynamics of the solar system* ed. Duncombe, R.L., D. Reidel, Dordrecht, 277–282

Weissman, P.R. (1980) Stellar perturbations of the cometary cloud. *Nature* **288** 242–243

Weissman, P.R. (1985a) The origin of comets: implications for planetary formation. In: *Protostars and planets II*, eds Black, D.C., & Matthews, M.S., Univ. Arizona Press, Tucson, 895–919

Weissman, P.R. (1985b) Cometary dynamics. *Space Sci. Revw.* **41** 299–349

Weissman, P.R. (1986a) Are cometary nuclei primordial rubble piles? *Nature* **288** 242–244

Weissman, P.R. (1986b) Post-perihelion brightening of Halley's Comet: A case of nuclear summer. In: *20th ESLAB Symposium on the Exploration of Halley's Comet, ESA SP-250*, **3** 517–522

Weissman, P.R. (1987) How typical is Halley's Comet? In: *Diversity and similarity of comets*, ESA SP-278, 31–36

Whipple, F.L. (1950) A comet model I: The acceleration of Comet Encke. *Astrophys. J.* **111** 375–394

Whipple, F.L. (1978) Cometary brightness variation and nucleus structure. *Moon & Planets* **18** 343–359

Whipple, F.L., & Lecar, M. (1976) Comet formation induced by the solar wind. In: *The study of comets*, NASA SP-393, 660–662

Yeomans, D.K., & Kiang, T. (1981) The long term motion of Comet Halley. *Mon. Not. Roy. Astron. Soc.* **197** 633–643

21

The ageing of Comet Halley and other periodic comets

Ľubor Kresák

1 INTRODUCTION

From the evolutionary point of view, comets are the most conspicuous objects in the solar system. On one hand, they are the most primordial products of the first phases of formation of the system, most of them having remained essentially unchanged for the following 4.5 thousand million years. On the other hand, they are the only objects in which physical ageing processes, as well as drastic changes of orbits, can be observed within a few decades to centuries. This apparent paradox is due to the enormous acceleration of the ageing of individual objects under specific conditions.

The physical ageing of comets is a rather complicated process, including sublimation of the ice cover, formation and blow-off of the dust mantle, repeated violation of the forming crust by jets from the inside, and disintegration into fragments of different size. The ageing rate depends primarily on two orbital elements which undergo changes by planetary perturbations. The first of them, the perihelion distance q, determines the extreme conditions experienced by the comet during its revolution around the Sun. So long as q is large—tens of AU and more—the comet can remain essentially unchanged for an unlimited period. The ageing process begins at the time of the first significant reduction of q, either by a star passing through (or close to) the Oort Cloud, or by an outer planet encountered near the comet's perihelion. Later,

the ageing rate tends to increase with each subsequent reduction of q by planetary perturbation, and to decrease with each subsequent increase of q.

The other decisive element is the revolution period P, which determines the frequency of the destructive passages near the Sun. This element alone defines the *dynamical* age of the comet. We call *new comets* those of $P > 10^6$ years (binding energy $1/a < 10^{-4}$ AU^{-1}), because most of them are approaching the Sun for the first time from the Oort Cloud (Oort 1951), and *old comets* all the others. The *periodic comets*—or *short-period comets*—conventionally defined by $P < 200$ years, are at the other end of the evolutionary sequence. This stage is reached only by a very small fraction of the comets entering the inner solar system (Everhart 1976). It seems that, at a given time, less than one comet in a thousand million is moving in a short period orbit (Kresák 1980).

For the investigation of the ageing processes, however, the short-period comets are of fundamental interest. Their appearance at a number of successive revolutions makes it possible to witness the progressive evolutionary changes. The failure of rediscovery at a predicted return—in spite of a reliable ephemeris and favourable Sun–Earth–comet configuration—indicates its deactivation below a level set by the efficiency of the telescopes used. If there is a succession of missed returns, total extinction or decay can be inferred. Furthermore, short-period comets leave a part of the products of their ageing process within the inner solar system, accessible to observation as meteors and micrometeoric dust. For the long-period comets the situation is different because, owing to their low binding energy, all their finer débris is removed immediately by the solar radiation pressure.

Knowledge of the physical ageing rates and survival times of short-period comets is essential for the investigation of the nature and properties of these objects and of the physicochemical processes going on in them. It is also of crucial importance for understanding the overall structure and evolution of the huge system of comets, surrounding the planetary system to a 1000-fold distance of the outermost planet, only a minute fraction of which is accessible to observation. Our knowledge of the Oort Cloud is obtained by extrapolation and numerical modelling based on a sample of entirely atypical (observable) objects, which have undergone strong perturbations directing their orbits towards the Sun. Such extrapolations necessarily become incorrect when the ageing processes are not taken into account. The limited active lifetimes of comets introduce a definite asymmetry into the occurrence rate of the fundamental perturbing events, which is not borne out by numerical simulation of their past and future motion. At the moment when the time scale of physical ageing becomes shorter than that of the orbital evolution, the basic requirements of a random process fail: a comet whose disintegration was triggered by a decelerating encounter may not survive long enough to experience an analogous accelerating encounter. Captures into short-period orbits become more frequent than ejections from them; Jupiter's reflecting barrier becomes a partially absorbing barrier; and the equilibrium between captures and ejections changes into an equilibrium between captures and ejections plus physical decay. The proportion of the two latter processes can be determined only from the survival times of comets, and it is clear that these play a major role.

The ageing processes also introduce substantial complications into the extrapolation of orbital evolutions into the past. The erratic nongravitational effects in the motion of comets, produced by the jet effects of escaping gases, make it impossible to extrapolate the motion of individual objects far back. There is not a single case in which the present orbit of a periodic comet could be traced back into the Oort Cloud.

2 THE SUITABILITY OF INDIVIDUAL COMETS FOR INVESTIGATING THE AGEING PROCESSES

Up to February 1988, we know 146 short-period comets, 90 of which have already been observed at more than one apparition. The total number of 608 apparitions, with 254 missed returns in between, provides a good basis for statistical investigation of the ageing processes. However, if we wish to trace the physical evolution of individual comets, the situation becomes rather displeasing, and only a few well observed objects can be considered.

This is illustrated by Table 1 which lists 20 periodic comets, including the 10 objects with the greatest values of three tentative ageing criteria. The comet's name is followed by the dates of its first and last apparition, T_1 and T_2; the number of perihelion passages between the first and last apparition inclusive, N; and the mean absolute total magnitude $\bar{H}$. The insolation parameter I represents the integrated solar irradiation received by a unit area of the comet's surface during the whole period of observation, as a function of the varying perihelion distance q and eccentricity e of the comet's orbit. The unit is one year at $r = 1$ AU, corresponding to one revolution in the Earth's orbit. Then,

$$I = \int_{T_1 - P_1/2}^{T_2 + P_2/2} r^{-2}\, \mathrm{d}t = \sum_{T_1}^{T_2} (q + qe)^{-1/2} \qquad (1)$$

The activity parameter A represents the integrated activity, referred to a one-year stay of the given object at $r = 1$ AU, assuming that the activity varies with the inverse fourth power of the heliocentric distance. Then,

$$A = \int_{T_1 - P_1/2}^{T_2 + P_2/2} r^{-4}\, \mathrm{d}t = \sum_{T_1}^{T_2} (q + qe)^{-5/2}(2 + e^2)/2. \qquad (2)$$

Table 1. Periodic comets which should show best different signs of ageing

Comet	T_1	T_2	N	$\bar{H}$	I	A	L	Note
P/Encke	1786	1984	61	8.8	77.1	267.6	80	
P/Grigg–Skjellerup	1808	1982	36	11.7	30.0	18.5	0.4	
P/Halley	−239	1986	30	2.7	28.1	31.5	2600	H
P/Pons–Winnecke	1819	1983	29	9.0	23.0	12.1	3	
P/Tuttle–Giacobini–Kresák	1858	1978	23	10.5	16.8	5.8	0.4	
P/Tempel 2	1873	1983	22	7.7	15.2	4.0	3	
P/D'Arrest	1851	1982	21	7.8	14.5	3.9	3	
P/DeVico–Swift	1844	1965	21	8.7	14.0	3.3	1.1	
P/Faye	1843	1984	20	6.4	12.4	2.1	6	
P/Tempel 1	1867	1983	20	7.8	12.4	2.1	1.6	
P/Tempel–Tuttle	1366	1965	19	7.4	13.9	5.7	6	H
P/Peters–Hartley	1846	1982	18	8.7	11.4	2.1	0.7	
P/Denning–Fujikawa	1881	1978	12	13.1	10.2	7.3	0.1	
P/Beial	1772	1852	11	7.1	10.3	5.2	8	E
P/Honda–Mrkos–Pajdušáková	1948	1985	8	11.0	7.9	10.2	0.4	
P/Brorsen	1846	1879	7	8.3	6.6	7.1	3	E
P/Pons–Brooks	1812	1954	3	4.1	2.1	1.5	35	H
P/Olbers	1815	1956	3	3.9	2.0	0.5	15	H
P/Mellish	1917	1917	1	6.8	1.6	16.9	30	H
P/Swift–Tuttle	1862	1862	1	3.9	0.7	0.3	8	H

Notes: H = Halley type, $20 < P < 200$ years; E = already extinct.

The mass loss parameter L represents the estimated gas and dust loss during the whole observation period, in 10^{10} kg:

$$L = A \times 10^{3 - 2\bar{H}/5} \qquad (3)$$

For more details on the definition and determination of these parameters see Kresák & Kresáková (1987a,b). The peculiar comet P/Schwassmann–Wachmann 1, with $q > 5$ AU, is not included in the table, although it may rank among the ten objects with highest L.

The table illustrates very clearly the exceptional nature of two comets: P/Encke, with an orbit exposing it to far the strongest deterioration effects, and P/Halley, with far the longest observational history and far the largest mass loss during it. For P/Encke, owing to the shortest revolution period of all comets, there exist observations from 53 apparitions covering the last 200 years. For P/Halley only the last three returns fall within the same time span, and a great majority of observations refer to historical records of limited information content. However, during its 1986 apparition the first comet missions, artificial satellites, and unique groundbased campaigns using the whole range of modern observing techniques have assembled a broad variety of measurements, not available for any other comet. This has made P/Halley an extraordinarily suitable object for the investigation of comet ageing. It must be kept in mind, however, that P/Halley is by no means a typical short-period comet, whether in terms of its past dynamical history (circumstances of its capture into the inner solar system), or its size, composition, structure, and surface properties. The same applies to P/Encke.

3 THE SECULAR BRIGHTNESS DECREASE OF COMETS

The simplest indicator of the ageing of comets is the progressive decrease of their gas and dust production. This is reflected by the decrease of their absolute brightness, i.e. of their apparent brightness reduced to a unit heliocentric and geocentric distance. A general secular fading of periodic comets was discovered by Vsekhsvyatskij (1958), based on an extensive collection and statistical analysis of their brightness records. This finding is usually interpreted by means of a simplified comet model, assuming that its nucleus is spherical, homogeneous, and shrinks at a constant rate, unless the planetary perturbations appreciably change the orbit—and hence also the insolation regime. The condition of orbital stability is well satisfied both for P/Encke and P/Halley, whose insolation parameters I have varied within $\pm 1\%$ since the first discovery.

In this model, the change of the absolute magnitude between two successive returns ΔH can be expressed in terms of the change of the radius ΔR and the change of mass ΔM as follows:

$$\Delta H = -\frac{5}{\ln 10}\frac{\Delta R}{R} = -\frac{5}{3\ln 10}\frac{\Delta M}{M} = -0.724\frac{\Delta M}{M} \quad (4)$$

The number of revolutions N_L left until a complete disintegration is, accordingly,

$$N_L = -\frac{R}{\Delta R} = \frac{5}{\Delta H \ln 10} = \frac{2.171}{\Delta H}. \quad (5)$$

Vsekhsvyatskij (1972) quotes $\Delta H = 0.3$ to 0.4 as a typical value. His listing of H for 41 comets of more than one apparition (Vsekhsvyatskij 1958) implies a mean value of $\Delta H = 0.34$. Accordingly, the mean survival time should be only six returns after the mean epoch; in 20% of cases more than ten returns, and in another 20% only one or two. The first attempts to predict the death dates of individual comets by fitting relation (4) to their long-term absolute lightcurves (Whipple 1964, Whipple & Douglas-Hamilton 1966) have led to unbelievably short lifetimes. The failure of these predictions became evident by continuing observations of a number of comets (P/Pons–Winnecke, P/Tuttle, P/Wolf, P/Kopff, P/Brooks 2, P/Faye, P/Whipple) long after their predicted death dates.

There are three reasons for these gross underestimates of comet lifetimes, even within the framework of the simplified model: (A) The irregular brightness fluctuations from one revolution to another favour the discovery at a higher activity level, which biases the observed trend towards a steeper brightness decrease. (B) If the comet is decelerated by Jupiter's perturbations into an orbit of appreciably smaller perihelion distance, the abrupt change of the insolation regime tends to make the comet absolutely brighter for the next one or two revolutions. Such captures are responsible for about 20% of discoveries of short-period comets (Kresák 1982). (C) The brightness estimates of comets are seriously liable to instrumental effects, increasing with the telescope size and, hence, also with time. The point is that large telescopes would only detect the nuclear condensation of the comet, while the outer parts of the coma are not included, or are even subtracted as a part of the background. In the statistics of absolute magnitudes this effect is the most serious one.

The effects (A)–(C) have been discussed in more detail by Kresák (1974, 1982). Svoreň (1979) attempted to remove (A) and (B) by omitting the first observed apparition of each comet, and the mean fading rate dropped indeed to $\Delta H = 0.22$. This would nearly double the predicted survival times—to about ten revolutions on the average—but even this is too little to comply with other evidence. As for (C), corrections depending on the telescope aperture were introduced by Bobrovnikoff (1941) and Öpik (1963), and refined by Morris (1973) and others. These corrections already imply a time-dependent trend with increasing telescope size. However, they do not seem sufficient, because a higher magnification applied in the same telescope would also make the comet fainter for a visual observer, and the appearance of each individual comet—in particular its total brightness and the degree of nuclear condensation—also plays an important role. For the intricate problems of magnitude corrections see Meisel & Morris (1976, 1982).

Empirical corrections, depending on the apparent brightness at the time of maximum light, were recently introduced by Kresák & Kresáková (1987b). Their application leaves only five comets with statistically significant values of ΔH, and for four of them (P/Faye, P/Brooks 2, P/Finlay, and P/Pons–Winnecke) this fading may be associated with a rapid orbital evolution controlled by Jupiter's perturbations. An exception is P/Encke, whose orbit predestines it to far the most rapid ageing, as shown in Table 1. When all the short-period comets with more than one apparition are taken together, the mean brightness decrease per revolution is $\Delta H = 0.028 \pm 0.037$. Omitting the five abovementioned comets, we have $\Delta H = 0.016 \pm 0.042$; and omitting the first apparitions, $\Delta H = 0.020 \pm 0.047$. According to equation (5), these fading rates correspond to mean survival times of 80, 140, and 110 revolutions, respectively—but these figures have little meaning, the mean errors of ΔH being larger than their numerical values themselves. What can be accepted with confidence is the lower limit of about 50 revolutions. The data for individual objects show that only 55% of them were brighter during the first apparition than during the last one, and only 52% during the first predicted return than during the last one. Thus the introduction of instrumental corrections, on one hand increases the predicted survival times by more than an order of magnitude, removing the disagreement with other evidence; and on the other hand, it makes the slow progressive fading vanish against irregular brightness fluctuations and observational errors, thus not allowing us to predict the death dates with confidence. The total active lifetimes should be, on the average, about twice as long as the predicted survival times.

There is still little consensus on the significance of the instrumental effects and on the best way of taking them into account, but there is ample evidence of their importance. For example, without taking them into account, the upper limit of absolute brightness of short-period comets should be currently 10 times (2.5 magnitude) lower than a century ago, and the mean exponent in the brightness law of long-period comets would have to increase by 50% during the same period (Kresák 1985). Dobrovol'skij et al. (1983) have shown that the brightness decrease of P/Kopff—by 3 magnitudes over 7 revolutions, according to Vsekhsvyatskij (1958)—can be fully explained by the changes of observational geometry. These are also reflected in spurious periodic changes of the brightness of P/Encke. When a short-period comet arrives into a very favourable configuration—and this is the circumstance under which many of them were discovered—it often surprises by its unexpected brightness. P/D'Arrest, P/Kopff, P/Wirtanen, and P/Borrelly were never seen brighter than during their most recent apparitions. Another example is P/Grigg–Skjellerup which was thought to fade by 5 to 7 magnitudes per century (Vsekhsvyatskij 1958, 1972, Svoreň 1979). Its identification with a comet observed by Pons in 1808 (Kresák 1987) has shown that the

brightness decrease is definitely at least 10–15 times slower, and essentially unrecognizable.

4 THE SECULAR BRIGHTNESS DECREASE OF P/ENCKE AND P/HALLEY

The two exceptional comets, P/Encke and P/Halley, are very little sensitive to instrumental effects on the magnitude estimates when the peaks of their light-curves in each apparition are considered. At that time, P/Halley is always a bright naked-eye object, and P/Encke close to the naked-eye threshold, mostly between apparent magnitude 5.0 and 7.5. The main problem is the asymmetry of their lightcurves, peaking for P/Encke several weeks before and for P/Halley several weeks after the perihelion passage. Depending on the Sun–Earth–comet configuration, the maximum would often escape observation owing to proximity to conjunction with the Sun. For P/Encke similar conditions recur after three revolutions, and for P/Halley the sequence is quite irregular. While for P/Encke the analysis can be based on direct magnitude estimates and on the statements of naked-eye visibility, for most of the apparitions of P/Halley the only source of information is the recorded date of the discovery and of the last sighting, combined with the assumption that the apparent brightness at these moments remained approximately constant in earlier history. According to Ferrín (1984), the fading of P/Halley, which was not found by this type of analysis by Broughton (1979) appears only when the twilight and moonlight effects are taken into consideration.

The upper parts of Tables 2 and 3 list, for P/Encke and P/Halley, the computed fading rates ΔH (in magnitudes per revolution), the epochs E to which they refer, the corresponding numbers of remaining revolutions N_L after that epoch, and the predicted death dates T_D. At the bottom of each table the values of N_L and T_D based on the relative mass–loss rates $\Delta M/M$ (to be discussed in sections 5 and 6) are added for comparison. The large discrepancies illustrate the untenability of the assumption of a constant shrinking rate, with no inactive core left at the end.

Table 2.   The ageing of P/Encke

Reference	ΔH $(\Delta M/M)$	E	N_L	T_D
Kresák (1965)	0.033 ± 0.010	1875	66	2090
Whipple & Douglas-Hamilton (1966)	0.075 ± 0.008	1900	29	2000
Sekanina (1969)	0.066 ± 0.010	1850	33	1960
Sekanina (1969)	0.092 ± 0.007	1900	23	1980
Sekanina (1969)	0.122 ± 0.010	1950	18	2010
Ferrín & Gill (1986)	0.097 ± 0.022	1986	22	2060
Kresák & Kresáková (1987b)	0.029 ± 0.006	1885	75	2130
Sekanina (1969)	(0.0010)	1950	3000	12 000
Whipple & Sekanina (1979)	(0.0016)	1800	1900	8000
Whipple & Sekanina (1979)	(0.0009)	1950	3300	13 000

Note: The N_L and T_D values are computed from ΔH, using equation (5), and from $\Delta M/M$, using equation (7). They are not always quoted explicitly by the authors.

Table 3.   The ageing of P/Halley

Reference	ΔH $(\Delta M/M)$	E	N_L	T_D
Vsekhsvyatskij (1958)	0.21 ± 0.05	1340	10	2100
Sekanina (1964)	0.33 ± 0.03	1900	7	2400
Broughton (1979)	0.00 ±	760	large	large
Hughes (1983)	0.020 ± 0.021	760	110	10 000
Bortle & Morris (1984)	0.00 ±	990	large	large
Wallis & Wickramasinghe (1985)	0.055 ± 0.015	760	55	5000
Ferrín & Gill (1986)	0.055 ± 0.015	760	40	3800
McIntosh & Hajduk (1983)	(0.0050)	1986	600	50 000
Hughes (1985)	(0.0013)	1986	2300	180 000
Hajduk (1986)	(0.0013)	1986	2300	180 000
Rickman (1986)	(0.0030)	1986	320	26 000
Sagdeev et al. (1987)	(0.0020)	1986	500	40 000

Note: The N_L and T_D values are computed from ΔH, using equation (5), except for that by Wallis & Wickramasinghe (1985) applying a different formula to the fading rate determined by Ferrín (1984). The first three values based on $\Delta M/M$ use equation (7), while Rickman (1986) and Sagdeev et al. (1987) assume a constant mass-loss rate, i.e., equation (6).

5 THE MASS-LOSS RATE AND AGEING

Another approach to the determination of comet lifetimes makes direct use of the relative mass-loss rate per revolution, i.e.,

$$N_L = \frac{M}{\Delta M} \qquad (6)$$

for a constant mass–loss rate, or

$$N_L = \frac{3M}{\Delta M} \qquad (7)$$

for a constant reduction of the linear size. It was only recently that reliable estimates of ΔM began to accumulate. Those referring to gas production are mainly based on UV spacecraft measurements, and those referring to dust production on the groundbased IR measurements. For the measuring techniques and results see other papers in the present work. In interpreting such data it must be kept in mind that spectrophotometry can cover only a part of the emissions, and that the infrared measurements cover only a part of the size spectrum of the solid ejecta. The former problem is alleviated by the prevalence of water ice in comets (see, e.g., Delsemme 1982,

Huebner *et al.* 1986). In spite of the necessary assumptions about the radiation mechanism, the current estimates of gas production appear more reliable than those of the total production of solids.

Even worse is the situation regarding the masses of cometary nuclei, where the uncertainties in the sizes and densities make M uncertain by more than plus/minus one order of magnitude. The only exception is P/Halley, for which the Vega and Giotto missions were able to measure the size directly, with an uncertainty of about $\pm 20\%$ in the volume. Its total mass has been estimated at 2×10^{14} kg, with an uncertainty of a factor of three mainly due to the unknown bulk density, by Sagdeev *et al.* (1987). Combining this value with the estimated mass-loss rate $\Delta M = 5 \times 10^{11}$ kg (Whipple 1986), we obtain from equation (6) a survival time of 400 future revolutions, and from the preferable equation (7) one of 1200 revolutions. Whipple's estimate may seem rather conservative, in view of the hidden contribution of larger solid particles (McDonnell *et al.* 1986, Hajduk & Kapišinský 1987). This, and/or a lower mass and density, as suggested by Rickman (1986) and Rickman *et al.* (1987) would make the predicted survival time shorter—only a few hundred revolutions.

Another approach to the determination of the mass-loss rate is to compare the total mass of larger solid particles constituting the meteor stream of P/Halley with the time elapsed since the beginning of its formation. McIntosh & Hajduk (1983) have estimated the present mass of the stream at 5×10^{14} kg. With the same value for the present mass of the comet, Hajduk (1986) has estimated both the history of the comet since its capture into an orbit of small perihelion distance, and its future survival time, at 2300 revolutions. This agrees with the result of Hughes (1985) obtained with $M = 2.2 \times 10^{14}$ kg and $\Delta M = 2.8 \times 10^{11}$ kg, as estimated before the Halley mission encounters. Jones & McIntosh (1986) have recently questioned the long history of this meteor stream (see the paper by McIntosh in this volume), which would also have an impact on the predicted survival time. However, a substantial reduction would become inconsistent with the above value for the mass of the meteor stream, which already requires a very high proportion of the total mass production in the form of larger particles, and a very low rate of their progressive elimination.

Extensive laboratory experiments with icy samples of different chemical composition (Dobrovol'skij *et al.* 1983, 1986) have led to the conclusion that they would lose a surface layer 1.4 to 6.8 m thick during one revolution in the orbit of P/Halley. For pure water ice it is 2.0 m, and for other realistic compositions about 2.5 m, which would correspond to a survival time of some 2000 revolutions, starting from the present size of the comet.

6 DYNAMICAL EVIDENCE

The data on the motion of comets involve two sources of information on their ageing processes. One is the presence of nongravitational jet effects, which are indicative of the relative mass loss, and the other the influence of the finite lifetimes of comets on the maintenance of a steady population of active objects moving in short-period orbits.

The nongravitational effects are generally small, variable, and difficult to determine unless a number of apparitions are covered by accurate astrometrical observations (Marsden 1985; see also the paper by Yeomens in this volume). Their use for determining the relative mass-loss rate requires a complex model, including the shape and rotation of the nucleus, its thermal regime, ejection velocity, scatter of outflow directions, and perihelion asymmetry of the mass-loss rate. In an analysis of this type, Whipple & Sekanina (1979) have estimated the current relative mass loss of P/Encke at $\Delta M/M = 0.0009$, and that at the time of its discovery at $\Delta M/M = 0.0016$. With a constant absolute mass loss, these figures would imply a survival time of 1100 and 600 revolutions, respectively, and with a constant decrease of the comet's linear size three times as much. A recent analysis of P/Halley and P/Kopff by Rickman (1986) became the main argument for a low density of cometary nuclei. Rickman estimates the future survival time of P/Halley at < 180–460 revolutions, and that of P/Kopff at < 400 revolutions. For more details on the mass determination from nongravitational effects see the paper by Rickman in this volume.

Fernández (1985) has used the long-term integrations of the motion of all periodic comets by Carusi *et al.* (1985a, b) to determine the transfer rate of their perihelia from Jupiter's region to the region of terrestrial planets. He found that a steady-state population of this region would require an average lifetime of 1000 revolutions in a short-period orbit.

7 TRANSIENT AND ULTIMATE EXTINCTIONS OF PERIODIC COMETS

The most straightforward way to determine comet lifetimes is to compare the observational history of all periodic comets with the number of objects which have disappeared during this time span. From among the 146 known periodic comets, 34 remained unrecovered at the last perihelion passage, and 16 of them also during four or more preceding perihelion passages. There are various reasons for this. Mostly, it is a poorly-determined orbit, not allowing for ephemeris-aided searches. If this is combined with a very favourable geometry for observation at the first apparition, which is often the case, we have to wait for an independent rediscovery for a number of revolutions. In fact, 16 comets lost in this way have already been rediscovered after 5 to 18 missed returns, two of them (P/Tempel–Tuttle and P/Tuttle–Giacobini–Kresák) even twice. In other cases the rediscovery was prevented by planetary perturbations increasing the perihelion distance of the comet. For a listing of comet losses and rediscoveries see Belyaev *et al.* (1986); for a more detailed discussion of the currently lost comets see Kresák (1981b) and Spratt (1984). The analysis by Kresák has left only three or four objects indicative of a real extinction: P/Biela, P/Brorsen, P/Westphal, and probably P/Neujmin 2. This would imply a typical active lifetime of 150 to 300 revolutions, the uncertainty being due to the statistics of small numbers.

There is one important open question. Does a definite disappearance of a comet mean a total disintegration, a complete loss of volatiles, or a complete crusting? In the latter case the activity could be restored again after some time by the exposure of inner sources. Hence, it is essential to consider also those cases when a currently active comet was inactive in the past. A search for such objects has revealed seven suspect cases: P/Biela, P/Brorsen,

P/Finlay, P/Honda–Mrkos–Pajdušáková, P/Boethin, P/Hartley–IRAS, and P/Denning–Fujikawa (Kresák 1986). Interestingly enough, the first two of them are just two of the four currently extinct comets. It seems, therefore, that the final fading out is preceded by transient intermissions of activity, or at least by periods of its substantial reduction. The observed extinctions of comets, both of short and long periods, are in their final phase much more abrupt than predicted by equation (3) (Kresák 1984, Sekanina 1984). Under such conditions the active lifetime of a comet need not cover a single limited time span; it can also consist of several periods spaced by dormant phases of indeterminate duration. Accordingly, some of the asteroids moving in comet-like orbits (see Kresák 1985, Rickman 1985) might also change into active comets in the future.

8 JUMPS IN THE AGEING PROCESS

The physical evolution of comets is by no means a smooth process, which could be well approximated by a constant shrinking rate of their nuclei. In addition to the dormant phases, there are discrete events accelerating the ageing suddenly: splitting of cometary nuclei and their outbursts. For reviews of these phenomena see Hughes (1975), Whipple (1978), and Sekanina (1982).

Splitting of the nucleus into large fragments producing their own comas, and sometimes also tails, has so far been observed with certainty in five periodic comets: P/Biela in 1846, P/Brooks 2 in 1889, P/Giacobini in 1895, P/Taylor in 1916, and P/Du-Toit–Hartley in 1982. There are a few additional suspect cases (Golubev 1975, Sekanina 1977). Splitting of P/Brooks 2 occurred at the time of its close approach to Jupiter—the closest such approach for any comet ever recorded—so that the tidal forces could have played some role in it. Otherwise, there is no correlation with the position in the orbit or with any other external triggering event (Kresák 1981a). A collisional origin is contradicted by a deficiency of splits within the main asteroid belt (Pittich 1971). Of course, these conclusions refer to all comets, the sample of short-period comets being too small for statistical analysis.

The outcome apparently depends on the relative size of the fragments. The two comparable components of P/Biela have survived only two revolutions. Two smaller fragments of P/Brooks 2 have disappeared during their first perihelion passage, but the main component is still active after 14 additional revolutions, just a little fainter than in 1889. P/Giacobini and its small companion were never seen again, but the failure of rediscovery may well be due to its faintness and poorly determined orbit. P/Taylor was observed again in 1977 and 1984, after eight missed returns, as a single comet. P/DuToit–Hartley, split one revolution before its rediscovery in 1982, was recovered again as a single object in 1987.

The only example of persistent remnants indicative of an earlier splitting is the pair of comets P/Neujmin 3 and P/Van Biesbroeck (Carusi *et al.* 1985b). Both the components, widely separated by differential perturbations during a close encounter with Jupiter in 1850, have already survived 12 and 11 perihelion passages, respectively. P/Neujmin 3 was not recovered at its last predicted return in 1982, but this may have been due to the unfavourable observational geometry, much worse than at any previous apparition (Belyaev *et al.* 1986). A conspicuous family of inactive objects moving in similar orbits (Clube & Napier 1986) suggests that P/Encke could also be a remnant of a very large object disrupted long ago, like the Kreutz group of long-period Sun-grazing comets. This possibility is supported by the complex structure of meteor streams associated with it (Whipple & Hamid 1952, Štohl 1986).

The fact that the splits are not accompanied by strong brightness bursts, argues against any sudden exposure of a large inner reservoir, much richer in volatiles than the outer crust. It seems that we have to do with a mild process, in which a part of the irregularly shaped nucleus smoothly separates from the main body (Whipple 1963), and is driven away by the differential solar radiation pressure (Sekanina 1982). The statistics suggest one such event per 120 revolutions. With due corrections for the observability, it can be estimated at one per 90 revolutions. This is much less than the frequency of one per 12 revolutions found for the long-period comets which are apparently less resistive to disintegration (Kresák 1981a). However, a comparison with the probable lifetimes of short-period comets indicates that most of them must experience several splits during their active periods, which must have some impact on the ageing process.

Sudden outbursts, during which the total brightness of a comet may increase by a factor of 100 or 1000 within one day, have a clear tendency to appear repeatedly in a fraction of periodic comets. A special case is the peculiar object P/Schwassmann–Wachmann 1, with a number of outbursts of different strength during each revolution in its low-eccentricity orbit beyond Jupiter. The only outbursts comparable with those of P/Schwassmann–Wachmann 1, or even stronger, occurred in two otherwise normal comets, P/Holmes (1892/3) and P/Tuttle–Giacobini–Kresák (1973), in pairs separated by 6 and 9 weeks, respectively. There is, again, no indication of an external triggering event, and no concentration to the main asteroid belt which would imply a collisional origin (Pittich 1971, Andrienko & Vashchenko 1981). The recurrence cannot be explained by chance, anyway, and it became the basis for a tentative explanation by Whipple (1983, 1984). If some cometary nuclei are double, as suggested by Van Flandern (1981), the reduction of the minimum distance between the components by differential nongravitational forces can result in a grazing encounter (first burst), followed by a collision one revolution later (second burst). However artificial this scenario may seem, we still do not have a better one.

The lightcurves of the outbursts are very poorly determined, owing to their unpredictability. Even so, the mass loss involved is apparently, in the extreme cases, one order of magnitude higher than during a whole normal revolution, reducing the survival time appreciably. P/Holmes is still active 13 revolutions after its violent outbursts. P/Tuttle–Giacobini–Kresák was not recovered at its return in 1984, under bad observing conditions, after 23 revolutions since its first discovery. We have to wait for the next favourable return in 1990 to be sure about it. Its impending extinction is only indicated by erratically increasing nongravitational effects, resembling those experienced by P/Biela and P/Brorsen before their disappearance (Marsden 1985).

9 LESSONS FROM COMET HALLEY

The ageing processes in comets are governed by the periodic and progressive changes going on in the

surface layers of their nuclei. These include sublimation of the ice cover, escape of gas carrying with it
solid particles of various sizes, and formation and
blow-off of the dust mantle. Before the spacecraft
encounters with P/Halley, most of the research in this
direction was based on modelling, which had to start
with a number of theoretical assumptions and end
with a comparison with the limited sources of
observational evidence.

The close-up pictures of P/Halley from Vega-1 and -
2 and Giotto, and the direct measurements in their
environment, have added a substantial study of new
data (see the papers by Keller and Sagdeev & Szegö in
the present work). So it has been possible to check and
refine the earlier models (see papers by Houpis and
Weissman in this volume). The observed complex
morphology of the nuclear surface, resulting in
irregular variations of the mass loss and its composition, makes the problem of comet ageing even more
complicated than it appeared before the Halley
missions.

The variety of observations from a single apparition
of P/Halley has provided us with unique information
on one cycle of the ageing process, i.e., on the changes
taking place during one perihelion passage and
recurring periodically. However, what we need for
determining the survival times and eventual fates of
periodic comets, is the progressive evolution from one
return to another: in particular, the trends in the
relative size of the free sublimating area on the
comet's surface, in the resulting mass-loss rate, and in
the differentiation of the nucleus with depth. The
composition and structure could have been different
for individual comets since the beginning, and could
have changed in a different way during their past
orbital evolution. The interaction between the formation and blow-off of the insulating mantle may be a
rather complicated function of the perihelion distance, which itself varies because of planetary
perturbations. Some comets, in particular those of the
Jupiter family, have arrived in their present orbits by
a stepwise decrease of the perihelion distance, while
other comets—apparently including P/Halley—have
skipped over the phase of repeated very low activity,
in a single-stage capture process.

In spite of the detailed imaging of P/Halley by
Vega-1, -2 and Giotto, there is still little consensus

about the fraction of its surface active at any one time.
The suggested values range from a few per cent to
20%, and Whipple's (1986) summarizing estimate
puts it at about 15%. In fact, the pure percentage does
not express the actual complexity of the surface, on
which several strong local sources and considerable
differences between larger areas have been detected.
Ip & Rickman (1986) estimate that the relative
contribution of two strong sources of gas, four weaker
sources, and the remaining background was in a ratio
of 3:2:1 during the spacecraft encounters.

For the determination of the survival time, the
effective active fraction of the surface area, defined by
the gas production rate, is essential. It would be very
difficult to explain any systematic increase of it with
time. The simplest assumption of a constant fraction
would be, as to the mass loss variations, analogous to a
constant shrinking rate of the nucleus, as expressed by
equations (5) and (7). A more realistic progressive
decrease would admit the high fading rates and short
survival times of Tables 2 and 3, with, finally, a totally
crusted asteroid-like object (Sekanina 1969). However, this would not agree with other evidence. The
active lifetimes would become much too short, and
extinctions much more frequent than observed. Also,
when applied to the whole sample of known comets,
these fading rates would require something unbeliev-
able: that the periodic comets *in general* are much
fainter now than they were a century ago, and that the
whole Jupiter family will disappear completely within
a few centuries (Kresák 1985).

It appears more reasonable to assume that the
active fraction of the nuclear surface varies quite
irregularly, but with a slowly decreasing trend. As
demonstrated by modelling experiments (Fernández
1985), this would imply erratic variations of the
absolute brightness, leaving essentially no information on the survival time over tens of revolutions. This
is consistent with the absolute magnitude data by
Kresák & Kresáková (1987b), giving for the Jupiter
family of comets a mean change of $+0.01 \pm 0.05$
magnitude per revolution. The presence of dormant
phases makes the problem still more complicated, the
more so as we have very little information on their
duration. The extinct comet P/Biela and the absolutely faintest periodic comet P/Denning–Fujikawa are
the only examples where observations preceding a

transient deactivation seem to exist (Kresák 1986).

The disproportion of three orders of magnitude between the mass-loss rate of P/Halley and typical comets of the Jupiter family under identical conditions offers two alternative explanations. Either they are much smaller—typically 0.5 to 1 km in diameter—or their active surface areas are effectively much smaller, say 0.1 to 0.2%. Any compromise between these two limiting explanations, when combined with a very low albedo, as found for P/Halley, would explain the old question, 'Where are the extinct comets?' Observations and recoveries of inactive objects such as P/Biela, P/Brorsen, 944 Hidalgo, 2212 Hephaistos, 3200 Phaethon, 5025 P-L, 1982 YA, 1983 SA, 1983 XF, 1984 BC, 1985 WA, and 1986 JK deserve special attention. A reactivation of any such object would represent a unique opportunity for a new insight into the evolution of comets.

A typical active lifetime of a periodic comet seems to be a few hundred revolutions, i.e. short compared with a typical dynamical lifetime of an object moving in an orbit of that type. However, with intermissions of the active periods by dormant phases, the total time span covered can become substantially longer. The fact that the active lifetime of P/Halley seems to be longer than that of typical periodic comets, in spite of its ~1000 times larger mass loss per revolution, implies its exceptionality by size, surface structure, or supply of volatiles. Thus most of the problems of comet ageing have to wait for rendezvous missions to other comets—not too long, let us hope.

REFERENCES

Andrienko, D.A., & Vashchenko, V.N. (1981) Outbursts of cometary brightness. *Probl. Kosm. Fiz.* **16** 6–20

Belyaev, N.A., Kresák, Ľ., Pittich, E.M., & Pushkarev, A.N. (1986) *Catalogue of short-period comets* Veda, Bratislava

Dobrovnikoff, N.T. (1941) Investigations of the brightness of comets. *Perkins Obs. Contr.* **2** 49–300

Bortle, J.E., & Morris, C.S. (1984) Brighter prospects for Halley's comet. *Sky Telesc.* **67** 9–12

Broughton, R.P. (1979) The visibility of Halley's Comet. *J. Roy. Astron. Soc. Canada* **73** 24–36

Carusi, A., Kresák, Ľ., Perozzi, E., & Valsecchi, G.B. (1985a) *The long-term evolution of short-period comets* A. Hilger, Bristol

Carusi, A., Kresák, Ľ., Perozzi, E., & Valsecchi, G.B. (1985b) First results of the integration of motion of short-period comets over 800 years. In: *Dynamics of comets: their origin and evolution*, eds Carusi, A., & Valsecchi, G.B., D. Reidel, Dordrecht, 319–340

Clube, S.V.M., & Napier, W.M. (1986) Giant comets and the Galaxy: implications of the terrestrial record. In: *The Galaxy and the solar system*, eds Smoluchowski, R., Bahcall, J.N., & Matthews, M.S. Univ. Arizona Press, Tucson, 260–285

Delsemme, A.H. (1982) Chemical composition of cometary nuclei. In: *Comets* ed. Wilkening, L.L. Univ. Arizona Press, Tucson, 85–130

Dobrovol'skij, O.V., Ibadinov, Kh.I., Aliev, S., & Gerasimenko, S.I. (1983) Lifetime of cometary ice nuclei and secular brightness decrease of periodic comets. *Dokl. AN Tadzh.SSR* **26** 25–29.

Dobrovol'skij, O.V., Ibadinov, Kh.I., Aliev, S., & Gerasimenko, S.I. (1986) Thermal regime and surface structure of periodic comet nuclei. *ESA SP-250* **II** 389–394

Everhart, E. (1976) *The evolution of comet orbits* NASA SP-393, 445–464

Fernández, J.A. (1985) Dynamical capture and physical decay of short-period comets. *Icarus* **64** 308–319

Ferrín, I. (1984) Aging of Comet Halley detected. *Astron. Astrophys.* **135** L 7–9

Ferrín, I., & Gill, C. (1986) Sizes, aging rates and extinction dates of comets Halley and Encke. *ESA SP-250* **II** 427–432

Golubev, V.A. (1975) Splitting of cometary nuclei—I. The catalogue of split comets. *Probl. Kosm. Fiz.* **10** 23–34

Hajduk, A. (1986) Meteoroids from Comet Halley and the comet's mass production and age. *ESA SP-250* **II** 239–243

Hajduk, A., & Kapišinský, I. (1987) The evolution of the mass distribution of cometary particles. *ESA SP-278*, 441–444

Huebner, W.F., Boice, D.C., Keady, J.J., Schmidt, H.U., & Wegman, R (1986) Model ion abundances for Comet Halley. *ESA SP-250* **I** 529–531

Hughes, D.W. (1975) Cometary outbursts: a brief survey. *Quart. J. Roy. Astron. Soc.* **16** 410–427

Hughes, D.W. (1983): Temporal variations of the absolute magnitude of Halley's Comet. *Mon. Not. Roy. Astron. Soc.* **204** 1291–1295

Hughes, D.W. (1985) The size, mass, mass loss and age of Halley's Comet. *Mon. Not. Roy. Astron. Soc.* **213** 103–109

Ip, W.-H., & Rickman, H. (1986) A comparison of nucleus surface models to space observations of Comet Halley. *ESA SP-249* 181–184.

Jones, J., & McIntosh, B.A. (1986) On the structure of the Halley Comet meteor stream. *ESA SP-250* **II** 233–237

Kresák, Ľ. (1965) On the secular variations in the absolute brightness of periodic comets. *Bull. Astron. Inst. Czechosl.* **16** 348–355

Kresák, Ľ. (1974) The aging and the brightness decrease of comets. *Bull. Astron. Inst. Czechosl.* **25** 87–112

Kresák, Ľ. (1980) Dynamics, interrelations and evolution of the systems of asteroids and comets. *Moon and Planets* **22** 83–98

Kresák, Ľ. (1981a) Evolutionary aspects of the splits of cometary nuclei. *Bull. Astron. Inst. Czechosl.* **32** 19–40

Kresák, Ľ. (1981b) The lifetimes and disappearance of periodic comets. *Bull. Astron. Inst. Czechosl.* **32** 321–339

Kresák, Ľ. (1982) Comet discoveries, statistics, and observational selection. In: *Comets*, ed. Wilkening, L.L. Univ. Arizona Press, Tucson, 56–82

Kresák, Ľ. (1984) The lifetimes and disappearance of long-period comets. *Bull. Astron. Inst. Czeckosl.* **35** 129–150

Kresák, Ľ. (1985) The aging and lifetimes of comets. In: *Dynamics of comets: their origin and evolution*, eds Carusi, A., & Valsecchi, G.B., D. Reidel, Dordrecht, 279–302

Kresák, Ľ. (1986) On the aging process of periodic comets. *ESA SP-250* **II** 433–438

Kresák, Ľ. (1987) The 1808 apparition and the long-term physical evolution of periodic Comet Grigg–Skjellerup. *Bull. Astron. Inst. Czechosl.* **38** 65–75

Kresák, Ľ., & Kresáková, M. (1987a) The mass loss rates of periodic comets. *ESA SP-278*, 739–744

Kresák, Ľ., & Kresáková, M. (1987b) The absolute total magnitudes of periodic comets and their variations. *ESA SP-278* 37–42

Marsden, B.G. (1985) Nongravitational forces on comets: the first fifteen years. In: *Dynamics of comets: their origin and evolution*, eds Carusi, A., & Valsecchi, G.B., D. Reidel, Dordrecht, 343–352

McDonnell, J.A.M., Kissel, J., Grün, E., Grard, R.J.L., Langevin, Y., Olearczyk, R.E., Perry, C.H., & Zarnecki, J.C. (1986) Giotto's dust impact detection system DIDSY and particulate impact analyser PIA: interim assessment of the dust distribution and properties within the coma. *ESA SP-250* **II** 25–38

McIntosh, B.A., & Hajduk, A. (1983) Comet Halley meteor stream: a new model. *Mon. Not. Roy. Astron. Soc.* **205** 931–943

Meisel, D.D., & Morris, C.S. (1976) Comet brightness parameters: definition, determination, and correlations. *NASA SP-393* 410–444

Meisel, D.D., & Morris, C.S. (1982) Comet head photometry: past, present and future. In: *Comets* ed. Wilkening, L.L. Univ. Arizona Press, Tucson, 413–432

Morris, C.S. (1973) On aperture corrections for comet magnitude estimates. *Publ. Astron. Soc. Pacific* **85** 470–473

Oort, J.H. (1951) The structure of the cometary cloud surrounding the solar system and a hypothesis concerning its origin. *Bull. Astron. Inst. Neth.* **11** 91–110

Öpik, E.J. (1963) Photometry, dimensions, and ablation rate of comets. *Irish Astron. J.* **6** 93–112

Pittich, E.M. (1971) Space distribution of the splitting and outbursts of comets. *Bull. Astron. Inst. Czechosl.* **22** 143–153.

Rickman, H. (1985) Interrelations between comets and asteroids. In: *Dynamics of comets: their origin and evolution*, eds Carusi, A., & Valsecchi, G.B., D. Reidel, Dordrecht, 149–172

Rickman, H. (1986) Masses and densities of Comets Halley and Kopff. *ESA SP-249* 195–205

Rickman, H., Kamél, L., Festou, M.C., & Froeschlé, C. (1987) Estimates of masses, volumes and densities of short-period comet nuclei *ESA SP-278* 471–481

Sagdeev, R.Z., Elyasberg, P.E., & Moroz, V.I. (1987) Is the nucleus of Comet Halley a low-density body? IKI preprint

Sekanina, Z. (1964) Secular variations in the absolute brightness of short-period comets. *Bull. Astron. Inst. Czechosl.* **15** 1–7

Sekanina, Z. (1969) Dynamical and evolutionary aspects of gradual deactivation and disintegration of short-period comets. *Astron. J.* **74** 1223–1234

Sekanina, Z. (1977) Relative motions of fragments of split comets. *Icarus* **30** 574–594

Sekanina, Z. (1982) The problem of split comets in review. In: *Comets* ed. Wilkening L.L., Univ. Arizona Press, Tucson, 251–287

Sekanina, Z. (1984) Disappearance and disintegration of comets. *Icarus* **58** 81–100

Spratt, C.E. (1984) The lost periodic comets of more than one appearance. *J. Roy. Astron. Soc. Canada* **78** 39–47

Štohl, J. (1986) On meteor contribution by short-period comets. *ESA SP-250* **II** 225–228

Svoreň, J. (1979) Secular variations in the absolute brightness of short-period comets. *Contr. Skalnaté Pleso Obs.* **8** 105–140

Van Flandern, T.C. (1981) Do comets have satellites? *Icarus* **47** 480–486

Vsekhsvyatskij, S.K. (1958) *Physical chracteristics of comets* Gos. Izd. Fiz.-Mat. Literatury, Moscow; translation NASA TT F-80.

Vsekhsvyatskij, S.K. (1972) Cometary observations and variations in cometary brightness. *IAU Symp.* **45** D. Reidel, Dordrecht, 9–15

Wallis, M.K., & Wickramasinghe, N.C. (1985) Halley's Comet: its size and decay rate. *Mon. Not. Roy. Astron. Soc.* **216** 453–458

Whipple, F.L. (1963) On the structure of the cometary nucleus. In: *The solar system IV*, eds, Middlehurst, B.M., & Kuiper, G.P., Univ. Chicago Press, 639–663

Whipple, F.L. (1964) Brightness changes in periodic comets. *Astron. J.* **69** 152

Whipple, F.L. (1978) Comets. In: *Cosmic dust*, ed. McDonnell, J.A.M., J. Wiley, Chichester, 1–73

Whipple, F.L. (1983) Comets: nature, evolution and decay. *Highlights Astron.* **6** 323–331

Whipple, F.L. (1984) Comet P/Holmes 1892 III: a case of duplicity? *Icarus* **60** 522–531

Whipple, F.L. (1986) The cometary nucleus: current concepts. *ESA SP-250* **II** 281–288

Whipple, F.L., & Douglas-Hamilton, D.H. (1966) Brightness changes in periodic comets. *Mém. Soc. Roy. Sci. Liège, Ser. 5* **12** 469–480

Whipple, F.L., & Hamid, S.E. (1952): On the origin of the Taurid meteor streams. *Helwan Obs. Bull.* No. 41

Whipple, F.L., & Sekanina, Z. (1979) Comet Encke: precession of the spin axis, nongravitational motion, and sublimation. *Astron. J.* **84** 1894–1909

Index

Index

Index

Index

Index